AF847428

Topics in Safety, Risk, Reliability and Quality

Volume 36

Series Editor

Adrian V. Gheorghe, Old Dominion University, Norfolk, VA, USA

Advisory Editors

Hirokazu Tatano, Kyoto University, Kyoto, Japan
Enrico Zio, Ecole Centrale Paris, France and Politecnico di Milano, Milan, Italy
Andres Sousa-Poza, Old Dominion University, Norfolk, VA, USA

More information about this series at http://www.springer.com/series/6653

Alexandru Georgescu · Adrian V. Gheorghe ·
Marius-Ioan Piso · Polinpapilinho F. Katina

Critical Space Infrastructures

Risk, Resilience and Complexity

Springer

Alexandru Georgescu
Romanian Association for Space Technology and Industry
Bucharest, Romania

Adrian V. Gheorghe
Engineering Management and Systems Engineering
Old Dominion University
Norfolk, VA, USA

Marius-Ioan Piso
Research and Development
Romanian Space Agency
Bucharest, Romania

Polinpapilinho F. Katina
Informatics and Engineering Systems
University of South Carolina Upstate
Spartanburg, SC, USA

ISSN 1566-0443 ISSN 2215-0285 (electronic)
Topics in Safety, Risk, Reliability and Quality
ISBN 978-3-030-12603-2 ISBN 978-3-030-12604-9 (eBook)
https://doi.org/10.1007/978-3-030-12604-9

Library of Congress Control Number: 2019930277

© Springer Nature Switzerland AG 2019
This work is subject to copyright. All rights are reserved by the Publisher, whether the whole or part of the material is concerned, specifically the rights of translation, reprinting, reuse of illustrations, recitation, broadcasting, reproduction on microfilms or in any other physical way, and transmission or information storage and retrieval, electronic adaptation, computer software, or by similar or dissimilar methodology now known or hereafter developed.
The use of general descriptive names, registered names, trademarks, service marks, etc. in this publication does not imply, even in the absence of a specific statement, that such names are exempt from the relevant protective laws and regulations and therefore free for general use.
The publisher, the authors and the editors are safe to assume that the advice and information in this book are believed to be true and accurate at the date of publication. Neither the publisher nor the authors or the editors give a warranty, expressed or implied, with respect to the material contained herein or for any errors or omissions that may have been made. The publisher remains neutral with regard to jurisdictional claims in published maps and institutional affiliations.

This Springer imprint is published by the registered company Springer Nature Switzerland AG
The registered company address is: Gewerbestrasse 11, 6330 Cham, Switzerland

To my parents, Maria and Nicolae, and all the members of my family, on whose support I could always count.

—Alexandru Georgescu

To my children, Anastasia, Alexandra, and Paul

—Adrian V. Gheorghe

To my colleagues at RoSA: Romanian Space Agency

—Marius-Ioan Piso

To Shelley, Beatrice, Deborah, Crystal, Kelly, Percilla, Leticia, Rachel, Sally, Sandy, Terry, Veronica, Zawadi, and Elizabeth

—Polinpapilinho F. Katina

Foreword

The issue of safely exploring and exploiting outer space, sustainably and in such a manner that limits the impact of various vulnerabilities, is part of a much broader concept defined as the governance of space. According to United Nations (UN) documents of reference, outer space is humanity's most expansive commons, a global commons in fact, which has increasingly become more complex owing to the growing number of both governmental and non-governmental actors actively engaged with it, and to the emergence of new technologies and approaches such as public-private partnerships and private funding initiatives. In line with the dynamic developments in space activities, the ongoing task of the international community is to prove its response capacity and to react to current developments in order to conserve outer space as a global commons and use this otherwise limited resource responsibly for the benefit of humankind.

The *Space2030* agenda and the global governance of outer space activities are among the main current and future concerns of the sole UN committee specializing in outer space; COPUOS (The United Nations Committee on the Peaceful Uses of Outer Space). It provides a space for the intersection of multiple areas of global governance of outer space activities, space science and technology, policy and law, thus establishing a unique cross-domain approach which is essential in order to strengthen the contribution of space to the achievement of the goals and targets of the internationally defined development goals.

Among the constitutive elements of the concept of global governance of outer space activities are the problems related to managing space debris, utilizing nuclear power in outer space and, in a more broadly defined manner, the issue of Long-Term Sustainability of Outer Space Activities. The latter requires the development of cross-sector guidelines for UN Member States to diminish the risks, vulnerabilities, and threats to space systems and activities, as well as increase the confidence of states in the utilization of such systems. This is a very current issue which, in one way or another, and beyond the realm of the elaboration of purely technical solutions, involves assessments and recommendations regarding the criticality of space infrastructures. One such approach is provided by the Space

Situations Awareness dedicated structures within the various space agencies and organizations which address specific vulnerabilities of space systems.

It is emerging as a foregone conclusion that space systems need to be viewed and treated as critical infrastructures, but this also creates challenges for decision makers at the highest level through the very act of designation. For example, every type of critical infrastructure which has been designated on the basis of the National or European framework for critical infrastructure protection must be accompanied by an Operator Security Plan (with periodic revisions) in which the owner/operator/administrator of the critical infrastructure must identify key assets and resources, must perform a scenario-based risk assessment while highlighting vulnerabilities and threats, must define, select and prioritize countermeasures and their define their implementation, and so on.

The general approach for decision makers contemplating the vulnerability of space systems is to view them through the lens of specific and concrete threats, either deliberate and terrestrial, such as cyber-attacks, or originating in the cosmos such as the space weather phenomena. However, viewing space systems as critical space infrastructures and the application of the existing multidisciplinary framework for critical infrastructure protection imparts a new complexity to the perspective of the decision makers and involves nuances and side issues which are still to be identified and explored.

The present book, *Critical Space Infrastructures*, will prove useful to a wide variety of readers who are interested in the emerging field of space systems as critical infrastructures. *Critical Space Infrastructures* conjures up a vision of assets, resources, systems, and networks/interconnections, both physical and virtual, which are so vital to the globalized and globalizing world we inhabit that the disruption or destruction would have a damaging and destabilizing effect on its safety and security, on the national economies, on public safety, and many other subdomains of thriving and advanced societies. When considering these elements, various questions take form: What are the methods, instruments, and management techniques which can be employed in this emerging field?

The authors utilize a wide array of concepts, such as risk, vulnerability, reliability, resilience, robustness, fragility, and anti-fragility as basic building blocks for the description and evaluation of the criticality of space systems, while also describing in detail the specificities of space systems which set them apart from the terrestrial systems which are the mainstay of Critical Infrastructure Protection as a field. This also clarifies the types of interactions, interconnections, and dynamics with which we are faced and which influence the security outcomes. A taxonomy of space systems is given, but the study of criticality focuses on a few key categories —Earth Observation satellites, communication satellites, Global Navigation Satellite Systems, accompanied, also as a critical infrastructure, by the actual legislative and administrative framework which provides the current governance of human activities in space. The terrestrial critical infrastructures which feature dependencies on space systems, some of them critical, are varied and include energy-, transport-, industry-, and food-related critical infrastructures. The authors address the National and even global implications of the disruption of space systems and underline the necessity of protection also against threats which are specific to space systems, such as space debris collisions, space weather phenomena, electromagnetic pulse usage, and other anti-satellite weaponry. The authors also provide various theoretical inputs for assessing the criticality of space systems, such as a generic model for crisis and emergency situation management which is dependent on space systems through a three-phase paradigm, outlining the contribution of space systems to Critical Infrastructure Protection. Other contributions concern the use of "serious gaming" to close gaps between the methods of analysis for conventional problems and those involving complex systems, which feature new characteristics and behaviors stemming from the interactions of their individual components, as well as their environment.

A wide variety of topics are included which are pertinent to space system contributions and future development, from the nuclear industry to the role of emerging technologies such as blockchain in governance for critical terrestrial and space infrastructures. These are some of the novel contributions of this work.

The book also analyzes the applicability of several governance model proposals for the space environment, one based on Complex System Governance theory and the others derived from the observed patterns and behavior of the main space actors, and utilizing multi-criteria decision analysis techniques (fuzzy sets, evidentiary reasoning, etc.) to compare the defined models. The application of Complex System Governance theory to the critical space infrastructure system-of-systems is novel, and the authors highlight the advantages, disadvantages, and challenges of its use, as well as highlighting avenues of future research.

In conclusion, the present book, *Critical Space Infrastructures*, is a valuable contribution to the field of space security governance, as well as Critical Infrastructure Protection, and fits into a wider trend of conceptualizing, defining, and addressing a very important and specific challenge of our times, whose relevance for the future is also assured.

Bucharest, Romania
November 2018

Cosmonaut Dumitru-Dorin Prunariu, Ph.D.
Former President of the Romanian Space Agency
Former Chairman of UN COPUOS
Member and Trustee of the International Academy of Astronautics
Honorary Member of the Romanian Academy

Preface

Critical infrastructure is the body of systems, networks, and assets that are so essential that their continued operation is required to ensure the security of a given nation, its economy, and the public's health and/or safety. A naïve person would argue that such systems are limited to set of essential systems such as agriculture, water, healthcare, transportation, security services, electricity, telecommunication, and the economy. Shrewdness, and perhaps a call for thinking differently, will lead to new paths and models for questioning and understanding. On this quest and as highlighted by recent events and developments, for example, a call for "space army" in the US, this research extends the traditional view of critical infrastructure to include those in the "space-cloud". It is not always safe in the cloud as rain comes from the clouds.

Critical Space Infrastructures may prove to be a useful read for a variety of readers interested in the emerging domain of space as a critical infrastructure. On one extreme, "*Critical space infrastructures*" involves assets, systems, and networks, whether physical or virtual, so vital to the "world" that their incapacitation or destruction would have a debilitating effect on security, national economic security, national public health or safety, or any combination thereof. On the other extreme, and contingent on the first extreme, is the question: *What are the management methods, tools, and technologies that can be used in this emerging domain?* In the end, we need to be able to address real-life problems and eliminate (or at least reduce) the possible debilitating effect on the society. The navigation of such foggy waters is left to a variety of stakeholders—industry leaders, researchers, as well as students—most of whom are addressed in this text. With this audience in mind, the following chapters have been developed:

Chapter 1 provides fundamental notions of critical infrastructures, key resources, and key assets along with core concepts of interest: risk, vulnerability, reliability, resilience, robustness, rapidity, fragility, and anti-fragility. Finally, space as a critical infrastructure is announced.

Chapter 2 provides the basics of *Critical space infrastructure* including elements of key resources and key assets. In this chapter, a careful classification of orbits is undertaken to describe different key resources and assets related to space. An attack on these systems could produce, in the worst case, significant loss of life and/or public health and safety consequences.

Chapter 3 provides a taxonomy for critical space infrastructure. This taxonomy is based on five main categories of Remote Sensing, Communication, Meteorological, Global Navigation Satellite Systems (GNSS), and Administrative and Legislation Framework.

Chapter 4 provides an elaboration of on a key theme of critical space infrastructure: interdependency—relationships to other systems. These relationships are examined in relation to critical systems including water, energy, chemical, transportation, and food supply.

Chapter 5 explores failures associated with "critical space infrastructure" with as emphasis at the national and global scale effects.

Chapter 6 explores the need for protection of "critical space infrastructures." Emphasis is placed on need for protection measures against special threats such as space weather, space debris, electromagnetic pulse, and anti-satellite attacks.

Chapter 7 explores a generic model of crisis and emergency situation management. A three-phase space-based paradigm with a number of processes has been identified and described for potential contribution to space systems.

Chapter 8 puts forward "serious gaming" as a viable approach for understanding infrastructure systems. This approach and the developed model attempts to close a gap associated with conventional strategy making methods espoused by practitioners and management scholars dealing with complex systems.

Chapter 9 attempts to address the role of the Nuclear Industry from the perspective of "critical space infrastructure." This chapter discusses required approval of new reactors, security of the nuclear industry's various processes, permanent security for the waste disposal sites, and non-proliferation efforts as well as support of decision different stakeholders.

Chapter 10 addresses the role of emerging technologies in governance of space with implication of earthly systems. Specifically, blockchain technology is seen as positioned to addressing pressing issue related to space systems.

Chapter 11 offers an initial perspective on "critical space infrastructure" from the Complex System Governance perspective using multi-criteria decision analysis techniques of Fuzzy Sets-based decision analysis, Evidential Reasoning, the Analytic Hierarchy Process, and Logical Decisions for Windows to determine the efficacy of two proposed governance scenarios for the space domain.

Chapter 12 expounds on Chap. 11 by providing analysis of a governance scenario grounded in Complex System Governance. Harking to words commonly attributed to Albert Einstein: *we cannot solve our problems with the same thinking we used when we created them,* this chapter provides an emerging and exciting approach that would provide insights into the problem of governing "critical space infrastructure."

Chapter 13 provides a forward-looking view of governance for "critical space infrastructure." This chapter explores advantages, limitations, and challenges brought by CSG to practitioners and the practices for governing complex systems as well as providing possible a venue of research. The book closes with a Postface by Marius-Ioan Piso, the current President of the Romanian Space Agency, outline the role, prominence, and further venues of research.

Acknowledgements

Beyond the authors' toiling, this book is a measurable expression of their intense intellectual interaction and cross-fertilization of ideas with several distinguished colleagues and partners-in-mind from the Academia and the World of Business of many denominations. Most especially our gratitude goes to the following:

Prof. Wolfgang Kröger—ETH Zürich, Switzerland; Dr. Ioannis Papazoglou—National Center for Scientific Research 'DEMOKRITOS', Greece; Adolf Dörig—Dörig + Partner AG, Switzerland; Prof. Radu Cornel—University Politehnica of Bucharest, Romania; Jürg Birchmeier—ETH Zürich, Switzerland; Prof. Charles Keating—National Centers for System of Systems Engineering, Old Dominion University, USA; Dan Vamanu—Horia Hulubei National Institute of Physics and Nuclear Engineering, IFIN-HH Bucharest, Romania; Marcelo Masera—Directorate for Energy, Transport, and Climate, Institute for Energy and Transport, Joint Research Centre of the European Commission, the Netherlands; Prof. Liviu Muresan—EURISC Foundation, Romania; Prof. Jeannie M. Chapman, College of Science and Technology, University of South Carolina Upstate, USA; Iulia Jivănescu of the Romanian Space Agency Bucharest, Romania; Ştefan Popa and Ştefan Arseni of the Military Equipment and Technologies Research Agency (METRA) works within the Ministry of National Defence, Bucharest, Romania.

The authors are also grateful to graduate students and young researchers at Horia Hulubei National Institute of Physics and Nuclear Engineering, IFIN-HH Bucharest, Romania, ETH Zürich, Laboratorium fur Siecherheistanalytik, Zürich, Switzerland, and Engineering

Management and Systems Engineering, Old Dominion University, Norfolk, Virginia; for many mutually seminal discussions and participative enthusiasm.

Bucharest, Romania	Alexandru Georgescu
Norfolk, USA	Adrian V. Gheorghe
Bucharest, Romania	Marius-Ioan Piso
Spartanburg, USA	Polinpapilinho F. Katina

Contents

About the Authors

Alexandru Georgescu serves as a Researcher for the Romanian Association of Space Technology and Industry, from which position he is involved in research projects funded by, among others, the European Space Agency. He is also a Research Assistant with the Department of Cybersecurity and Critical Infrastructure Protection of the National Institute for Research and Development in Informatics (ICI Bucharest). He is also affiliated with the EURISC Foundation, the European Centre for Excellence for Blockchain, Eurodefense Romania, and the Atlantic Treaty Association's Romanian chapter. He holds a BA in Economics and an MA in Geopolitics from Bucharest University of Economic Studies (Bucharest, Romania). Having developed as a security governance professional, he deepened his knowledge of Critical Infrastructure Protection and now holds a Ph.D. in Industrial Engineering from the Politehnica University of Bucharest (Bucharest, Romania). He has published widely, both in peer-reviewed journals and conferences proceedings. He has significant international experience in places such as China, having been a Visiting Fellow of the Shanghai Institutes for International Studies. He is also present as a security analyst in the popular media, writing in the press and making media appearances, along with participations in conferences and fora.

Adrian V. Gheorghe currently serves as Professor of Engineering Management and Systems Engineering and is the Batten Endowed Chair on System of Systems Engineering with the Department of Engineering Management and Systems Engineering at Old Dominion University (Norfolk, Virginia, USA). He holds an M.Sc. in Electrical Engineering from the Faculty of Power Engineering, Bucharest Polytechnic Institute (Bucharest, Romania), an MBA from Academy of Economic Studies (Bucharest, Romania), an M.Sc. Engineering-Economics, Bucharest Polytechnic Institute, (Bucharest, Romania), and a Ph.D. in Systems Science/ Systems Engineering from City University (London, U.K.). His profile includes editorship for several international scientific journals and scientific board memberships. He has published several books, including "*Emergency Planning Knowledge*" (VdF Verlag, 1996), "*Critical infrastructures at risk: Securing the*

European electric power system" (Springer, 2006), and "*Critical infrastructures: Risk and vulnerability assessment in transportation of dangerous goods - Transportation by road and rail*" (Springer, 2016).

Marius-Ioan Piso is a Romanian scientist serving as the Head of the Romanian Space Agency (ROSA). Since 1998, he is Research Professor in the Institute of Space Science Bucharest. In 1994, he undertook studies in Doctor in Physics from the "Alexandru Ioan Cuza" University of Iasi, Iași, Romania. He is a graduate of the "Babes-Bolyai" University of Cluj with a degree in Physics. He is also a 1999 graduate of the National Defense College in Bucharest, Romania. Since 1990, he worked for the establishment of the Romanian Space Agency and became its first Executive Director in 1995 and appointed President and Chief Executive officer in 2004. He did early work in gravitational radiation and space-time physics. He also performed experimental work in motion measurement and developed a class of inertial sensors and actuators with nanomagnetic fluids, authoring more than 20 patents and 40 papers. Since the late 1990s, he involved in studies and decision-making for utilizing space technology in disaster management and planetary defense. For more than two decades, Marius-Ioan Piso has acted as a high-level representative for international bodies including: *United Nations Committee of the Peaceful Uses of Outer Space* (UNCOPUOS), *North Atlantic Treaty Organization* (NATO), and national delegate for the *European Space Agency* (ESA). He is a member of several national and international research bodies on space research and development and security research including the *International Academy of Astronautics* (IAA) serving as an elected trustee and chairman of section. He established, within the IAA, a study group on "Space Systems as Critical Infrastructures" in 2006 and organized a series of conferences on these topics and studies on planetary defense. He has received several hours including "*Doctor Honoris Causa*" and knighthoods. An asteroid discovered in 1981 bears his name.

Polinpapilinho F. Katina serves as an Assistant Professor of Advanced Manufacturing Management in the Department of Informatics and Engineering Systems at University of South Carolina Upstate, Spartanburg, South Carolina. He previously served as a Postdoctoral Researcher at the National Centers for System of Systems Engineering (NCSOSE) at Old Dominion University, Norfolk, Virginia; Adjunct Assistant Professor in the Department of Engineering Management and Systems Engineering at Old Dominion University, Norfolk, Virginia; and Adjunct Assistant Professor in the Department of Engineering and Technology, Embry–Riddle Aeronautical University—the Worldwide Campus. He holds a B.Sc. in Engineering Technology, an MEng in Systems Engineering, and a Ph.D. in Engineering Management and Systems Engineering from the Old Dominion University and received additional training from among others, Politecnico di Milano, Milan, Italy. He is a founding board member of the *International Society for Systems Pathology*, Claremont, California. He received a 2018 Publons Peer Review Award (Top 1% in Field: Multidisciplinary). His profile includes more than

80 peer-reviewed journal articles, conference (papers) proceedings, and books. His research interests include complex system governance, critical infrastructure systems, decision analysis, engineering management, manufacturing systems, system-of-systems, system pathology, systems engineering, and systems science.

Chapter 1
Critical Infrastructure

Abstract This introductory chapter articulates the topics of critical infrastructure, key resources, and key assets (CIKRKA). The aim is to offer the reader a clear definition of the notions, explain their relevance and the need to develop intelligible and robust models to diagnose and predict risks, vulnerability, resilience, fragility, and perception. The chapter reveals space, undersea, and belowground as three privileged sectors of human endeavor were critical infrastructures, resources and assets coexist in the guise of complex systems that tend to assume a leading position in the overall, global CIKRKA system of systems. This chapters sets the stage for the reminder of the book.

1.1 Infrastructures Systems

The quality of life in the modern society depends, to a large degree, on the quality of its infrastructure. This feeling is shared across governments, industry as well as academia (Gheorghe et al. 2018; Katina and Keating 2015). Allegedly, at the core of this opinion is the concern about the *wellbeing* of the people. In the end, there is a need for protection of such infrastructures. This protection is needed to enable continues operation of infrastructure systems. Table 1.1 provides a collection of formal definitions for the critical infrastructures. Failure of such systems—regardless of the source—anthropic or natural—can have an alarming impact and consequences on the public wellbeing, which extends to individuals, business, government as well as the environment. It comes as no surprise that some authors suggest that "failure of these infrastructures…is one of the most important vulnerabilities of modern society" (Thissen and Herder 2003, p. 1). As an example, it was pointed out, Hurricane Katrina, which made landfall in New Orleans (Louisiana) on August 29, 2005, was the costliest and one of the five deadliest hurricanes to ever strike the United States' with an estimated damage of over $100 billion dollars (Knabb et al. 2011; Townsend 2006).

Throughout human history, there were always systems that failed (McCormick 2014; Sandage 2006). However, there is a sense that failure of systems is increasing with a higher level of calamity (Ansoff 1984; Cohen and Ahearn 1980; Richardson

© Springer Nature Switzerland AG 2019

A. Georgescu et al., *Critical Space Infrastructures*,
Topics in Safety, Risk, Reliability and Quality 36,
https://doi.org/10.1007/978-3-030-12604-9_1

Table 1.1 Defining features for critical infrastructures

Author(s)	Perspectives on infrastructures
U.S. Congress (2001, p. 115 Stat. 401)	…systems and assets, whether physical or virtual, so vital to the United States that the incapacity or destruction of such systems and assets would have a debilitating impact on security, national economic security, national public health or safety, or any combination of those matters
European Council (2004, p. 3)	…consist of those physical and information technology facilities, networks, services and assets which, if disrupted or destroyed, would have a serious impact on the health, safety, security or economic well-being of citizens or the effective functioning of governments in the Member States
Germany Federal Ministry of the Interior, FRG (2009, p. 4)	…organizational and physical structures and facilities of such vital importance to a nation's society and economy that their failure or degradation would result in sustained supply shortages, significant disruption of public safety and security, or other dramatic consequences
Clinton (1996, p. 37347)	…are so vital that their incapacity or destruction would have a debilitating impact on the defense or economic security of the United States
Zio (2016, p. 3)	…large scale, man-made systems that function interdependently to produce and distribute essential goods (such as energy, water and data) and services (such as transportation, banking and health care). An infrastructure is termed critical if its incapacity or destruction has a significant impact on health, safety, security, economics and social well-being
Gheorghe et al. (2006, p. 5)	Infrastructures are critical because they provide services that are vital to one or more broad governmental or societal functions or attributes. This can be related to survivability of citizens as far as the safety of their life is concerned, or to their quality of life
Gheorghe and Katina (2014, p. 195)	The domain of critical infrastructures deals with engineering systems which are characterized by a high degree of technical complexity, social intricacy, and elaborate processes, aimed at fulfilling important functions in the society
Calida and Katina (2012, p. 87)	…current approach to CI protection and mitigation primarily focuses on large malicious and cataclysmic events of terrorism, cyber-attacks, and natural events… [there is] … need to understand the slow, evolving, and inane events that could accumulate into significant events over-time

1994; Weick 1988). Thus, the Rasmussen and Batstone (1989) assertion that: "the frequency and magnitude of organizational failures and the subsequent impacts are increasing at an alarming rate" (Rasmussen and Batstone 1989, p. ii) should stand true. This conjecture is supported by many researchers, including an annual analysis by Swiss Reinsurance.[1] This phenomenon has created, at least in the domain of critical infrastructures, a need for grading the operating state of an infrastructure system as 'under threat,' 'vulnerable,' 'operable,' 'inoperable', as well as a need to develop means to strengthen the security of such systems.

[1] More information can be found at http://www.swissre.com/.

At this point, one needs no reminder of the fact that our modern societies are highly dependent on the continuous operation of critical infrastructures that ensure the supply of crucial goods and services. These include amongst others power supply, drinking water supply, information and communication technologies or waste disposal. Disruptions may have rapid repercussions for the population and the basis of its livelihood and can affect other critical infrastructures through domi-no effects: for instance, a large-scale power blackout will also disrupt the water supply, telecommunications, and rail transport.

While Moteff et al. (2003) definition: "…so vital that their incapacitation or destruction would have a debilitating impact on defense or economic security" is enough, it only offers a limited perspective of critical infrastructures. In fact, the general welfare of our societies has come to rely upon many infrastructure systems. It is hard to think of an economic or societal activity that does not depend on infrastructure-related services, such as electricity and water supply, transportation, information and communication technology, health emergency response and others. A long record of effective operation has caused Western societies to take the availability of such services for granted. Moreover, while infrastructural systems have shown to be considerably trustworthy, they have failed in the past and may still fail in the future. The complexity of infrastructures has grown to overwhelming levels, making it more difficult to understand the potential causes and consequences of failures.

Furthermore, as infrastructure-related services pervade economy and society, the severity of potential failures increases. The ever-accelerated geographical expansion of the energy, transportation, and telecommunications infrastructures has resulted in the emergence of enormous networks that transcend national borders and even continental shores. Infrastructures are interconnected not only across national borders and continents, but also across infrastructure sectors. The energy sector is increasingly vulnerable to disruptions in telecom and information services, and vice versa. Recent studies on protection of infrastructure systems (e.g., Gheorghe et al. 2006) have shown that many of these interconnections are poorly understood, even within individual sectors. Interconnection between infrastructures adds one more challenge to dealing with the complexity and vulnerability of infrastructures.

Moreover, other definitions tend to highlight the fact that infrastructures do not exist in isolation—that is to say, increasing **interdependence** among infrastructure systems. Several facilities and services depend on each other. Airports and railways depend on electricity and communications. The power grid itself depends on communication among power plants and distribution nodes, and the telecommunications network depends on power supply for the transmission links and the exchange nodes. Disruption in any of the systems could jeopardize the continued operation of the entire infrastructure system (Bouchon 2006). In Gheorghe and Schläpfer (2006) research, the definition of the concept of interdependency entails a bi-directional relationship between two infrastructures, through which the state of each infrastructure influences or is correlated to the state of the other.

The consideration of interdependency leads, natural, to the need to consider a related concept and its implications: complexity. Complexity in this case conforms to Sussman's (2005) definition of a system composed of a group of related units (subsystems) for which the degree and nature of the relationship is imperfectly known. When a system is in this state, its behavior and structural patterns are always in constant flux making it difficult to understand and ascertain any useful knowledge (Sousa-Poza et al. 2008). Nonetheless, strides towards understanding 'relationships' are being tackled. The term 'relationship' is used to refer to 'mutually reliant relations between systems' and is often assimilated with interdependency in infrastructure research. Debatably, efforts to understand complexity in infrastructures begin with understanding interdependencies. In this case, Rinaldi et al. (2001) research may provide a good starting point. Six categories of interdependency have been suggested and elaborated elsewhere (e.g., see Dudenhoeffer, et al. 2006; Gheorghe et al. 2018; Katina et al. 2014; Masys 2015). Table 1.2 is adapted from Katina et al. (2014) and provides a summary of types of infrastructure interdependencies describing system relationships.

Because of the interdependencies and interrelations within the system-of-systems, it is no longer possible for an organization to manage its own risks with the presumption of consistent success. The individual critical infrastructures may end up forming, for all intents and purposes, a single system, with them as individual components. Managers end up integrating concepts of interdependencies, contingency transfers and domino effects or cascading disruptions into their thinking (Gheorghe et al. 2006). Robert et al. (2007) developed a common information exchange framework they termed "cooperative space", allowing managers in different organizations to establish interpersonal relationships and chart their dependencies on others and vice versa. Facilitated by technology or state-mediated institutional and legislative frameworks, the cooperative spaces pool knowledge and information and benefit every stakeholder involved. Such a management method corresponds to a risks, vulnerability and threat governance paradigm in continuous and proactive change, with visible effects especially at the level of planning for crisis and emergency situations.

Given the nature of infrastructure systems, it may stand to reason that the analysis of critical infrastructures requires a '**system-of-systems**' perspective. This is especially the case when one is interested in system failure (more appropriately system failure type) which might be related to type(s) of interdependencies—relationships. Rinaldi et al. (2001) suggests the following types of failures:

- **Cascading failures**: Failure in one infrastructure causes disturbances in another infrastructure. In this situation, there is a relationship between two or more infrastructures characterized as functional interdependency.
- **Escalating failures**: Failure in one infrastructure worsens an independent disturbance in another infrastructure. For example, a breakdown in the metro is significantly worse if a main road is unavailable due to a fire in a tunnel. The term "impact interdependency" is used to describe the relationship between these infrastructures.
- **Common cause failures**: A situation in which two or more infrastructure systems are disrupted at the same time due to a common cause, which is described as a geographical dependency because one or several elements of the infrastructures

Table 1.2 Type of infrastructure interdependencies

Interdependency type	Relevant themes	Implications for infrastructure development
Physical interdependency	This is a relationship that "arises from the physical linkage between the inputs and outputs of two agents [where the] commodity produced or modified by one infrastructure (an output) is required by another infrastructure for it to operate (an input)" (Rinaldi et al. 2001, p. 15) such as drinking water and electricity	A consideration of the direct and physical influence of external systems including outputs, product, goods, and services to a system of interest. For example, an operator of a water system should be concerned with risks in electrical grid, since availability of clean drinking water is physically dependents on electricity used in water treatment
Cyber interdependency	A relationship based on ubiquitous and pervasive use of information and communications technologies (ICT). Many critical systems provide essential goods and services with the help of control systems such as supervisory control and data acquisition (SCADA) systems that remotely monitor, and control operations used coded signals over ICT communication channels (Katina et al. 2016; Rinaldi et al. 2001)	A consideration of cyber interdependency could enable one to examine the nature of reliance of ICT within a given scenario. This analysis might include an overview of a cyber aspects a system including an articulation of relation between internal and extern systems, processes monitored and controlled, types of SCADA architectures deployed (i.e., 1st generation, 2nd generation, 3rd generation, 4th generation), and cyber-related risks as well as countermeasures
Geographical interdependency	This is a relationship that exists when different infrastructure systems share the same environment such as electrical power lines that share the same corridor with a bridge (DiSera and Brooks 2009; Katina et al. 2014)	This involves a consideration of geographical interdependencies associated with a need for common environment that typically enables coupling of infrastructure systems. Coupling creates a situation in which an attack on one is an attack to all. For example, a destruction of a bridge affects electricity transmission, if there is a shared corridor
Logical interdependency	A logical interdependency exists between infrastructures if the state of each infrastructure depends on the state of the other via some mechanism that is neither physical, cyber, nor geographical (Rinaldi 2004) such as power deregulation policy	An exploration into 'other mechanisms' beyond physical, cyber, and geography. Other mechanism could involve the role of time, space, perception, and geo-politics

(continued)

Table 1.2 (continued)

Interdependency type	Relevant themes	Implications for infrastructure development
Policy and/or procedural interdependency	This is a 'hidden' and not-so-obvious relationship that only becomes apparent after a change, in form of a policy and/procedure that takes effect in one part of the system. For example, several regulations that were issued in the wake of 9/11 attacks affected all air transport systems, changing the flying experience (Mendonca and Wallace 2006)	Attempts to feedforward and development of scenarios that might offer insights into how quality of goods and services could be influenced by changes in policy at national, state, regional, and local level. The intent of such efforts is the discovery of possible direct effects of changes as well as 'unintended' consequences on critical infrastructures
Societal interdependency	Societal interdependency is a situation in which infrastructure operations are affected by public opinion. For example, after 9/11 attacks, air traffic was reduced due to the public's evaluation of travel safety, resulting in job cuts and bankruptcies (Dudenhoeffer, et al. 2006; Katina et al. 2014)	This analysis involves examination of public opinion on critical topics as they relate to infrastructure goods, services, and operations. The intent of such efforts is an attempt to understand impact of infrastructure operations. This might include, for example, understanding public perception of emerging concepts, for example, Smart Grids

are in close proximity so that external threats may knock out several infrastructures at the same time.

Safeguards against single point failures generally depend on the proper functioning of the rest of the national infrastructure, a plausible assumption for high-reliability infrastructure systems when they experience random, uncorrelated single point failures. Single point failures are anticipated in the design of the systems and engineering solutions of various kinds, including redundancy, rapid repair, replacement, and operational rerouting. Planning for multiple failures, particularly when they are closely correlated in time, is much less common. It is safe to say that no one has planned for, and few have even imagined, a scenario with the loss of hundreds or even thousands of nodes across all the critical national infrastructures, all simultaneously.

Particularly, difficult to anticipate are situations in which the occurrence of simultaneous failures can bring into play dormant and previously hidden interaction pathways in which a destructively synergistic amplification of failure, normally locally contained, may be propagated through the network at large (Foster et al. 2008). Charles Perrow has drawn attention to these types of failures, which he has termed *normal accidents*, and which are posited as an inherent property of any tightly coupled system once a threshold of complexity has been passed (Perrow 1999).

The multitude and variety of nodes and links in these networks, and of the operations and services deployed, as well as the hosts of owners, operators, suppliers and users involved, have created enormously convoluted constructs. The intricacy of infrastructures limits the understanding of their behavior and, consequently, the options to effectively control and steer the processes involved. There is an urgent need to generate more systematic knowledge on these complex systems, if one is to succeed in adequately handling the many threats and vulnerabilities. In this case, one might argue in support of Arbesman (2016, p. 2):

> *Our technologies—from websites and trading systems to urban infrastructures, scientific models, and even the supply chains and logistics that power large businesses—have become hopelessly interconnected and overcomplicated, such that in many cases even those who build and maintain then on a daily basis can't fully understand them any longer.*

The above statement should not come as a surprise given the increasing number of failures associated with such systems.[2] Thus, the protection of such systems is paramount. Critical infrastructure protection (CIP) then, has emerged as a means to provides a comprehensive framework for managing such systems. However, CIP is plagued by several challenges (Hammerli and Renda 2010):

- The Public-Private Dichotomy—most critical infrastructures are owned/operated/administered by private entities pursuing profit, while the main agents of CIP at national and international levels are institutional actors and state authorities with a security mandate. Public-Private Partnerships become necessary under these conditions;
- The absence of borders—especially when it comes to informational infrastructures (ITC or not), but increasingly for other distributed infrastructure systems (transport corridors, electricity distribution, pipelines) the political or geographic borders are either irrelevant or an aggravating actor for CIP coordination. Attributing malicious activities and assigning the responsibility for implementing security policies become more difficult, especially for information systems. With regards to physical infrastructures passing through several jurisdictions, the gaps between what each state entity considers its jurisdiction, to the exclusion of coordination and cooperation with the others, is a space where new risks, vulnerabilities and threats are formed. Major breaches in security occur wherever the CIP activities are fragmented along national political lines;
- Powerful connectivity—critical infrastructures generate a vast and dynamic environment with significant interconnections which grow as the various stakeholders require;
- Complexity—certain facets of risk formation and manifestation are neglected by simplistic problem reduction models, leaving gaps in risk and vulnerability gover-

[2]This statement should be viewed in the context. Here, there the researchers make no distinction of the nature (and cause) of failure. For example, authors do not make any attributions to malicious, technical, or natural (e.g., see Gheorghe et al. 2018). The fact remains, these systems, despite being society's lifelines, they are failing, at large scales.

nance. A preferred habit of decomposing problems into smaller components to be addressed individually is not workable since "divide and conquer" produces too many other risks for it to be a viable option;

- The human factor—critical infrastructure operation/exploitation and CIP activities depend on human decision-making and suffer from uncertainties and expertise difficulties because of the limitless combinations of risk factors leading to potential crisis situations. Moving outside of the realm of the ordinary threat may prove challenging to the CIP human resources. With issues such as "enemy within" scenarios or "social engineering" for cyber intrusions, the human resources itself becomes a vector for threats;
- Omnipresent vulnerabilities—various attackers may exploit persistent or temporary vulnerabilities, gaining experience and insight into critical infrastructure functioning, especially through the repeated exposure possible in the cyber realm.

Thus, CIP needs to involve a number of elements, including, stakeholder models, agent-principal models, decision making models, public-private partnerships with attendant legislation, lines of communication and feedback, security standards, technical systems, security culture, modeling and simulations capabilities and much more. CIP concepts work with a number of core concepts, the most important of which are (Hokstad et al. 2012):

- ***Risk analysis***—a process of identifying potential hazards based on severity of consequence and likelihood of occurrence. The intent is to sort potential hazards and prioritize them for action based on objective criteria. One method is to grade likelihood and consequence on various scales
- ***Vulnerability***—a vulnerable system is open to losing its design functions. Vulnerability can then be taken as a degree to which a system, subsystem, or component is in situations where it is exposed to those specific hazards that would be harmful or damaging to the system. The focus in a vulnerability analysis moves away from the possibility that adverse events occur, to system properties determining how easy it is to eliminate major system functions
- ***Reliability***—reliability points to a system or a components ability to perform the predefined required functions. Reliability is measured in terms of the probability that a system or a component is able to perform its required function at a given point of time, or over a given period of time for a given set of conditions which may be at the extreme end of specified limits
- ***Resilience*** —the ability of a system to react and recover from unanticipated disturbances and events. The concept has gained in popularity in recent years and is now viewed as the ultimate goal of CIP processes, implying the minimization of damages and the rapid restoration of normal functions
- ***Robustness***—closely linked to resilience, it is a system's ability to withstand a certain amount of stress with respect to the loss of function of the system
- ***Rapidity***—another closely linked concept to resilience, it is a system's ability to recover from an undesired event as measured by the speed of recovery.

Johnson and Gheorghe (2013) expand on these concepts by adding two concepts on their models, based on work by Nassim Taleb:

- ***Fragility***—fragility and vulnerability are similar, but have critical differences, the first being endogenous, while the second is exogenous. Vulnerable systems fail because of their degree of exposure to stress of a specific nature, while fragile systems fail because they are easily broken regardless of the nature of stress they are exposed to
- ***Antifragility***—a system's ability to withstand stressors is, to a certain degree, a function of previous exposures to manageable stresses. A clear example can be found in the area of terrestrial and space infrastructure systems, where the various damages caused by recurring high levels of "space weather" activity has served to highlight the need for robustness and redundancy. The incremental improvement in their systemic resilience will be invaluable in the face of expected peak events which would, otherwise, be guaranteed to have devastating consequences.

Out of these concepts, resilience has become a very commonly utilized concept (featuring also in the present book) encapsulating complex notions related to safety, security and desired states. Johnsen (2010) describes seven principles of resilience:

- Graceful and controlled degradation—signifies a number of organizational capacities and competencies that diminish the speed which a cascading disruption will propagate. For instance, a facility may have a procedure for an immediate total or partial shutdown of activities or severance of links to the outside at a moment's notice, in order to avoid significant damage and enable quicker restoration of near normal functioning. Impact analyses and the identification of risky behaviors and policies are key, as well as the rapid mobilization of emergency capabilities and implementation of procedures;
- The management of margins—using proactive indicators, CI owners/operators/administrators ensure that the limits of the specifications for normal exploitation are not exceeded. Systems should be controlled and tested for the capacity to remain within safe operating parameters, especially as they increase in age or are hybridized with new-generation components such as late-model industrial control systems. Margins are eroded by long-term use, human error or disaster situations, and the tradeoff between system performance and system security can become substantial, and initially expressed as the narrowing of the interval between optimal performance parameters and safety maintenance parameters. Risk awareness is just as important as risk assessment;
- Common mental modes—promotes communication and cooperation between different systems, organizations, or distributed CI components. The commonality of mental modes is crucial to managing incidents of deviation from normal functioning and to promote a resumption of normality. Accident prevention and properly integrating feedback to improve security are also contingent on common mental modes. At its most basic, we could view the security culture of stakeholders as one such element;
- Redundancy—the entity must have alternative means to perform a given function and the definition scales up all the way to the level of system-of-systems. Alternatives can mean a great deal of things—other organizations, other technical means, other identical units, other procedures etc. Having systems waiting as back-up

or on parallel functioning is a good way to ensure resilience. Redundancies are also a new element of complexity and, therefore a risk, promoting vulnerability to common cause failures, for instance. This is why an alternative to redundancy is diversity as a function of flexibility;

- Flexibility—entails a diversity of systems and methods to perform the same function. Flexible systems are open to improvisations and incremental improvements, as well as featuring a tolerance for errors, so long as they are of the type that is easily observable and remediated;
- Complexity reduction—this can be achieved in a number of ways:
  - By segregating systems, even at the cost of certain synergies and efficiencies;
  - By going from common connectors and support systems to dedicated ones;
  - Promoting light substitution, as opposed to limited substitution;
  - Reducing the number of "feedback loops", which also increase the potency of escalating failures;
  - By redesigning a system so that multiple interactive control lines become instead dedicated, segregated and unique;
  - Indirect information systems should become direct (for every type of communication, including human one within the organization or outside of it);
  - Improving the understanding of the system-of-system through dynamic mapping;
  - The complexity of organization should be reduced, especially when they have extensive hierarchies with diffuse responsibilities and weak channels of communication;
- Coupling reduction—this hinders or prevents cascading disruption phenomena. Systems with this feature allow for delays in procedure, for flexible sequencing of operations, for flexible operating methods and flexibility in resources that are used. Such systems feature redundant capacities and substitutable systems, components and resources (Perrow 1999).

Vugrin et al. (2010) ties the qualitative analysis of a system-of-system's resilience to three capacities:

- Absorptive capacity—and endogenous system quality that allows is to absorbs perturbances in the system-of-systems with minimal consequences to its proper functioning and the production of critical goods and services. This capacity is affected by traits such as system properties such as redundancy and robustness. For instance, a chemical refinery may go offline for a while, but may have stocks of finished product to continue deliveries and absorb production shocks;
- Restorative capacity—the ability of a system to be easily maintained and repaired dynamically. Changes performed under a restoration process are permanent, as opposed to an adaptive one, which are temporary while environmental conditions warrant. The same chemical refinery can choose to invest in new technologies during the repairs following a negative event materializing, thereby profiting the most from enforced inactivity;

- Adaptive capacity—an endogenous system quality to reorganize when faced with a shock, so that it reaches its past performances to the fullest possible extent. Adaptive systems are very resilient, because, for instance, the absence of a resources may be initially compensated by substituting for another, at least for part of the total consumption requirement. System substitutability is an adaptive trait. For instance, clients of electricity companies may nonetheless choose to have reserve power generation capabilities based on rapid firing natural gas turbines.

Complicating the matter of critical infrastructure is the nature of the phenomenon itself: 'critical,' according to whom? This observation, made explicit by Katina and Keating's (2015) research in comparing the Patriot Act of 2002 and the PCCIP of 1996, suggests that we might be on a course of not granting the due attention to infrastructure systems essential to public well-being. Avoiding such a misconduct would call upon us to embrace a broadening of what constitutes critical infrastructures (Gheorghe et al. 2018). Beyond this discussion, there is a matter of key resources and key assets—and the main dish of ***'critical space infrastructure'***!

1.2 Key Resources and Key Assets

Gheorghe et al. (2018) suggests that the general concepts of critical infrastructures have penetrated and pervasive in the main streams. However, the newcomers include: 'key assets' and 'key resources.' At a fundamental level, one needs to know that both key assets and key resources are worth protecting, similarly to 'critical infrastructures.' Following the US Patriot Act of 2001 (US Congress 2001) a 'key resource' is defined as "a publicly or privately controlled resource essential to the minimal operations of the economy and government" (USDHS 2002, p. 116 STAT. 2141). The Homeland Security Act of 2002 calls for a comprehensive assessment of vulnerabilities in key resources including risk assessment involving risk classification, probability of attack occurrence, feasibility as well as efficacy of possible countermeasures. While this act did not offer any specific examples of key resources (Moteff and Parfomak 2004), Bennett's (2007, p. 54) research looks at the relevance of key resources but suggesting that the "destruction of a key resource would not endanger vital systems but could cause large-scale injury, death, or destruction of property and/or profound damage to our national prestige and confidence." This definition underscores the importance of understanding and protecting key resources.

Nearly a year after the Homeland Security Act of 2002, the *National Strategy for the Physical protection of Critical Infrastructures and Key Assets* introduced the concept of 'key assets.' These are defined as "individual targets whose destruction would not endanger vital systems but could create local disaster or profoundly damage our Nation's morale or confidence" (Bush 2003, p. 7). Three categories associated with key assets are described in Table 1.3. Similar to 'key resources,' 'key assets' can be distinguished from 'critical infrastructures' based on a 'scale of damage.' In the *National Strategy for the Physical protection of Critical Infrastructures and Key*

Table 1.3 Three main categories of key assets

Key asset category	Definition	Examples
Category I	This category comprises the diverse array of national monuments, symbols, and icons that represent our Nation's heritage, traditions and values, and political power. Such asset attracts large number of tourists and frequent media attention	Prominent historical attractions, monuments (e.g., African Burial Ground), cultural icons, and centers of government and commerce
Category II	This category includes facilities and structures that represent national economic power and technological advancement. These assets tend to house significant amounts of hazardous materials, fuels, and chemical catalysts that enable important production and processing functions	Specialized medical and pharmaceutical Laboratories, nuclear power and chemical plants and hydro electrical dams. Failures in such systems can have significant impact on public health and confidence, and the economy
Category III	This category includes such structures as prominent commercial centers, office buildings, and sports stadiums, where large numbers of people regularly congregate to conduct business or personal transactions, shop, or enjoy a recreational pastime	Entertainment and Media (e.g., motion picture studios, broadcast media), public assembly (e.g., arenas, and stadiums), and sports leagues (e.g., professional sports leagues)

Assets, it is noted that key assets, while "alone may not be vital to the continuity of critical services on a national scale, but an attack on any one of them could produce, in the worst case, significant loss of life and/or public health and safety consequences" (Bush 2003, p. 7). The importance of key assets (e.g., historical attractions and monuments) relates to the fact that such assets are seen as being part of a heritage of a given society and as such, an attack on these, is also an attack on the society's psyche and lifestyle.

An instantaneous salient feature of 'critical infrastructures,' 'key resources,' and 'key assets' First and foremost, incapacitating/crippling/destroying 'critical infrastructures, 'key assets', and 'key resources,' might as well be the significance of such systems on the public health and safety, public confidence, and economy (Katina and Keating 2015). To this end, there is vast interest in in "robust approaches for dealing with infrastructure systems" (Gheorghe et al. 2018, p. 13). In the case of risk management for infrastructures, the *National Infrastructure Protection Plan* (NIPP) developed a framework, originally introduced in 2006 and modified in 2013. It is not meant as a prescriptive approach for managing risk; rather, it is meant to be an 'organizing construct' for thinking and doing risk management in critical infrastructures,

Table 1.4 Phases and objectives of NIPP risk management framework, adapted from USDHS (2013)

Phase	Objectives
Set infrastructure goals and objectives	At the national level, this phase calls for the establishment of a set of broad national goals for critical infrastructure security and resilience. Owners and operators of critical infrastructure at regional entities, identify objectives and priorities for critical infrastructure consistent with national priorities, national goals, and sector objectives, tailored and scaled to their operational and risk environments and available resources
Identify infrastructure	To manage critical infrastructure risks effectively, assets, systems, and networks that are essential for continued operation must be identified, considering associated dependencies and interdependencies. It was observed that not all actors involved in this process (governments, industry, academia) view infrastructure systems from the same perspective (Katina and Keating 2015). This includes differences in appraising infrastructure 'criticality' at the national, regional, and sector levels (USDHS 2013)
Assess and analyze risks	Critical infrastructure risks are assessed in terms of threat, vulnerability, and consequence. In the context: • 'Threats' are agents and/or circumstances having the potential to harm life, information, operations, the environment, and/or property • 'Vulnerability' represents a physical feature or operational attribute that renders an entity open to exploitation or susceptible to a given hazard • 'Consequence' is an effect of an event, incident, or occurrence There is a broad range of methodologies that can be used in risk assessment enabling informed decision making
Implement risk management activities	This phase is concerned with prioritizing activities to manage critical infrastructure risk based on the criticality of the affected infrastructure, costs, and potential for risk reduction. Implementation activities include: • Identify, deter, detect, disrupt, and prepare for threats and hazards • Reduction of vulnerabilities • Mitigating the consequences
Measure effectiveness	Once risk management activities are implemented, the next phase involves an evaluation of the effectiveness of risk management efforts within sectors and at national, State, local, and regional levels by developing metrics for both direct and indirect indicator measurement. This phase serves, among others, as the basis for re-articulation of vision and national and regional goals and assessing progress

key resources, and key assets (USDHS 2013). It should be noted that the solution proposed considers three elements of critical infrastructures: the physical, cyber, and human aspects, that are dealt with in a sequence of five key phases as described in Table 1.4.

In coming to grip with the reality critical infrastructures, assets resources, and key assets: their increasing failures and potential consequences, one of the ways that might offer comfort is the need to the boundaries on the 'traditional' viewpoints on infrastructure systems. One might look at elements of spaces and time. For example, one could look at local, national, regional, or global level infrastructure systems and how such systems can be protected across time. Yet another view could be 'expanding' the scope of infrastructure systems. For example, introducing new infrastructures and addressing them within the paradigm of critical infrastructures. This includes the concepts of undersea (e.g., submarine cables), belowground (e.g., pipelines), and space systems (e.g., satellites). This book attempts to contribute to the later by proposing space as a critical infrastructure. While this is an emerging area of research, the path forward is grounded in the tenets of the 'traditional' themes of critical infrastructures, spearheaded by recent developments in research (e.g., see Francis 2010; Gheorghe and Yuchnovicz 2015; Gheorghe et al. 2018; Hesse and Hornung 2015; Schmieer 2015).

Applying 'criticality' of infrastructure systems to space is a complex, not only because 'space' is large, but because of our current state of knowledge regarding space. To this end, the approach we have undertaken is deliberate, adapted to specific situations, traits and diversity, while also integrating these efforts in the wider sphere of the protection and resilience of the system (and system-of-systems) of space. While allowing for divergence from the current CIP orthodoxy to meet the needs of space systems, CIP for space assets will involve much of the same components, only with the added burden of having to work in a global setting, with numerous stakeholders, in which a lack of consensus will prejudice protection efforts and allow vulnerabilities to go unchecked.

Various actors, mostly governmental, have pursued an independent development in this field, generating solutions of a technical, organizational and strategic nature. Great advances can be made through cooperation and experience-sharing, bringing together these isolated developments and constructing a comprehensive image of the complexities involved. Rapid development of the sector has enhanced our understanding of the depth to which critical infrastructures are interconnected and, therefore, the potential dangers that disruptions (intentional or unintentional, natural or man-made, sudden or gradual) in a single area pose to the entire system-of-systems emerging in modern societies. Today's critical infrastructures are large-scale sociotechnical systems, comprised of multiple components, involving various stakeholders, technologies, policies and social factors.

In recent years, numerous sociotechnical systems have started to undergo a series of transitions. These are long-term fundamental changes (irreversible, high impact and of high-magnitude) in the cultures (mental maps, perceptions), structures (institutions, infrastructures and markets) and practices (use of resources) of a societal system. In other words, the transition includes a structural change in both technical and social subsystems. The planning and implementation phases of such large-scale infrastructure transitions require close monitoring of performance parameters like safety, efficiency, and sustainability. Ensuring that infrastructure evolution reveals a safer and more sustainable system becomes a major challenge for society. Therefore, it is vital to maintain the operability of the critical infrastructures while enabling an evolution that is sustainable within the context of the security concerns generated by our increasing dependences. The ultimate goals are ensuring quality of life and business continuity in the most adverse conditions (Gheorghe et al. 2006).

For its case, space is already present among them as an area of emerging systems which has received attention through its close association with terrestrial infrastructure systems. There have been important advances in the area of strategic documentation and legislation. For example:

- *European Security Research and Innovation Forum*—ESRIF was established in September 2007 aiming at developing a mid and long-term strategy for civil security research and innovation through public private dialogue by 2009. Since then, there has been developments in European security. These efforts are not limited to space systems[3]
- *Directive 114/2008*—This directive creates a treaty for the identification and designation of European critical infrastructures and the assessment of the need to improve their protection[4]
- *The Lund Declaration, July 2009*—A declaration calling for European research that must focus on the Grand Challenges of the 21st century moving beyond the current rigid thematic approaches[5]
- *Com (2009) 149*—Protecting Europe from large scale cyber-attacks and disruptions: enhancing preparedness, security and resilience[6]
- *Com (2011) 163*—Critical information infrastructure protection—achievements and next steps: towards global cyber-security[7]
- *Com (2011) 152*—Towards a space strategy for the European Union that benefits its citizens, that lays down, in detail, EU priorities in this area[8]
- *Com (2012) 417*—Policy calling for a greater focus on Security R&D as well as greater cooperation to improve competitiveness and create an EU brand that can adequately rival that of the US, with its harmonized framework and robust internal market Action Plan for an innovative and competitive security industry[9]
- *Com (2011) 808 (Horizon 2020)*—The biggest EU Research and Innovation programme ever with nearly €80 billion of funding available over 7 years (2014–2020)—in addition to the private investment that this money will attract. It promises more breakthroughs, discoveries and world-firsts by taking great ideas from the lab to the market[10]
- *SWD (2013) 318*—A working documents on new approaches for enhancing security level of European infrastructures[11]

---

[3]http://ec.europa.eu/enterprise/policies/security/files/esrif_final_report_en.pdf.

[4]http://eur-lex.europa.eu/LexUriServ/LexUriServ.do?uri=OJ:L:2008:345:0075:0082:EN:PDF.

[5]http://www.vr.se/download/18.7dac901212646d84fd38000336/.

[6]http://eur-lex.europa.eu/LexUriServ/LexUriServ.do?uri=COM:2009:0149:FIN:EN:PDF.

[7]http://eur-lex.europa.eu/LexUriServ/LexUriServ.do?uri=COM:2011:0163:FIN:EN:PDF.

[8]http://ec.europa.eu/enterprise/policies/space/files/policy/comm_pdf_com_2011_0152_f_communication_en.pdf.

[9]http://eur-lex.europa.eu/LexUriServ/LexUriServ.do?uri=COM:2012:0417:FIN:EN:PDF.

[10]http://eur-lex.europa.eu/LexUriServ/LexUriServ.do?uri=COM:2011:0808:FIN:en:PDF.

[11]http://ec.europa.eu/energy/infrastructure/doc/critical/20130828_epcip_commission_staff_working_document.pdf.

- *Com (2013) 711*—A policy on long-term vision of European infrastructures and infrastructures of European importance from beyond the EU borders.[12]

The tangential approach to critical space infrastructures will, eventually, lead to a greater focus on their protection. However, the peculiar traits we have previously ascribed to 'critical space infrastructures,' particularly their complicated stakeholder model and significant risk profiles, would suggest that important avenues for the propagation of risks, vulnerabilities and failures are being insufficiently regarded. 'critical space infrastructure' protection efforts should consider the unique characteristics and circumstances of these systems and define a new paradigm which acknowledges the reality that space systems are literally the 'upper' component of most other critical infrastructure systems, rather than shunt it aside in favor of sector-centric perspectives on space criticality.

Final Remarks

The domain of critical infrastructure describes the physical and cyber systems and assets and resources that are so vital to the United States (or any nation and region) that their incapacity or destruction would have a debilitating impact on our physical or economic security or public health or safety. In effect, a nation's critical infrastructure provides the essential services underpinning that society. Critical infrastructure must be secure and able to withstand and rapidly recover from all threats and hazards. This requires proactive and coordinated efforts. Clearly, such efforts are shared responsibility among different entities (e.g., federal, state, local, tribal, and territorial entities, and public and private owners and operators) of critical infrastructure systems. Having an effective strategic guidance to public and private partners, can promote national unified efforts. Such efforts could then yield applied research that promotes aspects of secure, robust, and resilience critical infrastructure at local, regional, national, and international scale. However, present efforts are segregated (at best) and lacking (at worst). Along these lines there remains a need for:

- Investing and investigative physical and cyber risk management products and plans
- Educating the masses (not only employees) about critical infrastructure security and resilience
- Plan for business continuity
- Sharing threats and incident information
- Reporting and documenting suspicious activity.

As researchers, we hope that this book contributes to these elements. Along with this hope, we invite all the masses to join us on a journey that adds, yet another twist to critical infrastructure: *The 'space' aspect*.

---

[12]http://eur-lex.europa.eu/resource.html?uri=cellar:a2574790-34e9-11e3-806a-01aa75ed71a1.0007.01/DOC_1&format=PDF.

References

Ansoff, H. I. (1984). *Implanting strategic management*. Englewood Cliffs: Prentice-Hall.

Arbesman, S. (2016). *Overcomplicated: Technology at the limits of comprehension*. New York: Current.

Bennett, B. T. (2007). *Understanding, assessing, and responding to terrorism: Protecting critical infrastructure and personnel*. Hoboken, NJ: Wiley.

Bouchon, S. (2006). *The vulnerability of interdependent critical infrastructures systems: Epistemological and conceptual state-of-the-art* (No. EUR 22205 EN). Ispra, Italy: Institute for the Protection and Security of the Citizen, European Commission Joint Research Center. Retrieved from https://publications.europa.eu/en/publication-detail/-/publication/e3c8da0d-9d5a-408d-a098-38f6b624508d/language-en.

Bush, G. W. (2003). *The National strategy for the Physical Protection of Critical Infrastructures and Key Assests*. Washington, DC: The White House. Retrieved from https://www.dhs.gov/xlibrary/assets/Physical_Strategy.pdf.

Calida, B. Y., & Katina, P. F. (2012). Regional industries as critical infrastructures: A tale of two modern cities. *International Journal of Critical Infrastructures, 8*(1), 74–90.

Clinton, W. J. (1996). Executive order 13010: Critical infrastructure protection. *Federal Register, 61*(138), 37345–37350.

Cohen, R. E., & Ahearn, F. L. (1980). *Handbook for mental health care of disaster victims*. Hutchinson: Johns Hopkins University Press.

DiSera, D., & Brooks, T. (2009). The geospatial dimensions of critical infrastructure and emergency response. *Pipeline and Gas Journal, 236*(9), 1–4.

Dudenhoeffer, D., Permann, M. R., & Manic, M. (2006). CIMS: A framework for infrastructure interdependency modeling and analysis. In *Proceedings of the 38th conference on Winter simulation* (pp. 478–485). Monterey, California: Winter Simulation Conference. https://doi.org/10.1109/WSC.2006.323119.

European Council. (2004). *Communication from the Commission to the Council and the European Parliament: Critical infrastructure protection in the fight against terrorism* (pp. 1–11). Brussels, Belgium: Commission of the European Communities. Retrieved from http://eur-lex.europa.eu/LexUriServ/LexUriServ.do?uri=CELEX:52004DC0702:EN:NOT.

Foster, J. S., Gjelde, E., Graham, W. R., Hermann, R. J., Kluepfel, H. M., Lawson, R. L., et al. (2008). *Report of the commission to assess the threat to the United States from electromagnetic pulse (EMP) attack* (No. A2473—EMP). Washington D.C.: EMP Commission. Retrieved from http://www.empcommission.org/docs/A2473-EMP_Commission-7MB.pdf.

Francis, M. S. (2010). UAS uses, capabilities, grand challenges. In *Encyclopedia of Aerospace Engineering* (pp. 1–11). New York: NY: Wiley, Ltd.

FRG. (2009). *National strategy for critical infrastructure protection* (pp. 1–18). Berlin, Germany: Federal Ministry of the Interior. Retrieved from http://www.bmi.bund.de/cae/servlet/contentblob/598732/publicationFile/34423/kritis_englisch.pdf.

Gheorghe, A. V., & Katina, P. F. (2014). Editorial: Resiliency and engineering systems—Research trends and challenges. *International Journal of Critical Infrastructures, 10*(3/4), 193–199.

Gheorghe, A. V., & Schlapfer, M. (2006). Ubiquity of digitalization and risks of interdependent critical infrastructures. In *2006 IEEE International Conference on Systems, Man and Cybernetics* (Vol. 1, pp. 580–584).

Gheorghe, A. V., & Yuchnovicz, D. (2015). The space infrastructure vulnerability cadastre: Orbital debris critical loads. *International Journal of Disaster Risk Science, 6*(4), 359–371.

Gheorghe, A. V., Masera, M., Weijnen, M. P. C., & De Vries, J. L. (Eds.). (2006). *Critical infrastructures at risk: Securing the European electric power system (Vol. 9)*. Dordrecht: Springer.

Gheorghe, A. V., Vamanu, D. V., Katina, P. F., & Pulfer, R. (2018). *Critical infrastructures, key resources, and key assets*. Cham, Switzerland: Springer International Publishing.

Hammerli, B., & Renda, A. (2010). *CEPS Task Force Report—Protecting critical infrastructures in the EU*, European Center for Policy Studies, p. 23, Brussels, ISBN: 978-94-6138-070-8.

Hesse, M., & Hornung, M. (2015). Space as a critical infrastructure. In K.-U. Schrogl, P. L. Hays, J. Robinson, D. Moura, & C. Giannopapa (Eds.), *Handbook of space security* (pp. 187–201). New York: Springer New York.

Hokstad, P., Utne, I. B., & Vatn, J. (Eds.). (2012). *Risk and interdependencies in critical infrastructures: A guideline for analysis*. London: Springer.

Johnsen, S. (2010). Resilience in risk analysis and risk assessment. In T. Moore & S. Shenoi (Eds.), *Critical infrastructure protection IV—Fourth annual IFIP WG 11.10 international conference on critical infrastructure protection*, p. 211–227, Washington DC, US: 15–17 March 2010, IFIP Advances in Information and Communication Technology 311, Springer 2010. ISBN 13 978-3-642-16805-5.

Johnson, J., & Gheorghe, A. V. (2013). Antifragility analysis and measurement framework for systems of systems. *International Journal of Disaster Risk Science, 4*(4), 159–168.

Katina, P. F., & Keating, C. B. (2015). Critical infrastructures: A perspective from systems of systems. *International Journal of Critical Infrastructures, 11*(4), 316–344.

Katina, P. F., Pinto, C. A., Bradley, J. M., & Hester, P. T. (2014). Interdependency-induced risk with applications to healthcare. *International Journal of Critical Infrastructure Protection, 7*(1), 12–26.

Katina, P. F., Keating, C. B., & Gheorghe, A. V. (2016). Cyber-physical systems: Complex system governance as an integrating construct. In H. Yang, Z. Kong, & M. D. Sarder (Eds.), *Proceedings of the 2016 Industrial and Systems Engineering Research Conference*. Anaheim, CA: IISE.

Knabb, R. D., Rhome, J. R., & Brown, D. P. (2011). *Tropical cyclone report Hurricane Katrina 23–30 August 2005* (pp. 1–43). Washington, DC: National Oceanic and Atmospheric Administration. Retrieved from http://www.nhc.noaa.gov/data/tcr/AL122005_Katrina.pdf.

Masys, A. (Ed.). (2015). *Disaster management: Enabling resilience*. New York, NY: Springer International Publishing.

McCormick, R. (2014). *Croatia under ante pavelic: America, the ustase and croatian genocide*. New York: I.B. Tauris.

Mendonca, D., & Wallace, W. A. (2006). Impacts of the 2001 World trade center attack on New York city critical infrastructures. *Journal of Infrastructure Systems, 12*(4), 260–270.

Moteff, J., & Parfomak, P. (2004). *Critical infrastructure and key assets: Definition and identification*. Washington: DC: Congressional Research Service. Retrieved from http://www.fas.org/sgp/crs/RL32631.pdf.

Moteff, J. D., Copeland, C., & Fischer, J. (2003). *Critical infrastructures: What makes an infrastructure critical?* (pp. 1–17). Washington, DC: The Library of Congress.

Perrow, C. (1999). *Normal accidents: Living with high risk technologies*. New Jersey, NY: Princeton University Press.

Rasmussen, J., & Batstone, R. (1989). *Why do complex organisational systems fail?* World Bank Environmental Working Paper, No. 20.

Richardson, B. (1994). Socio-technical disasters: Profile and prevalence. *Disaster Prevention and Management: An International Journal, 3*(4), 41–69.

Rinaldi, S. M. (2004). Modeling and simulating critical infrastructures and their interdependencies. In *Proceedings of the 37th Hawaii International Conference on System Sciences* (pp. 1–8). Big Island, Hawaii. https://doi.org/10.1109/HICSS.2004.1265180.

Rinaldi, S. M., Peerenboom, J., & Kelly, T. K. (2001). Identifying, understanding, and analyzing critical infrastructure interdependencies. *IEEE Control Systems, 21*(6), 11–25.

Robert, B., Morabito, L., & Quenneville, O. (2007). The preventive approach to risks related to interdependent infrastructures. *International Journal of Emergency Management, 4*(2), 166–182.

Sandage, S. A. (2006). Born losers: *A history of failure in America*. Cambridge: Harvard University Press.

Schmieer, M. (2015). *Inventory of space based assets (infrastructures and services) to be classified as part of a critical infrastructure*. Ispra: Institute for the Protection and Security of the Citizen, European Commission Joint Research Center.

Sousa-Poza, A. A., Kovacic, S., & Keating, C. B. (2008). System of systems engineering: An emerging multidiscipline. *International Journal of System of Systems Engineering, 1*(1/2), 1–17.
Sussman, J. M. (2005). *Perspectives on intelligent transportation systems*. New York, NY: Springer.
Thissen, W. A., & Herder, P. M. (2003). *Critical infrastructures: State of the art in research and application*. Boston, MA: Kluwer Academic Publishers.
Townsend, F. F. (2006). *The federal response to Hurricane Katrina: Lessons learned*. Washington, DC: US Government Printing Office.
US Congress. (2001). *Uniting and strengthening America by providing appropriate tools required to intercept and obstruct terrorism (USA PATRIOT ACT) Act of 2001* (No. 147) (p. 115 Stat. 271–402). Washington, DC: 107th Congress. Retrieved from http://www.gpo.gov/fdsys/pkg/PLAW-107publ56/content-detail.html.
USDHS. (2002). *Homeland security act of 2002* (No. public law 107–296) (p. 116 Stat. 2135–2321). Washington, DC: 107th Congress. Retrieved from https://www.dhs.gov/homeland-security-act-2002.
USDHS. (2013). *NIPP 2013: Partnering for critical infrastructure security and resilience*. Washington, D.C.: U.S. Dept. of Homeland Security. Retrieved from http://www.dhs.gov/xlibrary/assets/nipp-ssp-national-monuments-icons.pdf
Vugrin, E., Wahren, D., & Ehlen, M. (2010), *A resilience assessment framework for infrastructure and economic systems: Quantitative and qualitative resilience analysis of petrochemical supply chains to a Hurricane*. Sandia National Labs: 6th Global Congress of Process Security organized by the American Institute for Chemical Engineers, March 2010.
Weick, K. E. (1988). Enacted sensemaking in crisis situations. *Journal of Management Studies, 25*(4), 305–317. https://doi.org/10.1111/j.1467-6486.1988.tb00039.x.
Zio, E. (2016). Critical infrastructures vulnerability and risk analysis. *European Journal for Security Research*, 1–18.

Chapter 2
Critical Space Infrastructures

Abstract This chapter provides the basics of space as a critical infrastructure including elements of key resources and assets. Critical space infrastructure (CSI) is presented as a set of interdependent system-of-systems encompassing workforce, environment, facilities and multidirectional interactions essential for the maintenance of vital societal functions, health, safety, security, economic or social well-being of people, whose destruction or disruption would have a significant impact in a given state. Topics of orbits are also discussed in the context of critical infrastructures.

2.1 The Basics of CSI

Space systems have become key enablers for a wide variety of commercial, scientific and military applications. The rapid growth of their capabilities has offset some of the size of the required investment and new developments promise an even greater reduction in the cost of space infrastructure. As such, some of the extant space systems have become deeply embedded in the functioning of advanced societies, supporting economies, lifestyles and governance processes. The increasing dependence on certain space systems places them firmly in the area of critical infrastructure, whose disruption or destruction would generate lasting damage. This inclusion into critical infrastructure theory is even more warranted as space systems have become a technological backbone for existing recognized critical infrastructures, such as energy, transportation, administration and others. Specifically, Booz and Company suggest that 'satellites will and must be an integral part of the future communications ecosystem' (Acker et al. 2011). The increasing use of satellites is also depicted in Fig. 2.1.

The reliance of infrastructure systems-of-systems on space-based command, coordination and control capabilities during normal functioning, but especially during emergency and crisis situation management processes, means that space systems fulfil the requirements for critical status. We propose to include space systems into the existing critical infrastructure protection framework, while also underlining the key differences between "terrestrial" infrastructures and extra-atmospheric ones. The geographic and economic realities of space systems also call for a different approach

© Springer Nature Switzerland AG 2019

A. Georgescu et al., *Critical Space Infrastructures*,
Topics in Safety, Risk, Reliability and Quality 36,
https://doi.org/10.1007/978-3-030-12604-9_2

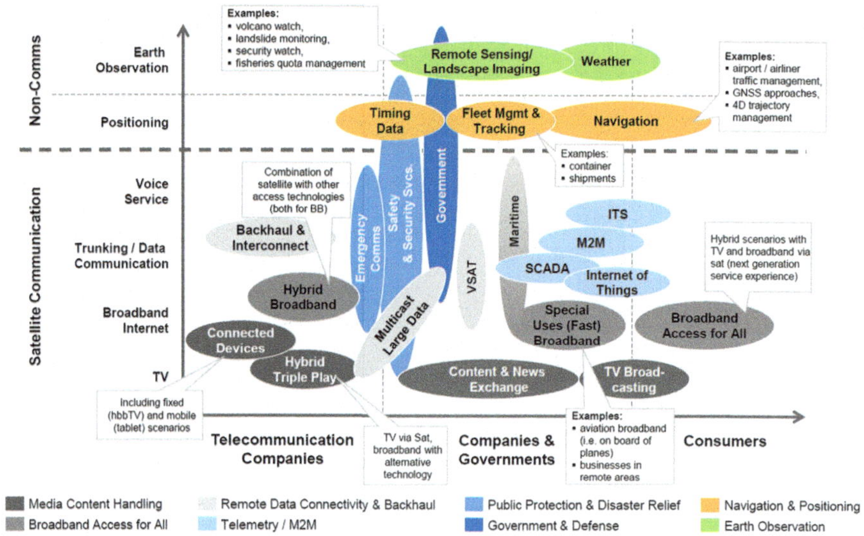

Fig. 2.1 The increasing satellite use cases. Figure adapted from Acker et al. (2011, p. 5)

to protection efforts, based on international cooperation and collective action. Finally, 'critical space infrastructures' are also faced with significant specific risks, such as orbital debris and the effects of 'space weather' phenomena which have attracted scientific attention, but limited awareness from political decision makers.

Most of the existing CSI are located in close proximity to Earth, where they provide critical services to earth-based consumers. Only a few of the systems not present in the immediate vicinity of Earth may be considered critical, and they will be detailed in the appropriate heading. In time, more CSI will be constructed in positions other than Earth-centric, and may even begin to provide critical material goods, in addition to services. For the time being, however, CSI constitute a limited pool of significant assets concentrating valuable services for a multitude of users. While most of them are not run with a monetary benefit in mind, it can be considered that, given the current state of fact, some of the existing CSI are the greatest creators of added value for their users in history, which serves to enhance the perception of their criticality.

Their central importance to communications, coordination (and so on) have made us wholly dependent on the operational capacity and resilience of a limited pool of assets in the most hostile environment known to men. It is so hostile, that it features two unique threats to infrastructure operation—space debris and space weather, the latter of which can also impact terrestrial systems. Space systems are, in many respects, unlike their terrestrial counterparts, and these particular traits must be taken into account before successful Critical Infrastructure Protection activities are undertaken. The following is a non-exhaustive account of the vast differences between critical space systems and terrestrial infrastructures:

- Infrastructure development in space face a significant economic barrier of entry, relating to the cost of the system itself and that of its placement in the desired vector. The consequences of this reality are manifold
- The marginal productivity of a new space system must be very high to justify the cost, meaning that the number of satellite units in space is often the minimum required to meet demand
- The number of active satellites is around 1,800 (with a big increase in recent years as marked in industry analyses because of the interest shown in cubesats, a platform of limited capabilities but affordable cost) which is a vanishingly small inventory, by numbers, volume and weight, which provides the entirety of the global community with services such as communications, positioning, navigation, timing, weather forecasting, disaster monitoring, early warning, emergency response coordination, search and rescue and myriad other services. While this means that every satellite is an extremely useful asset, it is also true that their critical nature is heightened, as well as the vulnerability of the provision of critical services, since replacing a defective system in a timely manner is also very difficult
- The weight limitations of space launch systems mean that every system must be tailor-made to suit its unique mission profile. This limits the ability for other satellites, outside the mission specific units, to serve as substitutes or extra capacity to increase the resilience of the system-of-systems. In turn, this lack of compatibility has also been a hindrance in the development of technical standards and standardized components, which would have created economies of scale that would reduce the price of individual systems. Another consequence is the current lack of refueling capability, which limits the useful lifetime of satellites even when the components themselves are still operational
- Space systems are expected to be continuously active in one of the most hostile environments known to man, replete with kinetic and electromagnetic threats which occur on a daily basis
- Even in the absence of a particular threat, the numerous stressors of the environment and the intricacy of the components of space systems ensure that sudden and unexplainable malfunctions occur often
- While being thought of as a vast place, the majority of space systems are crowded in a number of orbital bands, valuable because they pass directly above important markets or areas of scientific and military interest. This means that accidental collisions, not just with pieces of debris, but also between satellites, are possible. The company that maintains the Iridium satellite constellation reports that their units pass within 5 km of another satellite on a weekly basis, and the inactive Russian satellite which destroyed the Iridium 33 unit was anticipated to pass within 500 m of the satellites, at speeds of thousands of kilometers an hour
- The orbital mechanics of space systems also mean that space is a very international environment. Countries with no space program depend directly on space systems which are under the sovereignty of another country and are the property of a foreign company operating under foreign laws. Even though such a situation may be encountered in transcontinental infrastructures on Earth, such as pipelines, the level of disconnect between stakeholder authority and dependence on a certain

Table 2.1 Satellite quick facts (Figures include launches through 4/30/2018. *Source The Union of Concerned Scientists*: https://www.ucsusa.org/)

Total numbers of operating satellites: 1,886			
USA: 859	China: 250	Russia: 146	Other: 631
LEO: 1,186	MEO: 112	Elliptical: 40	GEO: 548
Total Number of U.S. satellites: 859			
Civil: 20	Commercial: 495	Government: 178	Military: 166

system is taken to its maximum when it comes to space assets. This complicates the Critical Infrastructure Protection activities and vulnerability analyses of countries without space development capabilities, which is an impetus both for international cooperation in CIP, but also for national scientific development to reduce some of the gaps

- A majority of space systems and their underlying technologies are dual use, both civilian and military. The most widely used GNSS system, the American GPS, became a huge success and innovation in the private sector when the military, which operates the network of satellites, decided to improve the quality of signals to civilian users. The military still retains the authority to disrupt service even to allies in case of national emergency. On the other hand, this means that the military, in addition to its own systems, is a significant consumer of services from private operators of space infrastructures. This dependency breeds new vulnerabilities.

The Union of Concerned Scientists[1] maintains an open source database of all satellites in Earth's orbit, though the current numbers are just a fraction of the number of satellites launched throughout the last half century (see Table 2.1). Space systems (mainly satellites) are the latest critical infrastructure to be defined, a development owing partially to groundbreaking work in Romania, with the project "Space Systems as Critical Infrastructure" (2011–2016) lead by the Romanian Space Agency (RoSA), with the EURISC Foundation and the Military Equipment and Technologies Research Agency. While this new domain has grown in scope, future development of these concepts in a way that services both identified Romanian needs in this domain and the accessing of opportunities to generate new capabilities, new products and new services is now being undertaken by the "Critical Space Infrastructure Protection at ROSA" project which will create a center of excellence in this field.

Crucially, from the perspective of resilience (the ability of a society to withstand shocks and negative events with minimal losses and rapid resumption of normal operations), space systems have become an upper layer of command, control and co-ordination capabilities which are increasing, due to technical prowess, the efficiency and cost effectiveness of many critical-infrastructure systems. This engenders a critical dependency on these capabilities, suggesting that space systems are themselves becoming critical infrastructure and, therefore, are not just part of a solution, but also part of the problem. Both the EU and national authorities (overwhelmingly those of the most developed states, but also the strongest of the emerging world),

[1] https://www.ucsusa.org/.

Fig. 2.2 Member companies of the Satellite Industry Association in the USA (https://www.sia.org/wp-content/uploads/2017/03/SSIR-2016-update.pdf)

which are the main drivers of critical infrastructure protection efforts and research, are increasingly coming around to this development and scrambling to update their mental modes, organizations and documents of reference to reflect this new variable.[2]

The yearly State of the Satellite Industry[3] report underlines the rapid development and growth of the satellite service industry and the extent to which it is a private industry, something we have termed CSI, operated and administered by companies operating under normal economic incentives, as opposed to the government run basic science missions that people normally envision in orbit. Large-scale private ownership is a reality (see Fig. 2.2) also for terrestrial critical infrastructures, like power plants, ports, farms and other important assets, which will affect considerations regarding security governance and security resources.

The 2016 report, which only covers private industry, not state activities, highlights that, in 2015, as in every year before that, the industry's growth has outpaced global or US growth, reaching 208 billion dollars, almost twice what it was in 2006 (see Figs. 2.3 and 2.4). The number of satellites which were launched stayed roughly the same as in 2014, being over 200, but this ignores the fact that the number doubled after 2013, due to large scale introduction of CubeSats, very small satellites that are more affordable and can be launched in large groups, lowering the financial cost of access to space tremendously for marginal operators (universities, small countries etc.). Fifty three percent of satellites launched in 2015 were CubeSats,

[2]These comments are based on a presentation by Martin Nix http://www.acser-archive.unsw.edu.au/snap/1-03-MNix.pdf.

[3]http://www.sia.org/wp-content/uploads/2017/03/SSIR-2016-update.pdf.

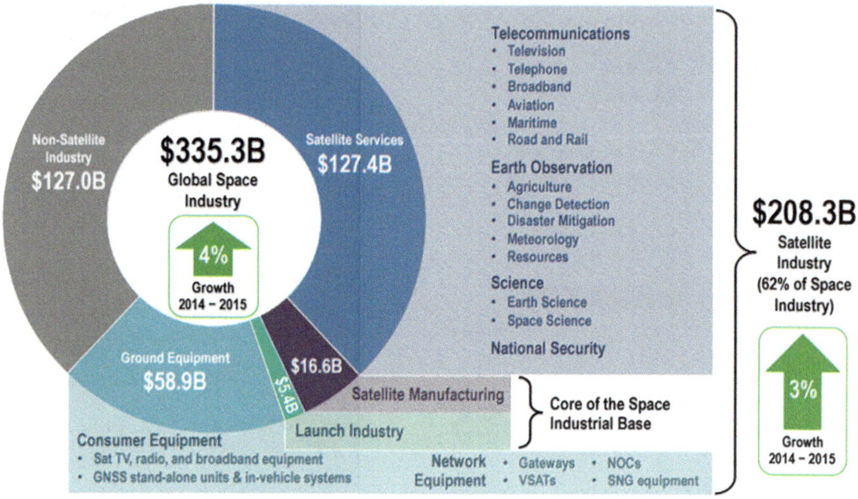

Fig. 2.3 Breakdown of revenue of the global satellite industry (https://www.sia.org/wp-content/uploads/2017/03/SSIR-2016-update.pdf)

and the area has significant potential for growth, in addition to the more financially and technologically intensive large satellite systems.

The 2018 report on the State of the Satellite Industry (Fig. 2.4) emphasizes the dynamics of growth in the hardware and services sectors of the satellite industry, outstripping global growth and not reacting to downturns. Figure 2.5 shows countries with satellites and space launching countries with satellites.

As mentioned before, the risk profile of space systems is unique. Consequently, those systems which appropriate methodologies have identified as CSI in order for them to be officially designated as such by competent authorities and included in protection efforts, feature an assortment of challenges which require collective protection efforts by emphasizing the limitations of individual national protection initiatives.

2.2 Orbits—A Key Asset and Resource

While most of these space systems are custom built to the specifications of their owner and, as such, are unique except within their family of units, the greatest distinguishing feature of most CSI in close proximity to Earth is their orbit, which is a description of their trajectory around the Earth and movement through space. Figure 2.6 provides satellite capability applications (Acker et al. 2011).

The common orbits include geostationary, low-earth, circular medium, polar, molniya, tundra, sun-synchronous, and Lagrange points.

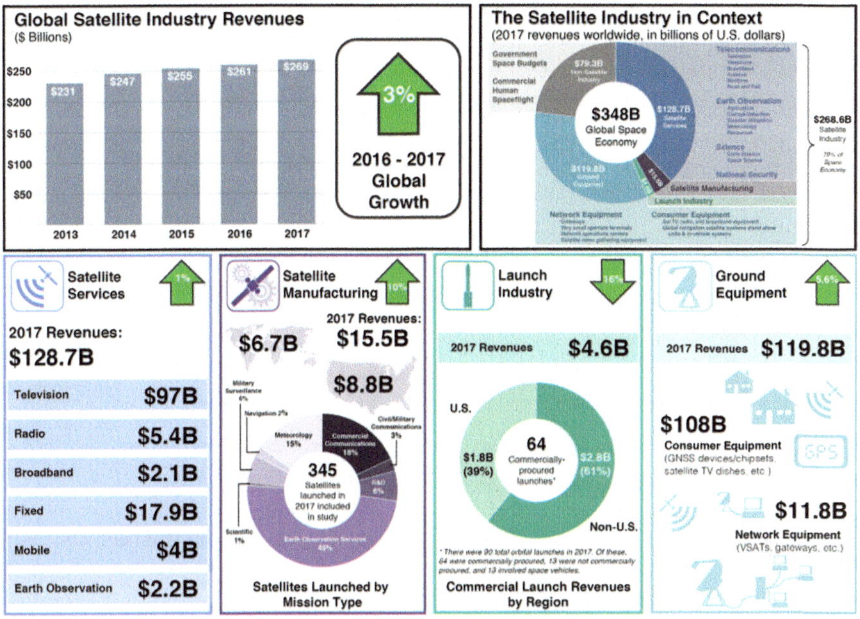

Fig. 2.4 State of the Satellite Industry 2018 Report vital statistics (Bryce 2018)

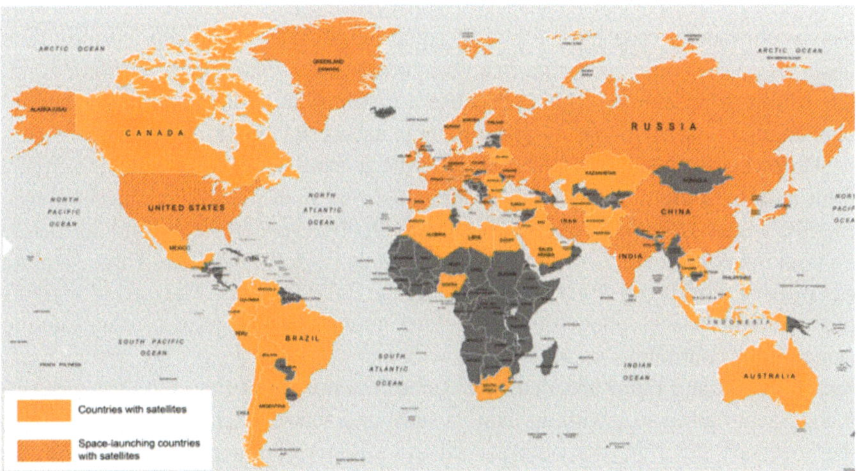

Fig. 2.5 List of countries which own satellites (*Source* https://www.ucsusa.org/)

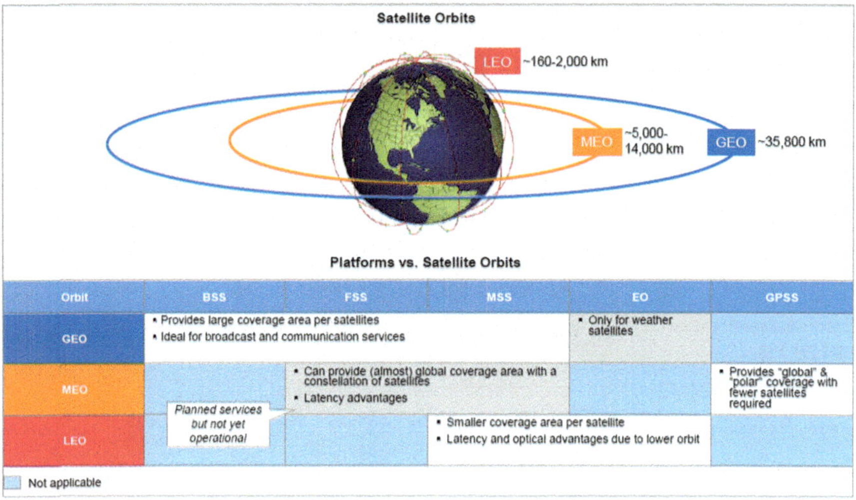

Fig. 2.6 Main applications for satellite capabilities, depending on orbit. Adapted from Acker et al. (2011)

2.2.1 Geostationary Orbits

Geosynchronous orbits have a period equal to the Earth's rotation period. To an observer on the earth, a satellite in a geosynchronous orbit appears motionless, in a fixed position in the sky. This is because it revolves around the earth at the earth's own angular velocity (360° every 24 h, in an equatorial orbit).

The most useful geosynchronous orbit is the geostationary orbit, which is a circular orbit at an altitude of 35,786 km in the equatorial plane. Because a geostationary satellite appears as a fixed point in the sky to all observers on the ground, users need no tracking equipment to send or receive signals from the satellite. Three satellites can provide worldwide coverage, excluding the Polar Regions. The area of visibility of the satellite is large, about 43% of the Earth's surface. Thus, geostationary satellites can provide continuous service over a wide geographical area. This is very useful for television and radio broadcasting, since it permits real-time data transfer over a wide geographic area without using a store-and-forward scheme. It also provides the necessary flexibility for commercial and military communications, which need to support users from widely different, non-predetermined locations. Geostationary satellites operate outside the densest regions of the Van Allen belt, but they are subject to infrequent bursts of high-energy particles from the Sun that can damage or degrade them. Table 2.2 discusses advantages and disadvantages associated with geosynchronous orbits when it comes to the current discussion.

Table 2.2 Utility associated with geosynchronous orbits

Advantages	Disadvantages
Ground antennas can be aimed at the satellite in a geostationary orbit without their having to track the satellite's motion. This is relatively inexpensive. In applications that require a large number of ground antennas, such as Direct TV distribution, the savings in ground equipment can more than outweigh the cost and complexity of placing a satellite into orbit	The main drawback of a geostationary orbit is that a satellite cannot service extreme northern and southern areas of the world. Another drawback is the height of the orbit, which usually requires more powerful transmitters, larger-than-normal (usually dish) antennas, and higher-sensitivity receivers on the earth. The large distance also introduces a significant delay, of approximately 0.25 s, into communications

2.2.2 *Low-Earth Orbits*

A Low Earth Orbit (LEO) typically is a circular orbit about 400 km above the Earth's surface and, correspondingly, a period (time to revolve around the earth) of about 90 min. Satellites at orbital heights of a few thousand kilometers could also be said to be in low earth orbits, but few satellites populate this part of space because of the large amount of radiation there. Because of their low altitude, these satellites are only visible from within a radius of roughly 1000 km from the sub-satellite point. In addition, satellites in low earth orbit change their position relative to the ground position quickly. So even for local applications, a large number of satellites are needed if the mission requires uninterrupted connectivity. On the other hand, space at these altitudes is mostly free from high radiation and charged particles.

Low earth orbiting satellites are less expensive to launch into orbit than geostationary satellites and, due to proximity to the ground, do not require as high signal strength (since signal strength falls off as the square of the distance from the source). Thus, there is a trade-off between the number of satellites and their cost. In addition, there are important differences in the on-board and ground equipment needed to support the two types of missions.

Since a satellite in LEO cannot see a large ground area and since it moves relative to the Earth's surface, LEO may not seem to be useful for missions such as communications. However, a network that contains enough LEO satellites to see all regions of the Earth and that can relay signals between the satellites can provide continuous worldwide coverage. A group of satellites working in concert is known as a satellite constellation. If such a network includes polar or near-polar orbits, it can also provide coverage of polar and high latitude regions, as geostationary satellites cannot. Because they are in low orbits, the round-trip transmission time from these satellites is relatively short (0.005 s to and from the ground), eliminating the need for echo control or other special treatment. (The time required for signals transmitted over long distances around the Earth, when relayed through multiple satellites, is dominated by the distance along the Earth rather than the altitude of the satellite: transmission halfway around the Earth—20,000 km—requires at least 0.067 s.[4])

Moreover, if some of the satellites are on highly inclined orbits, observers at high latitudes can see the satellites at high elevation angles, which reduces interference with the signals by buildings and other objects. These qualities make LEO orbits useful for personal communications systems. The disadvantage of using LEO satellites for this purpose is that the network requires many satellites. Recall that any observer sees a satellite passing overhead for roughly 10 min out of its 90-min orbit, so nine satellites would be required to provide continuous coverage of a single band on the Earth around its ground track (for an orbital altitude of 500 km, the width of this band is roughly 3,000 km). For broader coverage, considerably more satellites would be needed. For example, the Iridium constellation, which is used for a variety of military and commercial purposes, has 66 satellites distributed in six different orbits with an altitude of 780 km. The six orbits are in six different orbital planes, each at an inclination angle of 86.4°.[5] Another LEO satellite constellation is Global star, with 60 satellites. An interesting case is the network known as Teledesic, started with backing from Microsoft entrepreneur Paul Allen, which was to have over 840 satellites. This was later scaled back to 288 and ultimately ended up only launching one test satellite.

It is also possible to offer discontinuous coverage using a low Earth orbit satellite capable of storing data received while passing over one part of Earth and transmitting it later while passing over another part. This arrangement is known as 'store-and-forward.' This will be the case with the CASCADE system of Canada's CASSIOPE communications satellite. Another system using this store and forward method is Orbcomm. For missions that are not time critical, the motion of the LEO satellites relative to the Earth means that a single satellite in polar orbit can cover the entire Earth. If the orbital period is chosen so that the ground coverage areas on successive orbits lie next to each other, a satellite in a polar orbit can see any spot-on Earth twice a day. Some missions require low orbits. Earth observation and reconnaissance satellites intended to take high-resolution images of the Earth must be close to the Earth to get such resolution. For example, the U.S. Keyhole satellites, which took optical photographs for intelligence purposes, were usually deployed in elliptical orbits with an apogee and perigee at 1,000 and 300 km, respectively.[6] These have been replaced by a new generation of imaging satellites in similar orbits. Since these satellites move with respect to the Earth, they cannot offer continuous coverage of a particular area.

---

[4]The relay process, which requires each relay satellite to receive and retransmit the signal, also adds to the transmission time.

[5]Iridium Satellite Solutions: http://www.iridium.com, accessed January 15, 2005.

[6]Federation of American Scientists, "KH-11 Kennan/Crystal Satellites," http://www.fas.org/spp/military/program/imint/kh-11.htm, accessed January 15, 2005.

2.2.3 Circular Medium Orbits

Satellites in circular medium earth orbits (MEO), also termed intermediate circular orbits (ICO), have altitudes between those of low earth orbits and geosynchronous orbits: from roughly 1,500–36,000 km. A common orbit is one with an altitude of roughly 10,000 km and an orbital period of about 6 h. Continuous worldwide real-time coverage can be obtained with fewer satellites than are needed for a constellation of satellites in low earth orbits. For example, the ICO communications satellite system under construction will consist of 10 satellites in 2 orbits at an altitude of 10,390 km. The two orbital planes will be at 45° inclination, rotated 180° around the Earth's axis with respect to one another.[7]

Satellites in such medium earth orbits are relatively slow moving as seen from the Earth, thus requiring fewer and simpler handover arrangements than a LEO system. The round-trip transmission time to these satellites from the ground is longer than to a satellite in low earth orbit: the ICO transmission time is 0.069 s, whereas for the Iridium system it is 0.0052 s. This longer transmission time is less of an issue for communications over long distances (a signal traveling halfway around the world would along the Earth's surface require a minimum of 0.067 s, comparable to the time it takes for a round trip to the ICO satellite) and using higher altitude satellites reduces the number of satellites the signal must be relayed between to cover long distances. However, satellites in MEO orbits must employ radiation-hardened components (particularly to protect their computer systems) to survive long term.

A special type of medium earth orbit is the semi-synchronous orbit, which has a period of 12 h and an altitude of roughly 20,000 km. Both the U.S. NAVSTAR Global Positioning System (GPS) and Russian GLONASS navigational satellites use these orbits. A navigational system needs at least four satellites within view of the user at all times, where a continuous communications system needs only one. Thus, a navigational system requires more satellites than does a communications system deployed at the same altitude: both GPS and GLONASS (when fully deployed) use 24 satellites. The GPS satellites are in six orbital planes at an inclination angle of 55°; GLONASS is designed to use three orbital planes at an inclination angle of 65°.

2.2.4 Polar Orbits

Polar orbiting satellites circle the Earth at a typical altitude of 850 km (530 miles) in a north to south (or vice versa) path, passing over the poles in their continuous flight. Polar satellites are in sun-synchronous orbits, which means they are able to observe any place on Earth and will view every location twice each day with the same general lighting conditions due to the near-constant local solar time. Polar orbiting

---

[7]ICO Satellite Wireless Services, http://www.ico.com, accessed January 15, 2005, and "Lloyd's Satellite Constellations," http://www.ee.surrey.ac.uk/Personal/L.Wood/constellations/ico.html, accessed January 15, 2005.

weather satellites offer a much better resolution than their geostationary counterparts due their closeness to the Earth.

2.2.5 Molniya Orbits

Geostationary satellites must operate above the equator and will therefore appear lower on the horizon as the receiver gets the farther from the equator. This will cause problems for extreme northerly latitudes, affecting connectivity and causing multipath (interference caused by signals reflecting off the ground and into the ground antenna). For areas close to the North (and South) Pole, a geostationary satellite may appear below the horizon. Therefore, Molniya orbit satellites have been launched, mainly in Russia, to solve this problem. The Soviet Union first used this type of orbit for its Molniya satellite system, hence the name. Molniya orbits are highly elliptical, with a period of 12 h and an inclination of 63.4°.

At this inclination, the apogee remains over the same latitude in the northern (or southern) hemisphere. The Molniya orbit is highly inclined, guaranteeing good elevation over selected positions during the northern portion of the orbit. (Elevation is the extent of the satellite's position above the horizon. Thus, a satellite at the horizon has zero elevation and a satellite directly overhead has elevation of 90°). They are sometimes referred to as highly elliptical orbits (HEO). A satellite in a highly elliptical orbit with the apogee over the northern hemisphere covers Earth's high-latitude regions for a large fraction of its orbital period. The speed of a satellite is not constant on an elliptical orbit. The satellite has a high speed as it traverses the orbit near perigee and moves slowly near apogee—thus spending most of its time in the sky over the northern hemisphere. Its period is one half day, so that the satellite is available for operation over the targeted region for six to nine hours every second revolution. In this way a constellation of three Molniya satellites (plus in-orbit spares), with their major axes oriented at 120° with respect to each other, can provide uninterrupted coverage.

The first satellite of the Molniya series was launched on April 23, 1965 and was used for experimental transmission of TV signal from a Moscow uplink station to downlink stations located in Siberia and the Russian Far East, in Norilsk, Khabarovsk, Magadan and Vladivostok. In November 1967 Soviet engineers created a unique system of national TV network of satellite television, called Orbita, which was based on Molniya satellites. These satellites were in orbits with an apogee of roughly 40,000 km and a perigee of roughly 1,000 km (or an eccentricity of 0.75). For eight of their 12-h periods, each satellite remains visible to the regions under the apogee, with elevation angles above 70°.

Molniya orbits are also used by U.S. intelligence satellites that monitor Russia and by Russian early warning satellites that watch for U.S. missile launches.

2.2.6 Tundra Orbits

Like Molniya orbits, Tundra orbits have an inclination of 63.4°, so their apogees remain over one hemisphere. They are typically used to provide coverage of high latitude areas, with their apogee over the northern hemisphere. However, they are not as highly elliptical as are the Molniya orbits, and their period is 24 hour rather than 12. Satellites in Tundra orbits are visible to the regions under the apogee for 12 of their 24-hour periods. Thus, it is possible to obtain continuous coverage of this region with only two satellites whose orbits are rotated 180° with respect to each other. The Russian Tundra system uses two satellites in orbits with an apogee and perigee of roughly 54,000 and 18,000 km, respectively.

There are a further two orbits to be considered, as they are or will be popular destination for space missions, even though they are not in the vicinity of Earth.

2.2.7 Sun-Synchronous Orbits

Satellites in sun-synchronous orbits pass over a given part of the Earth at roughly the same local time of day (though not necessarily every day). That is, whenever the satellite observes a given ground location, the Sun is always in the same location in the sky. Such orbits are particularly useful for missions that take images of the Earth, because shadows from objects at a given location on the Earth's surface are always cast from the same angle. This simplifies the comparison of images taken on different days to detect changes. Satellites in these orbits are often placed at low altitudes (with short periods) so that they provide complete coverage of the Earth's surface at least once per day. The inclination of sun-synchronous orbits is chosen so that the precession of the orbital plane around the Earth due to gravitational irregularities keeps the plane at a constant angle with respect to a line between the Earth and Sun throughout the year. The precise inclination that produces this effect depends on the orbit's altitude and eccentricity; it is typically 96°–98°, making the orbits slightly retrograde. Figure 2.7 illustrates how a non-precessing orbit differs from an orbit that precesses synchronously with the Sun. The left-side panel shows a case in which the satellite's orbit does not precess and remains in a fixed orientation with respect to space. Thus, a satellite that is directly above a location on the Earth when the local time is midnight and noon, would four months later observe this location when the local time is 6 am and 6 pm. The right-side panel shows a sun-synchronous orbit. The orbit is in a plane chosen to precess at a rate synchronized with the Earth's trip around the Sun, so that the plane maintains a constant angle throughout the year with respect to a line between the Earth and Sun. As a result, during the entire year, this satellite observes a point on the Earth at the same local time.

In a special sun-synchronous orbit, called a dawn-to-dusk orbit, the satellite's orbital plane coincides with the plane that divides the half of the Earth that is illuminated by the Sun from the half that is dark. If the plane were aligned slightly

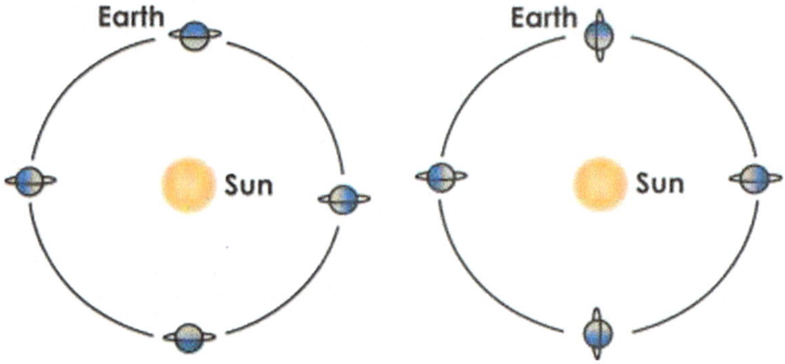

Fig. 2.7 Both panels show the Earth at four positions in its yearly orbit around the Sun, and the orbital plane of the same satellite in each case

differently, the satellite would spend half of its time in full sunlight and half in shadow, but a dawn-to-dusk orbit allows the satellite to always have its solar panels illuminated by the Sun. For example, the Canadian Radarsat Earth observation satellites use such a dawn-to-dusk orbit to keep their solar panels facing the Sun almost constantly, so they can rely primarily on solar power and not on batteries.[8]

2.2.8 Lagrange Points

There are five special orbits in which satellites orbit not the Earth but the Sun, and do so in a way that they maintain a fixed position relative to the Earth as it orbits the Sun. These fixed locations are called Lagrange points; there are five such points, one corresponding to each of the five orbits (see Fig. 2.8). A satellite orbiting the Sun closer than the Earth does has a shorter orbital period than the Earth's. However, such a satellite is pulled by the Earth's gravitational field as well as by that of the Sun. This effect is negligible if the satellite is far from the Earth but must be taken into account for a satellite close to Earth. For a satellite directly between the Earth and Sun, the direction of the Earth's pull is exactly opposite that from the Sun, effectively cancelling some of the Sun's gravitational pull.

At the first Lagrange point (L_1), the net gravitational force on the satellite is the same as the Sun's gravitational force on the Earth, so that the satellite orbits the Sun with the same orbital period as the Earth. There are analogous Lagrange points for the Earth-Moon system. These points are near the Moon and stationary with respect to it. The Lagrange points discussed here are all in the Earth-Sun system. The L_3 point, which lies on the other side of the Sun, directly opposite the Earth, is not very useful for satellites. The L_4 and L_5 points are along the Earth's orbit but precede

[8]Additional information on Radarsat can be found at the Radarsat International website http://www.radarsat2.info/, accessed January 15, 2005.

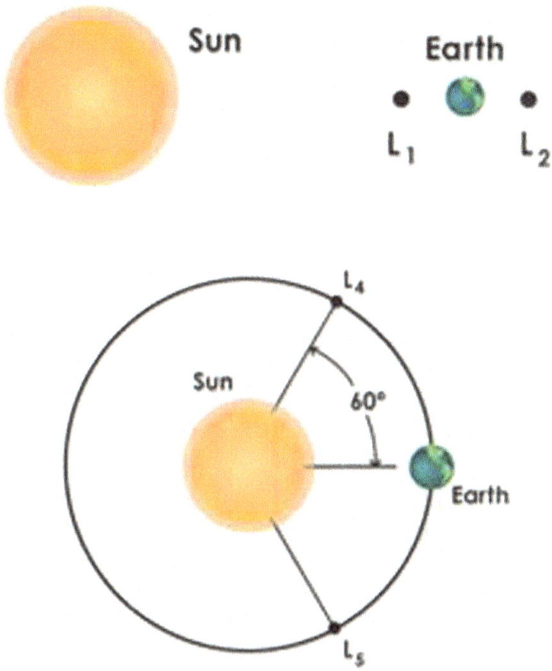

Fig. 2.8 The L_1, L_2, L_3, L_4 and L_5 Lagrange points for the Earth-Sun system

and lag it. They are 60° away from the Earth-Sun line. Some suggest the L_2 will be strategically interesting for space exploration or for space militarization.

A satellite in this position stays with the Earth throughout its journey around the Sun. The L_1 point is about four times more distant from the Earth than the Moon is. The L_1 point is particularly useful for scientific missions that study the Sun, and satellites positioned there can give early warning of increased solar winds. A second Lagrange point is located the same distance from the Earth but on the other side, directly away from the Sun. In this case, the Earth's gravitational pull adds to that of the Sun, increasing the orbital speed required for the satellite to stay in orbit. In this case, the satellite keeps up with the Earth in its orbit, while it would normally fall behind. Scientific missions are positioned there as well, allowing the satellite to be maximally far from the Earth (to minimize interference), but maintain constant contact. NASA plans to place its Next Generation Space Telescope (NGST), the successor to the orbiting Hubble telescope at or near L_2.[9]

Since craft at L_2 are in a stable position and need little fuel to remain there for extended periods of time, L_2 could be used as a place to assemble other spacecraft from parts lifted piece by piece. Such a scheme could be more energy efficient than trying to assemble large structures on the Moon and more feasible than assembling

[9]For example, see James Oberg's '*Will China's Space Plan Skip the Moon?*' at http://www.jamesoberg.com/05242004skipmoon_chi.html.

them on Earth and then launching them. Objects at L_2 are also out of easy observation by the Earth, being quite distant.

Final Remarks

Globally, the growth of space-related industry has outpaced world economies reaching 208 billion dollars, almost twice what it was in 2006. Failure to recognize critical space infrastructure can only delay the needed research along emerging risks and vulnerabilities along with approaches for analysis and assessment, remediation, indicators and warning systems, mitigation, and incident response, and reconstruction. These issues along with the interdependent nature key resources and asserts, presents a fatal ground for space as a critical infrastructure; a set of interdependent system-of-systems encompassing workforce, environment, facilities and multidirectional interactions essential for the maintenance of vital societal functions, health, safety, security, economic or social well-being of people, whose destruction or disruption would have a significant impact in a given state. This vast space features a host of threat. Two unique threats, space debris and space weather are known to have a direct impact on terrestrial systems—some of with are critical. However, there remains unknown threats due to lack of research.

References

Acker, O., Pötscher, F., & Lefort, T. (2011). *Why satellites matter: The relevance of commercial satellites in the 21st century—A perspective 2012–2020*. New York City, NY: Booz & Company. Retrieved from https://www.esoa.net/Resources/Why-Satellites-Matter-Full-Report.pdf.

Bryce Space and Technology. (2018). *2018 State of the Satellite Industry Report*, Report prepared for the Satellite Industry Association. Retrieved from https://brycetech.com/downloads/SIA_SSIR_2018.pdf.

Chapter 3
Critical Space Infrastructure Taxonomy

Abstract This chapter is developed to provide categories of critical space infrastructures. This include remote sensing, communication, meteorological, Global Navigation Satellite Systems, and framework for administrative and legislation. These categories are discussed in this chapter as well as their implications for the development and exploration of critical space infrastructures.

3.1 CSI Taxonomy

Taxonomies are primarily concerned with classification, especially of organisms. In the present case, however, a CSI taxonomy is related with classification of space systems. Authors suggest that space systems and their associated infrastructures are divided into the five main categories: remote sensing, communication, meteorological, Global Navigation Satellite Systems (GNSS), and administrative and legislation framework. Figure 3.1 articulates the main categories as well as the sub-categories associated with remote sensing and GNSS.

3.1.1 Satellites

The world's first artificial satellite, the Sputnik 1, was launched by the Soviet Union in 1957. Since then, thousands of satellites have been launched into orbit around the Earth. Artificial satellites originate from more than 50 countries and have used the satellite launching capabilities of ten nations. A few hundred satellites are currently operational, while thousands of unused satellites and satellite fragments orbit the Earth as space debris.

Some of the intrinsic attributes of satellites make them vulnerable in ways that ground-based systems are not. Satellites in orbit move at high speeds, rendering collisions with even small objects disastrous. Satellites are nearly impossible to hide: just as satellites can view large swaths of the Earth, they are also visible to observers

© Springer Nature Switzerland AG 2019

A. Georgescu et al., *Critical Space Infrastructures*,
Topics in Safety, Risk, Reliability and Quality 36,
https://doi.org/10.1007/978-3-030-12604-9_3

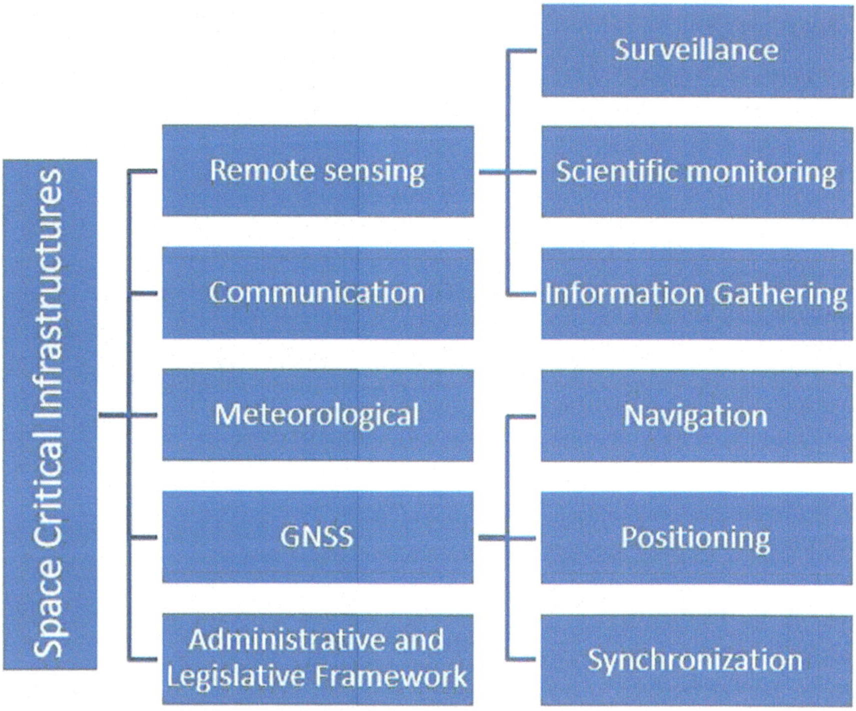

Fig. 3.1 Categories for critical space infrastructures

over large swaths of the Earth (see Fig. 3.2). The observable area is compared here for satellites at two different altitudes: the satellite at the lower altitude sees a much smaller area than the one at the higher altitude. Note that the observable area also describes the area on the Earth that can see the satellite.

Moreover, once in orbit, a satellite's motion is predictable, and it takes significant effort to appreciably change the orbit. Even small evasive maneuvers to escape an anti-satellite attack could add up to a prohibitively large effort, since an adversary can take multiple shots at the satellite. Satellites are also difficult to protect. Launch mass is at a premium, so armor and defensive measures come at some price. Some satellites, such as communications satellites, are designed to be easily accessed by users across the globe, a sensitivity that can be exploited to harm them or interfere with their operation. And essentially no satellite can now be repaired once damaged.

Satellite systems have a number of components, some of which make better targets than others. A satellite system comprises the satellite itself, the ground stations used to operate and control them, and the links between them. This section describes the components and their functions and how vulnerable and critical they are. We place particular emphasis on those elements that might be targeted and note that successful interference with a satellite system may not involve an attack on the satellite itself. Satellites vary greatly in size. For example, commercial communications satellites

Fig. 3.2 The size of the area of the Earth observable from a satellite depends on its orbital altitude

can be large. The body of a Boeing 702 communications satellite, which was first launched in 1999, is seven meters long, and its solar panels extend to a length of 48 meters. The average Boeing 702 weighs nearly 3 tons when launched (this mass includes its station keeping propellant).[1] Satellites can be small, as well. The SNAP "Nano" satellite, constructed by Surrey Satellite Technology Ltd., is only 0.33 m long, with a total mass of 6–12 kg, which includes a payload of up to 4 kg. This small satellite was placed in orbit in June 2000 and was able to maneuver, image, correctly keep attitude, and communicate with the ground.[2] All satellites have some basic elements, as outlined below and shown schematically in Fig. 3.3.

A structural subsystem or bus: The bus is a metal or composite frame on which the other elements are mounted. Because it bears the stresses of launch, the bus is generally resilient. It may be painted with reflective paint to limit the solar heat it absorbs, which could also provide some protection from laser attacks.

A thermal regulation subsystem: This system keeps the active parts of the satellite cool enough to work properly. Active satellite components such as the computer and receiver can generate a large amount of heat. Sunlight incident on the satellite's surface also generates heat, although the satellite's surface can be made highly reflective to minimize heat absorption.[3] Without an atmosphere, conduction and

[1] 'What Is a Satellite?' factsheet, http://www.boeing.com/defense-space/space/bss/what_is_a_satellite.pdf accessed December 15, 2004.

[2] For information on SNAP, see http://zenit.sstl.co.uk/index.php?loc=47, accessed February 3, 2005.

[3] The incident sunlight that is not reflected is absorbed. If the satellite is disguised by limiting its reflectivity (i.e., painting it black), then it has a higher heat load.

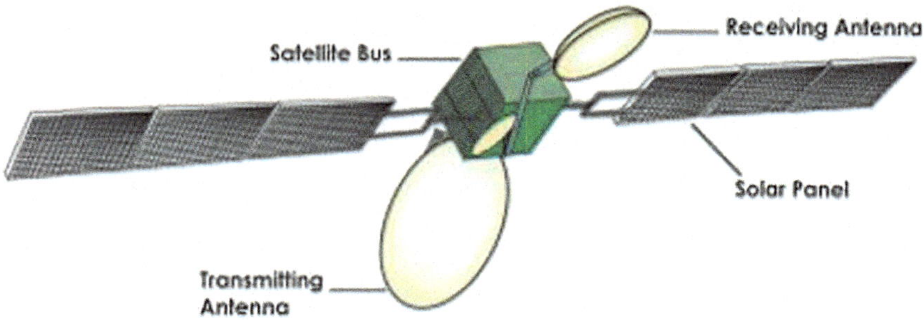

Fig. 3.3 A schematic illustration of a satellite, including the satellite bus, the receiving and transmitting antennae, and the solar panels

convection cannot remove heat from an object as they do on Earth, so the satellite must radiate the heat to eliminate it. In most cases, the thermal regulation system is passive: just a set of well-designed thermally conducting pathways (heat pipes) and radiators to radiate the heat away. However, some components, such as some infrared sensors, may need cryogenic cooling; in this case, loss of the coolant would dramatically degrade the system's performance. A large amount of heat introduced by an incident laser beam may be unmanageable: the internal electronics may fail if the bus conducts too much heat to them, or the structural integrity of the bus itself may be compromised.

A power source: Power is often supplied by arrays of solar cells ("solar panels") that generate electricity, which is stored in rechargeable batteries to ensure a power supply while the satellite is in shadow. For geostationary satellites, eclipses occur on 90 days a year, and last as long as 70 min (Pattan 1993). Technological improvements in battery technology have led to new battery types with high specific energy (energy stored per unit mass) and high reliability. Solar cells are mounted on the body of a satellite or on flat panels. Mounting the solar cells on the satellite's body results in a more compact configuration (which may be desirable if space and mass are limited, or the satellite is meant to be covert), but since not all cells will be illuminated by the Sun at any one time, the power generated is less than it would be from large panels made of solar cells that are continually positioned to face the Sun. Solar panels that are properly oriented toward the Sun can provide about 130 W/m^2 and 50 W/kg of power. Because solar cells mounted on the satellite's body will not, in general, be optimally oriented, they can typically provide 30–35 W/m^2 and 8–12 W/kg of power (Maral and Bousquet 2002).

The solar panels often have a large surface area compared with the rest of the satellite, so they sustain a relatively large number of collisions with debris particles. Solar panels are fragile and can be damaged easily, but partial damage to a solar panel may not disable the satellite. For example, the Telstar 14/Estrela do Sul communications satellite failed to fully deploy one of its solar panels.[4] Also, on March

[4]Loral Space and Communications reports that the satellite generates enough power to maintain satellite health and to operate 17 of its 41 Ku-band transponders ("Loral To Initiate Limited

26, 1996, a solar panel on Canada's Anik-E1 satellite was disconnected, causing a power shortage and safety shutdown of the satellite. The satellite was restarted and was able to transmit a reduced number of television programs.[5] Satellites often can continue to function with partially working solar panels, albeit with diminished capacity. However, if the solar panels fail to deploy or are torn off, a satellite without another power source would cease functioning fairly quickly. A malfunction of the power distribution system could also totally impair the satellite.

For example, in September 2003, Loral Space and Communications declared the Telstar 4 satellite a total loss after it experienced a short circuit on the primary power bus.[6] However, other sources of power are available. The Soviet Union reportedly used nuclear reactors to power energy-intensive missions such as orbiting radar systems, and the United States launched one reactor-powered satellite.[7] Currently, the United States is considering a project to develop a uranium fueled nuclear reactor to produce much higher levels of electric power in space. As early as 1994, there was discussion of reactor capable of producing 100 kW of electricity (Iannotta 2004). On-board chemical sources of power are currently not used for satellites, although newer satellite designs may use fuel cells, which produce electricity by combining chemicals such as hydrogen and oxygen. Generators that produce electricity from the heat released by radioactive materials. Radioisotope Thermoelectric Generators (RTGs), which use the heat of radioactive decay of plutonium-238 to produce electricity, can produce up to hundreds of watts. RTGs are currently used on deep space probes that move too far from the Sun to rely on solar panels.[8] RTGs have been used on earth-orbiting satellites in the past but are not normally used on these orbits.

A computer control system: The on-board computer monitors the state of the satellite subsystems, controls its actions, and processes data. High-value satellites may incorporate sophisticated anti-jamming hardware that is operated by the computer. If someone gained control of the satellite's computer, the satellite could be made useless to its owners. Computer systems are also sensitive to their electromagnetic environment and may shut down or reboot during solar storms or if barraged by high levels of electromagnetic radiation.

A communications system: Communications form the link between the satellite and its ground stations or other satellites. This system generally consists of a receiver, transmitter, and one or more radio antennae. The radio links between a satellite and

---

Service On Telstar 14/Estrela Do Sul In March," Loral press release, January 21, 2004, http://www.loralskynet.com/news_012104.asp, accessed December 15, 2004).

[5]Martyn Williams, "Galaxy IV Failure Highlights Reliance on Satellites," Government Computer News, May 20, 1998.

[6]More information is available at Loral press release, http://www.loral.com/inthenews/030922.html, December 17, 2004).

[7]Regina Hagen, "Nuclear Powered Space Missions—Past and Future," taken from Martin B. Kalinowski, ed., "Energy Supply for Deep Space Mission," IANUS 5/1998 Working Paper, http://www.globenet.free-online.co.uk/ianus/npsmfp.htm, December 15, 2004.

[8]See more information at US's Department of Energy Office of Space and Defense Power Systems, "Radioisotope Power Systems," http://www.ne.doe.gov/space/space-desc.html, accessed December 15, 2004.

the ground are one of the most critical and most vulnerable parts of a satellite system. All satellites require a link to and from the ground to perform "telemetry, tracking, and command" (TT&C) functions.[9] The TT&C system operates the satellite and evaluates the health of the satellite's other systems; it is therefore essential. The receivers on the satellite and on the ground can be overwhelmed by an intruding signal—called jamming—or confused by false signals—called spoofing. Although interfering with the TT&C channel could cause a great deal of damage, these channels are usually well protected with encryption and encoding. Generally, the more vulnerable piece of the communications system is that used for mission-specific communications. The TT&C system occupies only a small part of the satellite's total assigned bandwidth. Bandwidth is the width of the band of frequencies that the satellite is assigned to use—the difference between the highest and the lowest frequency. The amount of data that can be sent through a band is proportional to the bandwidth. A jamming attack would need to be mounted from the broadcast and reception area of the TT&C communications channel—the region from which a user can communicate with the satellite. Restricting the size of this area by increasing the antenna's directionality can help protect these channels from attack by reducing the region from which a jamming attack could take place. However, this may not be a viable solution for satellites that need to support users from a broad geographic area. Moreover, at a given frequency, improved directionality requires a bigger antenna.

An attitude control system: This system, which keeps the satellite pointed in the correct direction, may include gyroscopes, accelerometers, and visual guidance systems. Precise control is required to keep antennas pointed in the right direction for communication, and sensors pointed in the right direction for collecting data. If the attitude control system were not functioning, the satellite is unlikely to be usable.[10]

A propulsion subsystem: The satellite's propulsion system may include the engine that guides the spacecraft to its proper place in orbit once it has been launched, small thrusters used for station keeping and attitude control, and possibly larger thrusters for other types of maneuvering. If the propulsion system does not function, because of damage or lack of propellant, the satellite may still be functional. However, in orbits dense with other satellites, such as geostationary orbit, satellites must be able to maintain their position very accurately or they will be a danger to their neighbors and to themselves. Satellites in low-altitude orbits need to make regular station keeping adjustments, without which their orbits will decay.

Mission-specific equipment: In addition to the basic elements required for a satellite to operate, satellites also carry mission-specific equipment to carry out specific tasks. These may include:

---

[9]Telemetry refers to the information the satellite sends the control station about the status of its various components and how they are operating. Tracking refers to knowing where the satellite is; for example, the time for a signal to travel between the satellite and ground can be used to accurately determine the distance to the satellite. Command refers to the signals that are used to tell the satellite what to do.

[10]In October 1997, trading on Bombay's National Stock Exchange in India was halted for four days after the Insat-2D satellite lost attitude control and began spinning in space. The problem was blamed on a power failure and cost the exchange around US$2 billion in losses.

- **Radio receivers, transmitters, and transponders**: In addition to the communication equipment needed to operate the satellite, a satellite may carry similar equipment for other tasks. It may carry a radio antenna to collect radio signals, such as telephone or television signals, and to relay or rebroadcast them. The antenna serves to receive and transmit signals. It may be a parabolic dish (similar to satellite TV dishes), a feed horn (a conical or cowbell shaped structure), or a minimal metal construction (similar to a rooftop TV antennae). When a system is designed to automatically receive a transmission, amplify it, and send it back to Earth, possibly at a different frequency, it is called a transponder. A satellite-based radar system is also composed in part of transmitters and receivers used to send and then receive the radio waves. Receivers are also used by the military for signals intelligence—eavesdropping on military communications, detecting the operating frequencies of enemy radar, or collecting telemetry from ballistic missile tests. Similarly, a satellite may carry transmitters to send out radio signals, such as the navigation signals from the Global Positioning System. A satellite may be designed to transmit a signal to a specific receiver on the Earth, or to broadcast it over a large area.
- **Remote-sensing systems**: The satellite's mission may be to take detailed images of the Earth's surface or atmosphere or objects in space, or to collect other types of data about the Earth and the atmosphere. A satellite may therefore carry such devices as optical cameras, infrared sensors, spectrographs, and charge-coupled devices (CCDs). For civilian scientific missions, these payloads are often complex, unique, and the result of many years of development.[11]
- **Weapons systems**: A satellite may carry equipment to be used for attacking other satellites or targets on the ground or in the atmosphere. For example, it could carry a laser system and the fuel and mirrors needed to use it, or an explosive charge intended to destroy another satellite.

3.1.1.1 Remote Sensing Satellites (Mapping)

Remote sensing is the acquisition of information about an object or phenomenon without making physical contact with the object. In modern usage, the term generally refers to the use of aerial sensor technologies to detect and classify objects on Earth (both on the surface, and in the atmosphere and oceans) by means of propagated signals (e.g., electromagnetic radiation emitted from aircraft or satellites).

There are two main types of remote sensing: passive remote sensing and active remote sensing.[12] Passive sensors detect natural radiation that is emitted or reflected

[11] Satellites are also key assets as 'disruption of satellite services, whether intentional or not, can have a major adverse economic impact. One indication of the importance of satellite services was provided in 1998 by the failure of the Galaxy IV satellite, which disrupted 80–90% of 45 million pagers across the United States for 2–4 days and blocked credit card authorization at point-of-sale terminals (such as gasoline pumps)' see GAO (2002).

[12] For more information, see https://earthobservatory.nasa.gov/Features/RemoteSensing/remote_08.php.

by the object or surrounding areas. Reflected sunlight is the most common source of radiation measured by passive sensors. Examples of passive remote sensors include radiometer, imaging radiometer, spectrometer, and spectroradiometer. Active collection, on the other hand, emits energy in order to scan objects and areas whereupon a sensor then detects and measures the radiation that is reflected or backscattered from the target. RADAR and LiDAR are examples of active remote sensing where the time delay between emission and return is measured, establishing the location, speed and direction of an object. Remote sensing also replaces costly and slow data collection on the ground, ensuring in the process that areas or objects are not disturbed.

Satellite reconnaissance is used to perform numerous strategic and tactical military missions, including mapping terrain, gathering information on the military and industrial capabilities of other countries, monitoring one's own troop movements, choosing targets during a conflict, and assessing battle damage.

The United States has a number of dedicated military reconnaissance satellites: three optical imaging reconnaissance satellites, with ground resolution reported to be 12–15 cm; and three synthetic aperture radar satellites, with ground resolution reported to be roughly 1 m—there are a number of signals intelligence satellites, too—satellites that detect radio signals.[13] Many other countries operate reconnaissance satellites as well. Ownership of reconnaissance satellites is not restricted to governments. A few commercial satellites take optical and infrared images useful for intelligence work. For example, the French SPOT system of satellites takes images of the ground with up to 2.5-m resolution.[14] The EROS-A satellite can deliver images from 1- to 1.8-m resolution.[15] The Ikonos satellite provides images with up to 1-m resolution.[16] The U.S.-based Quickbird satellite provides images with resolution below 1 m.[17]

The United States is not just performing a gradual build-up of remote sensing capabilities. It is also investing in solutions to augment its capacity at a moment's notice, either to replace a hypothetical downed satellite or to provide extra coverage of a certain theatre of operations. For example, the United States is developing launch vehicles that could launch small payloads with a minimum of preparation time. Such launch vehicles could put into orbit imaging satellites that are smaller and less expensive than current reconnaissance satellites. These satellites could be in lower orbits than the current reconnaissance satellites to compensate for their lower power optics and thus provide adequate ground resolution.

---

[13]For more information, see http://www.nbcnews.com/id/3077885/ns/technology_and_science-space/t/spy-satellites-enter-new-dimension/#.WxwLbSAnaUk.

[14]SPOT 4 and SPOT 5 are currently in orbit. See Spot Image, http://www.spotimage.fr.

[15]EROS-A, launched on December 5, 2000, is owned ImageSat International N.V., a commercial endeavor founded by a team of engineers from the Israeli space program, http://www.imagesatintl.com/index.shtml.

[16]Space Imaging provides optical images obtained with the Ikonos satellite; see Space Imaging, http://www.spaceimaging.com.

[17]Digitalglobe provides images obtained with the Quickbird satellite; see DigitalGlobe, http://www.digitalglobe.com/.

It is important to note, however, that remote sensing satellites are vulnerable to blinding and dazzling attacks from ground-based lasers. Because they are usually in low earth orbits, they are also vulnerable to kinetic energy attacks launched by ballistic missiles. If remote sensing satellites are compromised, some of their functions can be provided by other systems, especially for regional or tactical use. It is also entirely possible to use conventional, low altitude reconnaissance and only use satellite-based systems to provide backup tactical reconnaissance capabilities.

3.1.1.2 Communication Satellites

A communications satellite or "comsat" is an artificial satellite used for telecommunications applications. Modern communications satellites use a variety of orbits including geostationary orbits, Molniya orbits, elliptical orbits and low (polar and non-polar Earth orbits). The general purpose of comsats is to provide a microwave radio relay technology complementary to that of communication cables, for fixed (point-to-point) services. They are use also for mobile applications such as communications to ships, vehicles, planes and hand-held terminals, and for TV and radio broadcasting. Comsats are usually composed of the following subsystems:

- Communication Payload, normally composed of transponders, antenna, and switching systems
- Engines used to bring the satellite to its desired orbit
- Station Keeping Tracking and stabilization subsystem used to keep the satellite in the right orbit, with its antennas pointed in the right direction, and its power system pointed towards the sun
- Power subsystem, used to power the Satellite systems, normally composed of solar cells, and batteries that maintain power during solar eclipse
- Command and Control subsystem, which maintains communications with ground control stations. The ground control earth stations monitor the satellite performance and control its functionality during various phases of its life-cycle.

The bandwidth available from a satellite depends upon the number of transponders provided by the satellite. Each service (TV, Voice, Internet, radio) requires a different amount of bandwidth for transmission. This is typically call known as link budgeting and a network simulator can be used to arrive at the exact value.

Communications is one of the first applications of satellites. Even Sputnik had a communications module that permitted it to transmit on 20.005 and 40.002 MHz which were monitored by amateur radio operators throughout the world. The signals continued for 22 days until the transmitter batteries ran out on 26 October. On the first orbit, the Telegraph Agency of the Soviet Union (TASS) transmitted 'as result of great, intense work of scientific institutes and design bureaus the first artificial Earth satellite has been built.'[18] The communications satellite idea appeared early

[18]For more information, https://www.mid-day.com/articles/interesting-facts-about-the-sputnik-i-satellite/15654192.

in the 20th century. Arthur C. Clarke proposed the concept of the geostationary communications satellite, based on work by Konstantin Tsiolkovsky and on the 1929 work of Herman Potočnik (writing as Herman Noordung) *Das Problem der Befahrung des Weltraums—der Raketen-motor*. Clarke published an article titled "Extra-terrestrial Relays" in the British magazine Wireless World (October 1945). The article described the fundamentals behind the deployment of artificial satellites in geostationary orbits for the purpose of relaying radio signals. Thus, Arthur C. Clarke is often quoted as being the inventor of the communications satellite.

The first geostationary satellite was Syncom 3, launched on August 19, 1964, and used for communication across the Pacific starting with television coverage of the 1964 Summer Olympics. Shortly after Syncom 3, Intelsat I (i.e., Early Bird) was launched on April 6, 1965 and placed in orbit at 28° west longitude. It was the first geostationary satellite for telecommunications over the Atlantic Ocean.

Telesat Canada lunched On November 9, 1972, its first geostationary satellite serving the continent, Anik A1. United States followed with the launch of Westar 1 by Western Union on April 13, 1974. And on May 30, 1974, the experimental satellite ATS-6 was launched, built for NASA the first geostationary communications satellite in the world to be three-axis stabilized.

After the launches of the Telstar through Westar 1 satellites, RCA Americom (later GE Americom, now SES Americom) launched Satcom 1 in 1975. It was Satcom 1 that was instrumental in helping early TV channels such as WTBS (now TBS Superstation), HBO, CBN (now ABC Family), and Channel become successful, because these channels distributed their programming to all of the local cable TV providers using the satellite. Additionally, it was the first satellite used by broadcast television networks in the United States, like ABC, NBC, and CBS, to distribute programming to their local affiliate stations. Satcom 1 was widely used because it had twice the communications capacity of the competing Westar 1 in America (24 transponders as opposed to the 12 of Westar 1), resulting in lower transponder-usage costs. Satellites in later decades tended to have even higher transponder numbers.

By 2000, Hughes Space and Communications (now Boeing Satellite Development Center) had built nearly 40% of the more than one hundred satellites in service worldwide. Other major satellite manufacturers include Space Systems/Loral, Orbital Sciences Corporation with the STAR Bus series, Indian Space Research Organization, Lockheed Martin (owns the former RCA Astro Electronics/GE Astro Space business), Northrop Grumman, Alcatel Space, now Thales Alenia Space, with the Spacebus series, and Astrium.

3.1.1.3 Applications

Telephone

The first and most important application, historically speaking, for communication satellites was in intercontinental long-distance telephony. The fixed Public Switched Telephone Network relays telephone calls from land line telephones to an earth station, where they are then transmitted to a geostationary satellite. The downlink follows an analogous path.

Improvements in submarine communications cables, through the use of fiber-optics, caused some decline in the use of satellites for fixed telephony in the late 20th century, but they still serve remote islands such as Ascension Island, Saint Helena, Diego Garcia, and Easter Island, where no submarine cables are in service. There are also regions of some continents and countries where landline telecommunications are rare to non-existent, for example large regions of South America, Africa, Canada, China, Russia, and Australia. Satellite communications also provide connection to the edges of Antarctica and Greenland.

Satellite phones connect directly to a constellation of either geostationary or low-earth-orbit satellites. Calls are then forwarded to a satellite teleport connected to the Public Switched Telephone Network.

Television

The demand for simultaneous delivery of relatively few signals of large bandwidth to many receivers grew significantly as television became the main market, its being a more precise match for the capabilities of geosynchronous comsats. Two satellite types are used for North American television and radio: Direct broadcast satellite (DBS), and Fixed Service Satellite (FSS).

The definitions of FSS and DBS satellites outside of North America, especially in Europe, are a bit more ambiguous. Most satellites used for direct-to-home television in Europe have the same high-power output as DBS-class satellites in North America but use the same linear polarization as FSS-class satellites. Examples of these are the Astra, Eutelsat, and Hotbird spacecraft in orbit over the European continent. Because of this, the terms FSS and DBS are more so used throughout the North American continent and are uncommon in Europe.

Fixed Service Satellites use the C band, and the lower portions of the Ku bands. They are normally used for broadcast feeds to and from television networks and local affiliate stations (such as program feeds for network and syndicated programming, live shots, and backhauls). FSS satellites are also used to distribute national cable channels to cable television providers.

FSS are the support channel for other governmental and commercial applications such as distance learning by schools and universities, business television (BTV), Videoconferencing, and general commercial telecommunications.

FSS satellites are also used for distribution of Free-to-air satellite TV channels, on in the Ku band. The Intelsat Americas 5, Galaxy 10R and AMC 3 satellites over North America provide a quite large amount of FTA channels on their Ku band transponders. The American Dish Network DBS service has also recently utilized FSS technology for their programming packages requiring their Super Dish antenna—Dish Network needing more capacity to carry local television stations per the FCC's 'must-carry' regulations. DBS used FSS also for more bandwidth to carry HDTV channels.

A direct broadcast satellite is a communications satellite that transmits to small DBS satellite dishes (usually 45–60 cm in diameter). Direct broadcast satellites generally operate in the upper portion of the microwave Ku band. DBS technology is used for DTH-oriented (Direct-To-Home) satellite TV services, such as DirecTV and DISH Network in the United States, Bell TV and Shaw Direct in Canada, Freesat and Sky Digital in the UK, the Republic of Ireland, and New Zealand and DSTV in South Africa.

Operating at lower frequency and lower power than DBS, FSS satellites require a much larger dish for reception (3–8 feet (1–2.5 m) in diameter for Ku band, and 12 feet (3.6 m) or larger for C band). They use linear polarization for each of the transponders' RF input and output (as opposed to circular polarization used by DBS satellites), but this is a minor technical difference that users do not notice. FSS satellite technology was also originally used for DTH satellite TV from the late 1970s to the early 1990s in the United States in the form of TVRO (Television Receive Only) receivers and dishes. It was also used in its Ku band form for the now-defunct Primestar satellite TV service.

Some satellites have transponders in the Ka band, such as DirecTV's SPACEWAY-1satellite, and Anik F2. NASA as well has launched experimental satellites using the Ka band recently. Some manufacturers have also introduced special antennas for mobile reception of DBS television. Using Global Positioning System (GPS) technology as a reference, these antennas automatically re-aim to the satellite no matter where or how the vehicle (on which the antenna is mounted) is situated. These mobile satellite antennas are popular with some recreational vehicle owners. Such mobile DBS antennas are also used by JetBlue Airways for DirecTV (supplied by LiveTV, a subsidiary of JetBlue), which passengers can view on-board on LCD screens mounted in the seats.

Digital Cinema

Realization and demonstration, on October 29, 2001, of the first digital cinema transmission by satellite in Europe of a feature film by Bernard Pauchon, Alain Lorentz, Raymond Melwig and Philippe Binant.

Radio

Satellite radio offers audio services in some countries, notably the United States. Mobile services allow listeners to roam a continent, listening to the same audio programming anywhere. A satellite radio or subscription radio (SR) is a digital radio signal that is broadcast by a communications satellite, which covers a much wider geographical range than terrestrial radio signals.

Satellite radio offers a meaningful alternative to ground-based radio services in some countries, notably the United States. Mobile services, such as Sirius, XM, and Worldspace, allow listeners to roam across an entire continent, listening to the same audio programming anywhere they go. Other services, such as Music Choice or Muzak's satellite-delivered content, require a fixed-location receiver and a dish antenna. In all cases, the antenna must have a clear view to the satellites. In areas where tall buildings, bridges, or even parking garages obscure the signal, repeaters can be placed to make the signal available to listeners.

Initially available for broadcast to stationary TV receivers, by 2004 popular mobile direct broadcast applications made their appearance with the arrival of two satellite radio systems in the United States: Sirius and XM Satellite Radio Holdings.

Radio services are usually provided by commercial ventures and are subscription-based. The various services are proprietary signals, requiring specialized hardware for decoding and playback. Providers usually carry a variety of news, weather, sports, and music channels, with the music channels generally being commercial-free.

In areas with a relatively high population density, it is easier and less expensive to reach the bulk of the population with terrestrial broadcasts. Thus, in the UK and some other countries, the contemporary evolution of radio services is focused on Digital Audio Broadcasting (DAB) services or HD Radio, rather than satellite radio.

Amateur radio operators have access to the amateur radio satellites that have been designed specifically to carry amateur radio traffic. Most such satellites operate as space borne repeaters and are generally accessed by amateurs equipped with UHF or VHF radio equipment and highly directional antennas such as Yagis or dish antennas. Due to launch costs, most current amateur satellites are launched into fairly low Earth orbits and are designed to deal with only a limited number of brief contacts at any given time. Some satellites also provide data-forwarding services using the X.25 or similar protocols.

Internet Access

Satellite communication technology has been used as a means to connect to the Internet via broadband data connections, more and more often, after the 1990s. This can be very useful for users who are located in remote areas, and cannot access a broadband connection, or require high availability of services.

3.1.1.4 Meteorological Satellites

The meteorological satellite is a type of satellite that is primarily used to monitor the weather and climate of the Earth. Satellites can be polar orbiting, covering the entire Earth asynchronously, or geostationary, hovering over the same spot on the equator.

Several geostationary meteorological spacecrafts are in operation. The United States has three in operation: GOES-12, GOES-13, and GOES-15. GOES-12 previously designated GOES-East and now used for South America, is located at 60° west.[19] GOES-13 took over the role of GOES-East on April 14, 2010 and is located at 75° west.[20] GOES-11 was GOES-West over the eastern Pacific Ocean until it was decommissioned December 2011 and replaced by GOES-15.[21] Russia's new-generation weather satellite Elektro-L 1 operates at 76° E over the Indian Ocean. The Japanese have one in operation; MTSAT-1R over the mid Pacific at 140° E. The Europeans have Meteosat-8 (3.5° W) and Meteosat-9 (0°) over the Atlantic Ocean

---

[19]NOAA Satellite and Information Service, http://www.oso.noaa.gov/goesstatus/spacecraftStatusSummary.asp?spacecraft=12.

[20]NOAA Satellite and Information Service, http://www.oso.noaa.gov/goesstatus/spacecraftStatusSummary.asp?spacecraft=13.

[21]NOAA Satellite and Information Service, http://www.oso.noaa.gov/goesstatus/spacecraftStatusSummary.asp?spacecraft=15.

and Meteosat-6 (63° E) and Meteosat-7 (57.5° E) over the Indian Ocean. India also operates geostationary satellites called INSAT which carry instruments for meteorological purposes. China operated the Feng-Yun geostationary satellites FY-2D at 86.5° E and FY-2E at 123.5° E, which are no longer in use.

The United States has the NOAA series of polar orbiting meteorological satellites, presently NOAA 17 and NOAA 18 as primary spacecraft, NOAA 15 and NOAA 16 as secondary spacecraft, NOAA 14 in standby, and NOAA 12. Europe has the Metop-A satellite. Russia has the Meteor and RESURS series of satellites. China has FY-1D and FY-3A. India has polar orbiting satellites as well.

The United States Department of Defense's Meteorological Satellite (DMSP) can see the best of all-weather vehicles with its ability to detect objects with size of a huge oil tanker. In addition, of all the weather satellites in orbit, only DMSP can see at night in the visual. Some of the most spectacular photos have been recorded by the night visual sensor; city lights, volcanoes, fires, lightning, meteors, oil field burn-offs, as well as the Aurora Borealis and Aurora Australis have been captured by this 450-mile-high space vehicle's low moonlight sensor.

3.1.1.5 Global Navigation Satellite Systems

Global Navigation Satellite Systems (GNSS) is the generic term for space-based systems that transmit signals that can be used to provide three services: Position, Navigation, and Timing—known collectively as PNT. Currently, the American NAVSTAR GPS system and the Russian GLONASS are the only fully functioning GNSS; however, several other systems are currently in development, most notably Galileo in Europe and Compass in China.

The Global Positioning System was developed by the U.S. Military in 1973 and became fully functional in 1995 with an original constellation of 24 satellites. It remains under military management and performs critical military missions, such as mission planning, guidance of precision munitions, and navigation for troops and vehicles on the ground and in the air. The GPS system has also become integrated deeply into the civil infrastructure. GPS signals are used for civil navigation, for air traffic management, and as a global time standard that synchronizes everything from cell phones to scientific experiments (Wright et al. 2005). The US Air Force manages the constellation to ensure the availability of at least 24 GPS satellites, 95% of the time. For the past several years, the Air Force has been flying 31 operational GPS satellites, plus 3–4 decommissioned satellites 'residuals' that can be reactivated if needed. The Air Force normally flies more than 24 GPS satellites to maintain coverage whenever the baseline satellites are serviced or decommissioned. The extra satellites may increase GPS performance but are not considered part of the core constellation. In June 2011, the Air Force successfully completed a GPS constellation expansion known as the 'Expandable 24' configuration. Three of the 24 slots were expanded, and six satellites were repositioned, so that three of the extra satellites became part of the constellation baseline. As a result, GPS now effectively operates

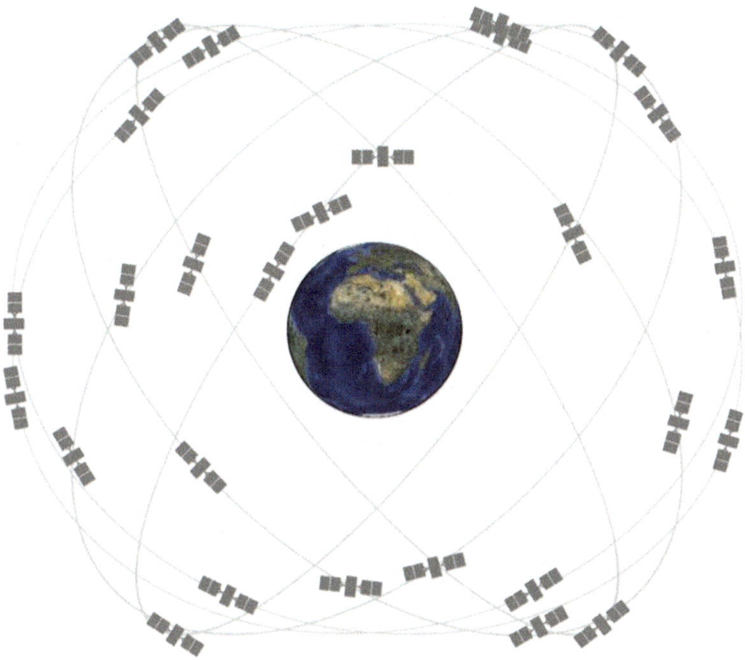

Fig. 3.4 The Enhanced 24 GPS satellite constellation

as a 27-slot constellation with improved coverage in most parts of the world as suggested by Fig. 3.4.[22]

Development of GLONASS began in 1976, with a goal of global coverage by 1991. Beginning on 12 October 1982, numerous launches added satellites to the system until the constellation was completed in 1995. With the collapse of the Russian economy GLONASS rapidly degraded, mainly due to the relatively short design life-time of the GLONASS satellites,[23] but the Russians fully restored the system in October of 2011 with 24 satellites in operation (21 operational and 3 spare satellites in orbit).[24] The constellation has eight satellites on each of its three orbital planes.

GLONASS constellation current status (Fig. 3.5)[25]:

- Operational satellites 23
- In maintenance 4 satellites
- Spares 3 satellites
- In flight tests phase 1 satellite.

[22]http://www.gps.gov/systems/gps/space/.

[23]http://www.positim.com/glonass_overview.html.

[24]http://gislounge.com/global-navigation-satellite-systems/.

[25]http://www.glonass-ianc.rsa.ru/en/GLONASS/.

Fig. 3.5 GLONASS satellite constellation (http://www.glonass-ianc.rsa.ru/en/GLONASS/)

Both the U.S. NAVSTAR Global Positioning System (GPS) and Russian GLONASS navigational satellites use a special type of medium earth orbit (the semi-synchronous orbit), which has a period of 12 h and an altitude of roughly 20,000 km. A navigational system needs at least four satellites within view of the user at all times, where a continuous communications system needs only one. Thus, a navigational system requires more satellites than does a communications system deployed at the same altitude: both GPS and GLONASS (when fully deployed) use 24 satellites. The GPS satellites are in six orbital planes at an inclination angle of 55°; GLONASS is designed to use three orbital planes at an inclination angle of 65° (Wright et al. 2005).

Galileo is an effort by European Union (EU) and European Space Agency (ESA). The Galileo constellation will consist of 30 satellites (27 active and 3 spares) organized into three orbital planes with an altitude of 23,222 km above the Earth and each satellite will orbit the Earth once every 14 h. This means the Galileo satellites are above the GPS and GLONASS satellites. Like the U.S. GPS, there will be four satellites visible from any given point on earth. Galileo will provide coverage reaching to the polar regions. Galileo should be fully functional by 2020 and will not only act as a rival to the American GPS, but also complement it. Future satellite navigation receivers will be able to receive both GPS and Galileo signal, and by using both sets of information, provide positioning that is even more precise.[26]

ESA's first two navigation satellites, GIOVE-A and -B, were launched in 2005 and 2008 respectively, reserving radio frequencies set aside for Galileo by the International Telecommunications Union and testing key Galileo technologies. Then on 21 October 2011 came the first two of four operational satellites designed to validate the Galileo concept in both space and on Earth. Two more will follow in 2012. Once this In-Orbit Validation (IOV) phase has been completed, additional satellites will be launched to reach Initial Operational Capability (IOC) around mid-decade.[27]

As an interesting new feature compared to GPS and GLONASS, Galileo will provide a global Search and Rescue (SAR) function, based on the operational COSPAS-SARSAT system. Each Galileo satellite will be equipped with a transponder that is able to transfer the distress signals from the user transmitters to the Rescue Coordination Centre, which will then initiate the rescue operation. At the same time, the system will provide a signal to the user, informing him that his situation has been detected and that help is under way. This latter feature is new and is considered a major upgrade compared to the existing system, which does not provide feedback to the user.[28] Figure 3.6 depicts Galileo's constellation.

The Galileo system was originally planned to have an operating frequency band that would overlap the U.S. encrypted military signal from GPS. This would have prevented the United States from jamming the Galileo signal without also jamming its own military signal. This was an intensely negotiated aspect of the Galileo system, and it appears that a compromise has been reached in which the Galileo system will not interfere with U.S. military operations.

The Europeans are not the only ones developing new GNSS. The Chinese are developing a new global system known as the Compass navigation system, which should also be fully functional by 2020. The BeiDou COMPASS Navigation Satellite System is composed of three parts: the space section, the ground section and the user section. The space section contains 5 geostationary orbit satellites and 30 non-geostationary orbit satellites. The ground section consists of a certain number of stations: including the main control stations, the injection stations and the monitoring stations. And the user section includes terminators of BeiDou system, and some compatible with other navigation satellite system.[29]

The BeiDou system is steadily accelerating the construction based on a "three-step" development strategy, with specifics as follows:

- Phase I began in 1994, when China started the construction of BeiDou Navigation Satellite Demonstration System. In 2000, two BeiDou navigation experiment satellites were launched, and the BeiDou Navigation Satellite Demonstration System was basically established. The third BeiDou navigation experiment satellite was launched in 2003 (Liu et al. 2014).

---

[26] http://gislounge.com/global-navigation-satellite-systems/.

[27] http://www.esa.int/esaNA/galileo.html.

[28] http://www.positim.com/galileo_overview.html.

[29] http://www.beidou.gov.cn/.

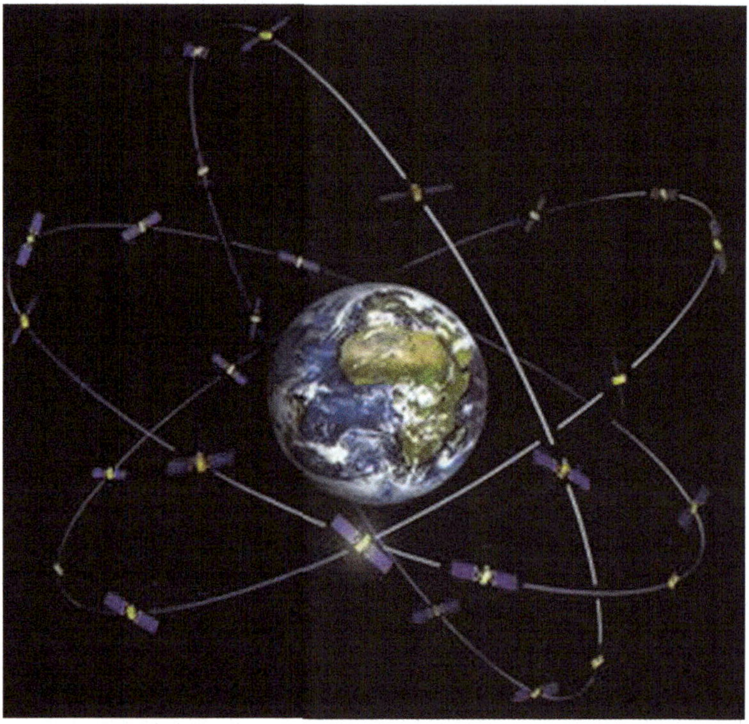

Fig. 3.6 Galileo's constellation, *Source European Space Agency* (https://www.esa.int/ESA)

- Phase II: BeiDou Navigation Satellite (regional) System. In 2004, China started construction of BeiDou Navigation Satellite System. In 2007, the first satellite, a round medium earth orbit satellite (COMPASS-M1) was launched. In 25 October 2012 was launched the sixteenth satellite constellation, which culminates the second phase. The system will provide regional Asia Pacific service by early 2013.[30]
- Phase III: BeiDou Navigation Satellite System with global coverage will completely be established by 2020.

The Indian Regional Navigation Satellite System (IRNSS) is an autonomous regional satellite navigation system being developed by the Indian Space Research Organisation (ISRO). The IRNSS will be composed of seven geostationary satellites to watch over the country from 36,000 km above the earth's surface. The first of seven satellites that will form the Indian Regional Navigation Satellite System, or IRNSS, is expected to be launched by December 2012, Indian Space Research Organisation (ISRO).[31]

[30] http://www.insidegnss.com/node/3246.

[31] http://www.indianexpress.com/news/scientists-excited-about-india-s-own-gps/951413.

The Quasi-Zenith Satellite System QZSS consists of a multiple number of satellites that fly in the orbit passing through the near zenith over Japan. By sharing almost, the same positioning signals for transmission with the currently operated GPS, the system enables to expand the areas and time duration of the positioning service provision in mountainous and urban regions in Japan.

Furthermore, the QZSS aims at improving positioning accuracy of one meter to the centimeter level compared to the conventional GPS error of tens of meters by transmitting support signals and through other means. In order to have at least one quasi-zenith satellite always flying near Japan's zenith, at least three satellites are necessary. The first quasi-zenith satellite 'MICHIBIKI,' launched on 11 September 2010, carries out technical and application verification of the satellite as the first phase. The verification results will be evaluated for moving to the second phase in which the QZ system verification will be performed with three QZ satellites.[32]

All of the systems work in almost the same way, so we will make a short description of the GPS in order to explain it. The GPS may be split in three areas: the ground, the space and user segments.

- The ground segment, also called the control segment, is used in order to send data to the satellites, to synchronize time in the entire constellation and to track the satellites to enable orbit and clock determination.
- The GPS space segment consists in satellites on six orbital planes. One full constellation (core constellation) is formed by 24 satellites. In order to identify a satellite into the orbit it is used a satellite code—this code is the main difference between GPS and GLONASS which differentiates satellites by frequency channel. GLONASS utilize a frequency division multiple access method (FDMA) and GPS and Galileo utilize a code division multiple access technique (CDMA).[33]
- The user segment is formed by receivers and associated antennas, utilized for receiving and decoding the signal for providing PNT information.

GPS is a system with three available carrier frequencies, which are multiples of fundamental frequency. The distance is derived from the measurement of the time difference between the transmission from the satellite and reception at the receiver of an encoded signal. This range is better known as being a pseudo-range because it is affected by a series of system unknowns including clock biases and propagation delays which must be solved or estimated. The signals carrier phase can also be used to derive the range, providing to a fix point a more accurate position, but with inherent ambiguity. In order to determine the position and time are needed ranges from four satellites (timing applications can function with one satellite in view but for verification reason, are preferred two).

The navigation message is transmitted from the satellite to the user and gives the satellite identifier along with information regarding the satellite health, predicted range accuracy, ionosphere and clock correction coefficients as well as orbital ephemeris for allowing the receiver to calculate the satellite position. The message

[32] http://www.jaxa.jp/projects/sat/qzss/index_e.html.

[33] http://www.positim.com/glonass_overview.html.

also contains an almanac which gives status, location and identifier information for all satellites in the constellation.

The main GPS system error sources are:

- Satellite orbit—Orbital biases occur within the ephemeris transmitted, mostly as a result of un-modelled gravitational forces.
- Satellite clock—The satellite clocks experience drift and noise which are modelled and included as part of the broadcast message, although residual error remains.
- Ionosphere and plasmasphere—The signals are delayed on the altitude above 80 km proportional to the free electrons number. The effect is lower when the satellite is at the zenith than when it is near the horizon and it is frequency dependent. This is the largest error source if it is uncorrected properly.
- Troposphere—The signal has delays because of the varying temperature and humidity levels at up to 12 km height. Basic models can correct up to 90%.
- Receiver noise—The inherent noise from the receiver, which causes signal interferences.
- Multipath—In addition to the direct satellite-to-receiver path, the signals are reflected from the ground and other objects. These cause multiple copies of the signal or a broadening of the signal arrival time both of which reduce precision.

In normal standalone operation, GPS gives a three-dimensional position accuracy of about 5–10 m, and also provides velocity of approximately 20 cm/s and time to within 1 μs. These accuracies are dependent on the user equipment, present error sources and on the tracked satellites configuration. For example, if the tracked satellites are all in the same area on the sky, the geometry is poor and attainable accuracy will be affected.

Different GPS applications demand varying degrees of positional accuracy. For example, in-car and personal navigation, need only the standard GPS positioning accuracy, whereas other more demanding applications require augmentation of the standard GPS data, seeing the integrity or correction information. In order to improve the GPS accuracy is used differential GPS (DGPS). In this case the corrections of pseudo-ranges (and/or carrier phases), are calculated in order to improve the positional accuracy of the user's GPS receiver. The DGPS corrections may be applied in post-processing stage or in real time. DGPS may improve the positional accuracy between a meter and a centimeter depending on the utilized signals, user equipment and the adopted methodologies. Some examples of DGPS services are:

- The European Geostationary Navigation Overlay Service (EGNOS) is a satellite-based augmentation system, developed by the European Space Agency, the European Commission and EUROCONTROL. This system is intended to supplement GPS, GLONASS and Galileo by providing integrity messages, corrections and additional ranging signals.
- The Wide Area Augmentation System (WAAS) was developed by the Federal Aviation Administration as an air navigation aid in order to improve the GPS accuracy, integrity and availability. WAAS is using a ground-based reference stations network for monitoring the GPS satellites signals, and geostationary satellites to transmit information to users.

- The General Lighthouse Authorities' DGPS is a network of 14 ground-based reference stations which provides corrections to GPS through MF radio transmissions for at least 50 nautical miles from the coast around the United Kingdom and the Republic of Ireland.
- Ordnance Survey runs the national OS Net GNSS infrastructure of 110 base stations. OS Net provides free GPS and GLONASS data products as well as a commercial high accuracy DGPS service.

Augmentation can also take the form of a data service enabling fast acquisition, where the receiver is sent orbital and timing information to enable almost immediate tracking when the receiver is switched on. This technique is one form of Assisted GPS. High-sensitivity GPS receivers exist and are used in difficult environments, for example to aid tracking of the very weak signals indoors.

In the following five years GNSS will beneficiate of a massive impulse, along with the satellites introduction up to four systems and the addiction of more powerful signals to GPS and GLONASS. Excepting the very low-cost applications, the GPS receivers will probably become a thing of the past. This will be partially determined by the plan of moving GLONASS from the frequency to code division multiple accesses, and partially driven by the Galileo system development and Compass system. Multi-GNSS tracking will deliver improved availability, accuracy and integrity. There are some concerns, however, in the fact that the increased number of signals will raise the noise level which will lead to a lower resilience against other interference sources (Royal Academy of Engineering 2011).

3.1.1.6 Nanosatellites

Large satellites are still affordable only to big national projects or extremely wealthy organizations. As such, emerging countries and small organizations are adopting smaller spacecraft as means to their space exploration endeavors by forcing the miniaturization age to the space industry.[34] Besides the cost issue, the main rationale for the use of miniaturized satellites is the opportunity to enable missions that a larger satellite could not accomplish, such as constellations for low data rate communications, using formations to gather data from multiple points, in-orbit inspection of larger satellites, university research.

Generally, the nanosatellite term designates satellites in the 1–10 kg mass range. However, the most representative for this class is the CubeSat which restricts developers to a volume of approximately $10 \times 10 \times 10\ \text{cm}^3$ (Barnhart et al. 2007). From the industry perspective, nanosatellites now represent an easy access to space for simple instruments or for test bed applications. Among the instruments best suited are the sensors for monitoring the radiation environment on LEO, the magnetic field and upper atmosphere phenomena.

One of the key applications of nanosatellites is as support in disaster management efforts. In these situations, low re-visitation periods are required to monitor major

[34] http://chaire-sirius.eu/wp-content/uploads/2015/07/Note-SIRIUS-Satellite-Miniaturization.pdf.

floods, fires or other large-scale natural disasters. For these types of conditions, rapid information delivery is more important than resolution as there is an immediate need to roughly identify the areas already affected and the ones most exposed to danger. Nanosatellites can therefore be used in conjunction with large spacecraft to identify precisely the locations where higher resolution images are required and request the specific areas to be monitored.

Several approaches have been proposed to address the problem of the size of the optical systems. Among them, worth mentioning are the use of complex deployable lens mounts and the use of multiple sensors. A nanosatellite that successfully demonstrated deployable optics is the 8 kg, 19 cm × 19 cm × 30 cm PRISM nanosatellite developed by the Intelligent Space Systems Laboratory (ISSL) of University of Tokyo.[35]

3.2 Space Stations and Transport Vehicles

A space station (i.e., an orbital station) is a spacecraft capable of supporting crewmembers, which is designed to remain in space (most commonly as an artificial satellite in low Earth orbit) for an extended period of time and for other spacecraft to dock. A space station is distinguished from other spacecraft used for human spaceflight by lack of major propulsion or landing systems. Instead, other vehicles transport people and cargo to and from the station. As of April 2018, two space stations are in Earth orbit: The International Space Station (operational and permanently inhabited) and China's Tiangong-2 (operational but not permanently inhabited). The International Space Station (ISS) is a habitable artificial satellite in low Earth orbit. It follows the Salyut, Almaz, Skylab and Mir stations as the ninth space station to be inhabited. The ISS is a modular structure whose first component was launched in 1998.

ISS components have been launched by American Space Shuttles as well as Russian Proton and Soyuz rockets. Budget constraints led to the merger of three space station projects with the Japanese Kibō module and Canadian robotics. In 1993 the partially built Soviet/Russian Mir-2, the proposed American Freedom, and the proposed European Columbus merged into a single multinational programme. The Russian Federal Space Agency (RSA/RKA) is using the ISS as a work site to assemble their next space station, called OPSEK. Modules and components for the new station began arriving on orbit in 2010, and the RSA plans to commission the new station before the remainder of the ISS is de-orbited.

The ISS serves as a microgravity and space environment research laboratory in which crew members conduct experiments in biology, human biology, physics,

[35]https://earth.esa.int/web/eoportal/satellite-missions/c-missions/cubesat-concept.

astronomy, meteorology and other fields.[36] The station is suited for the testing of spacecraft systems and equipment required for missions to the Moon and Mars.[37]

The station is serviced by Soyuz spacecraft, Progress spacecraft, the Automated Transfer Vehicle, the H-II Transfer Vehicle (Kitmacher 2010) and formerly the Space Shuttle. It has been visited by astronauts and cosmonauts from 15 different nations.[38] On 25 May 2012, Space Exploration Technologies Corporation (or SpaceX) became the world's first privately held company to send a cargo load, via the Dragon spacecraft, to the International Space Station.[39]

According to the original Memorandum of Understanding between NASA and RSA, the International Space Station was intended to be a laboratory, observatory and factory in space. It was also planned to provide transportation, maintenance, and act as a staging base for possible future missions to the Moon, Mars and asteroids.[40] In the 2010 United States National Space Policy, the ISS was given additional roles of serving commercial, diplomatic, and educational purposes.[41]

The ISS provides a platform to conduct scientific research that cannot be performed in any other way. While small unmanned spacecraft can provide platforms for zero gravity and exposure to space, space stations offer a long-term environment where studies can be performed potentially for decades, combined with ready access by human researchers over periods that exceed the capabilities of manned spacecraft.

The Ground Stations that serve the ISS belong to different nations, as follows:

- **United States of America—National Aeronautics and Space Administration (NASA)**

NASA Headquarters (HQ)

NASA headquarters, in Washington, D.C., exercises management over the NASA Field Centres, establishes management policies, and analyses all phases of the ISS program.

Johnson Space Centre (JSC)

Johnson Space Centre in Houston directs the station program. Mission control operates the U.S. on-orbit segment (USOS) and manages activities across the station in close coordination with the international partner control centers. JSC is the primary

---

[36] http://web.archive.org/web/20071208091537, http://pdlprod3.hosc.msfc.nasa.gov/B-gettingon board/index.html.

[37] http://issresearchproject.grc.nasa.gov/.

[38] https://www.nasa.gov/mission_pages/station/main/10th_anniversary.html.

[39] https://www.nytimes.com/2012/05/26/science/space/space-x-capsule-docks-at-space-station.html.

[40] Memorandum of Understanding Between the National Aeronautics and Space Administration of the United States of America and the Russian Space Agency Concerning Cooperation on the Civil International Space Station, http://www.nasa.gov/mission_pages/station/structure/elements/nasa_rsa.html.

[41] National Space Policy of the United States of America, http://www.whitehouse.gov/sites/default/files/national_space_policy_6-28-10.pdf.

center for spacecraft design, development, and mission integration. JSC is also the primary location for crew training.

Kennedy Space Centre (KSC)
Kennedy Space Centre, in Cape Canaveral, Fla., prepares the station modules and space shuttles for each mission, coordinates each countdown, and manages space shuttle launch and post-landing operations.

Marshall Space Flight Centre (MSFC)
Marshall Space Flight Centre's Payload Operations and Integration Centre (POIC) controls the operation of U.S. experiments and coordinates partner experiments aboard the station. MSFC oversaw development of most U.S. modules and the station's Environmental Control Life Support System.

Telescience Support Centres (TSCs)
Telescience Support Centres around the country are equipped to conduct science operations onboard the station. These TSCs are located at Marshall Space Flight Centre in Huntsville, Ala.; Ames Research Centre (ARC) in Moffett Field, Calif.; Glenn Research Centre (GRC) in Cleveland; and Johnson Space Centre in Houston.

- **Russia—ROSCOSMOS, the Russian Federal Space Agency**

  Roscosmos oversees all Russian human space flight activities.

Moscow Mission Control Centre (TsUP)
Moscow Mission Control Centre is the primary Russian facility for the control of Russian human spaceflight activities and operates the station's Russian segment. It is located in Korolev, outside of Moscow, at the Central Institute of Machine building (TsNIIMASH) of ROSCOSMOS.

Gagarin Cosmonaut Training Centre (GCTC)
The Gagarin Cosmonaut Training Centre (GCTC), at Zvezdny Gorodok (Star City), near Moscow, provides full-size trainers and simulators of all Russian station modules, a water pool used for spacewalk training, centrifuges to simulate g-forces during lift-off, and a planetarium used for celestial navigation.

Baikonur Cosmodrome
The Baikonur Cosmodrome in Kazakhstan is the chief launch centre for both piloted and unpiloted space vehicles. It supports the Soyuz and Proton launch vehicles and plays an essential role in the deployment and operation of the station.

- **Canada—Canadian Space Agency (CSA)**

Mobile Servicing System (MSS) Operations Complex (MOC)
Located in Saint Hubert, Quebec, the MSS Operations Complex is composed of the following facilities:

- Space Operations Support Centre (SOSC)
- MSS Operations and Training System (MOTS)

- Virtual Operations Training Environment (VOTE)
- Canadian MSS Training Facility (CMTF)

These facilities provide the resources, equipment and expertise for the engineering and monitoring of the MSS and provide crew training on Canadian systems.

Space Station Remote Manipulator System (SSRMS) Design and Development
The SSRMS was designed and built for the CSA by MDA of Brampton, Ontario.

Payload Telescience Operations Centre (PTOC)
The PTOC in Saint Hubert supports real time operations for Canadian Payloads on-board the station.

- **Europe—European Space Agency (ESA)**

European Space Research and Technology Centre (ESTEC)
The European Space Research and Technology Centre in Noordwijk, the Netherlands, is the largest ESA establishment, a test centre and hub for European space activities. It has responsibility for the technical preparation and management of ESA space projects and provides technical support to ESA's on-going satellite, space exploration, and human space activities.

Columbus Control Centre (COL-CC) and Automated Transfer Vehicle Control Center (ATV-CC)
Two ground control centres are responsible for controlling and operating the European contribution to the station program. These are the Columbus Control Centre and the Automated Transfer Vehicle (ATV) Control Centre. The COl-CC, located at the German Aerospace Centre (DLR), in Oberpfaffenhofen, near Munich, Germany, controls and operate the Columbus laboratory and coordinates the operation of the European experiments. The ATV-CC, located in Toulouse, France, on the premises of the French space agency, Centre national d'Etudes Spatiales (CNES), operates the European ATV during the active and docked mission phases of the ATV.

Guiana Space Centre (GSC)
Europe's Spaceport is situated in the northeast of South America in French Guiana. Initially created by CNES, it is jointly funded and used by both the French space agency and ESA as the launch site for the Ariane 5 vehicle.

European Astronaut Centre (EAC)
The European Astronaut Centre of the European Space Agency is situated in Cologne, Germany. It was established in 1990 and is the home base of the 13 European astronauts who are members of the European astronaut corps.

User Centres
User Support and Operation Centres (USOCs) are based in national centers distributed throughout Europe. These centres are responsible for the use and implementation of European payloads aboard the ISS .

- **Japan—Japan Aerospace Exploration Agency (JAXA)**

In addition to the JAXA headquarters in Tokyo and other field centres throughout the country, Tsukuba Space Centre and Tanegashima Launch Facility are JAXA's primary ISS facilities.

Tskuba Space Centre (TKSU)
JAXA's Tsukuba Space Centre (TKSU), located in Tsukuba Science City, opened its doors in 1972. The TKSC is a consolidated operations facility with world-class equipment, testing facilities, and crew training capabilities. The Japanese Experiment Module (JEM) or "Kibo," which translates in English as "Hope," was developed and tested at TKSC for the station. The Kibo Control Center plays an important role in control and tracking of the Japanese laboratory.

Tanegashima Space Centre (TNSC)
The Tanegashima Space Centre is the largest rocket launch complex in Japan and is located in the south of Kagoshima Prefecture, along the southeast coast of Tanegashima. The Yoshinobu launch complex is on site for H-IIA and H-IIB launch vehicles. There are also related developmental facilities for test firings of liquid- and solid-fuel rocket engines.

Research and Technology Facilities
Managing the international laboratory's scientific assets, as well as the time and space required to accommodate experiments and programs from a host of private, commercial, industry and government agencies nationwide, makes the job of coordinating space station research.

Teams of controllers and scientists on the ground continuously plan, monitor and remotely operate experiments from control centres around the globe. Controllers staff payload operations centres around the world, effectively providing for researchers and the station crew around the clock, seven days a week.

State-of-the-art computers and communications equipment deliver up-to-the-minute reports about experiment facilities and investigations between science outposts across the United States and around the world. The payload operations team also synchronizes the payload time lines among international partners, ensuring the best use of valuable resources and crew time.

The control centres of NASA and its partners include Payload Operations and Integration Centre (POIC), Marshall Space Flight Centre (Huntsville, Alabama), and Mission Control Centre (MCC) in Houston, Texas. Other centers and the associated organizations are listed below:

- ROSCOSMOS—Flight Control Centre (TsUP), Korolev, Russia;
- ROSCOSMOS—Transport Vehicle Control Room, Korolev, Russia;
- Japan Experiment Module Mission Control (JEMMC), Tsukubashi, Ibaraki, Japan;
- ESA—Columbus Control Centre (Col-CC), Oberpfaffenhofen, Germany (near Munich);
- ESA ATV—Control Centre, Toulouse, France;
- ESA—European User Support Operations Centres:

  - CADMOS, Toulouse, France
  - MARS, Naples, Italy
  - MUSC, Cologne, Germany
  - B-USOC, Brussels, Belgium
  - E-USOC, Trondheim, Norway
  - DAMEC, Odense, Denmark
  - BIOTESC, Zurich, Switzerland
  - ERASMUS, Noordwijk, The Netherlands

- CSA-Payloads Operations Telesciences Centre, St. Hubert, Quebec, Canada;
- Canadian Space Agency Mission Control Centre (CSA-MCC), Longueuil, Quebec, Canada.

NASA's Payload Operations Centre serves as a hub for coordinating much of the work related to delivery of research facilities and experiments to the space station as they are rotated in and out periodically when space shuttles or other vehicles make deliveries and return completed experiments and samples to Earth. The payload operations director leads the POIC's main flight control team, known as the 'cadre' and approves all science plans in coordination with Mission Control in Houston, the international partner control centres and the station crew.

3.3 Rockets

Placing satellites into orbit requires a tremendous amount of energy that must come from a launching vehicle or device. The satellite must reach an altitude of at least 200 km and a speed of over 29,000 km/h (8 km/s) to be successfully positioned on the orbit. Orbital launch vehicles are rockets and other non-rocket vehicles (planning stage) capable of placing payloads and/or satellites into or beyond Earth orbit. All satellites today get into orbit with the help of a rocket. Several countries have rocket launch capabilities, and satellites as large as several tons. For most satellite launches, at first, the launch rocket is aimed straight up. This gets the rocket through the thickest part of the atmosphere quickly and minimizes fuel consumption.

After a rocket launches straight up, the rocket control mechanism uses the inertial guidance system to do necessary adjustments to the rocket's nozzles to tilt the rocket to the established course. In most cases, the rocket will flight east direction because Earth rotates to the east, giving the rocket a free boost. The strength of the boost depends on the launch location (rotational velocity of Earth in that location). The boost is greatest at the equator, where the rotation of Earth is fastest.

At an altitude of about 193 km, once the rocket is in an extremely thin part of the atmosphere, the rocket's navigational system fires some small rockets, just enough to put the it into a horizontal position. Then the satellite is released.

Rockets have a long and colorful history.[42] Today, the rocket is an indispensable tool in the exploration of space. Starting with the ancient Chinese, the first to create rockets, these provided ceremonial and warfare uses for centuries. The rocket apparently made its debut on the pages of history as a fire arrow in 1232 AD used by the Chin Tartars for fighting off a Mongol assault on Kai-feng-fu.

There are reports of many rocket experiments all through the 13th to the 18th century. Giovanni de Fontana designed in 1420 a 'mechanical bird' that was powered by a rocket system, used for measurements of buildings. Kazimierz Siemienowicz, a Polish artillery specialist and pioneer of rocketry, published, in 1650, a book called Artis Magnae Artilleriae, which contains a large chapter on the caliber, construction, production and properties of rockets (for both military and civil purposes), including multistage rockets, batteries of rockets, and rockets with delta wing stabilizers. In 1696, Robert Anderson, an Englishman, published The Making of Rockets, a two-part treatise on how to make rocket molds, prepare the propellants, and perform the calculations.

The British used Congreve rockets against the United States in the War of 1812. William Hale, another British inventor, invented the stick-less rocket in 1846. The U.S. army used the Hale rocket more than one hundred years ago in the war with Mexico. Rockets were also used to a limited extent in the Civil War.

During the 19th century, rocket enthusiasts and inventors appeared in almost every country. Claude Ruggieri apparently propelled small animals into space as early as 1806. In 1821, sailors hunted whales using rocket-propelled harpoons. These rocket harpoons were launched form a shoulder-held tube equipped with a circular blast shield.

Skillful theorists, like Konstantin Tsiolkovsky in Russia, were beginning to consider the possibility of space travel. Persons particularly significant in the transition from the small rockets of the 19th century to the colossi of the space age were Konstantin Tsiolkovsky in Russia, Robert Goddard in the United States, Hermann Oberthand Wernher von Braun in Germany.

Launch vehicles can be classified in a number of ways:

- Expendable launch vehicles—designed for one-time use
- Reusable launch vehicles—designed to be recovered and used again.

Early rockets had a single engine. A better way to achieve great speed, however, is to place a small rocket on top of a big one and fire it after the first has burned out. Today almost every space rocket uses several stages, dropping each burned-out stage and continuing with a smaller booster. Explorer 1, the first artificial satellite of the US, launched in January 1958, used a 4-stage rocket. Even the space shuttle uses two large solid-fuel boosters which are dropped after they burn out.

The main purpose of rocket staging is to reduce the cost by reducing fuel required to send a payload into a desired orbit. The main reasons for rocket staging performance improvement by eliminating dead weight during flight as well as maintaining

---

[42] A brief history can be located here. https://www.thoughtco.com/invention-and-history-of-rockets-1992375.

Fig. 3.7 Images of series staged rockets

acceleration within acceptable limits by reducing the thrust in mid-flight.[43] Rocket staging can be performed either in series or in parallel. Series staging means that rocket motors are placed on top of one other to form a tower shape. The most commonly used rockets used to launch to LEO use series staging (see Fig. 3.7). Some examples include Explorer 1, Saturn Vand Falcon 9.

Parallel staging typically involves two or more solid propellant rockets strapped onto a central liquid rocket and are all simultaneously ignited at lift off (see Fig. 3.8). A few examples of parallel staged rockets Arethe Atlas, Titan IIIC and Delta IV rockets.

There are rockets that combine the two types of staging, known as mixed staging. Combining the series and parallel staging methods provides the advantages of both methods. Examples include the Delta II and the Soyuz rockets.

Rocket launch platforms are divided into three types, based on their host environment:

- Land—Spaceport and fixed missile silo
- Sea—fixed platform, mobile platform, submarine
- Air—aircraft.

The following (Table 3.1) classifies rockets based on their lift capacity to the most commonly used orbits, LEO and geostationary, as suggested by Darling (2002).

[43]http://blogs.discovermagazine.com/vintagespace/2018/02/17/the-how-and-why-of-rockets-staging/#.WxxKkCAnaUk.

Fig. 3.8 Images of parallel staged rockets

Table 3.1 Darling' (2002) rocket classification by capacity

Vehicle type	Capacity to LEO (kg)	Capacity to GTO (kg)
Medium lift launch vehicles (HLLV)	>10,000	>5,000
Large launch vehicles (LLV)	5,000–10,000	2,000–5,000
Medium launch vehicles (MLV)	2,000–5,000	1,000–2,000
Small launch vehicles (SLV)	<2,000	<1,000

LEO (Low Earth Orbit), GTO (Geostationary Transfer Orbit)

The following (Table 3.2) is a list of the best-known rockets, classified by family and nation that launched them.

The most important non-rocket launch vehicle systems proposed:

- **Static structures** (Space tower, Tensile structures, Skyhook, Space elevator, Hypersonic skyhook, SpaceShaft, Rotovators, Hypersonic rotovator, Endo-atmospheric tethers)
- **Dynamic structures** (Space fountain, Orbital ring, Launch loop, Pneumatic free-standing tower)
- **Projectile launchers** (Electrical, Chemical, Mechanical)
- **Reaction drives** (jets and unconventional rockets—Air launch, SpacePlanes, Laser propulsion)
- **Buoyant lifting** (Balloon, Buoyant space port)
- **Hybrid launch systems**.

Table 3.2 Classification of rockets and associated nations

Family	Nation
VLS-1	Brazil
KT-1	China
Long March 1	China
Long March 2-3-4	China
Long March 5	China
Long March 6	China
Ariane 1-2-3	Europe
Ariane 4	Europe
Ariane 5	Europe
Vega	Europe
Diamant	France
ASLV	India
GSLV Mk.I and II	India
GSLV Mk.III	India
PSLV	India
SLV	India
Safir	Iran
Simorgh	Iran
Shavit	Israel
Epsilon	Japan
H-I	Japan
H-II, IIA and IIB	Japan
J-I	Japan
L-4S	Japan
Mu 1-3-4	Japan
Mu 5	Japan
N-I and II	Japan
Naro	Korea
Unha	Noth Korea
Haas	Romania
Angara1.X	Russia
Angara A3	Russia
Angara A5	Russia
Angara A7	Russia
UR-100RokotStrela	Russia
R-29 ShtilVolna	Russia
Start-1	Russia
Black Arrow	UK

(continued)

Table 3.2 (continued)

Family	Nation
Dnepr	Ukraine
Athena I and II	USA
Atlas A-B-C-D-E-F-G	USA
Atlas I	USA
Atlas II	USA
Atlas III	USA
Atlas V	USA
Thor	USA
Delta	USA
Delta II	USA
Delta III	USA
Delta IV	USA
Falcon 1	USA
Falcon 9	USA
Falcon 9 v1.1	USA
Falcon Heavy	USA
Minotaur I	USA
Minotaur IV and V	USA
Pegasus	USA
Saturn I and IB	USA
Saturn V	USA
Scout	USA
SLS	USA
Space Shuttle	USA
Taurus	USA
Antares	USA
Titan I-II-3-IV	USA
Vanguard	USA
Energia	USSR
Kosmos	USSR
N1	USSR
UR-500 Proton	USSR Russia
R-7 SemyorkaSoyuz	USSR Russia
R-36 Tsyklon	USSR Ukraine
Zenit	USSR Ukraine

3.4 Space Probes and Other Planetary Satellites

A space probe is a scientific space exploration mission in which a spacecraft leaves Earth and explores space. It may approach the Moon; enter interplanetary space; flyby, orbit or land on other planetary bodies; or approach interstellar space. Space probes are a form of robotic spacecraft.

Some of the most famous space probes include Luna 16, the first unmanned robotic sample return probe from the Moon; Lunokhod 1, the first rover on Moon; Mariner 10, the first probe to Mercury; Venera 3, the Soviet Union probe was the first man-made spacecraft to impact on another planet (Venus); Venera 7, the first man-made spacecraft to successfully soft land on another planet (Venus) and to transmit data from there back to Earth; Mariner 9, the first space probe to maintain orbit around another planet; Mars 3, the first soft landing on Mars the planet, Sojourner, the first successful rover on Mars. Two of the most important space probes are the Mars Exploration Rovers, Spirit and Opportunity, launched in 2003. Communication with Spirit stopped on sol 2210 (March 22, 2010).[44] JPL continued to attempt to regain contact until May 24, 2011, when NASA announced that efforts to communicate with the unresponsive rover had ended. The last Martian rover that landed on the planet is the Mars Science Laboratory (Curiosity).

In terms of multinational interplanetary probes, we mention Halley Armada. The first probe to a comet is the ICE mission, the first solar wind sample return probe from sun-earth L_1 was the Genesis mission, while Stardust represents the first sample return probe from comet tail. Related to asteroids the NEAR Shoemaker was the first asteroid landing, while Hayabusa is the first sample return probe from an asteroid. Rosetta is the space probe that has flown by two asteroids and is aiming to rendezvous and explore comet 67P/Churyumov-Gerasimenko.[45]

Moreover, Pioneer 10 is the first probe to Jupiter, while its twin, Pioneer 11, is the first probe to flyby two planets and first probe to Saturn. Voyager 1 is a 733-kg probe launched September 5, 1977. It is currently still operational, making it the longest-lasting mission of the U.S. National Aeronautics and Space Administration (NASA). It visited Jupiter and Saturn and was the first probe to provide detailed images of the moons of these planets.

Voyager 1 is the farthest human-made object from Earth, traveling away from both the Earth and the Sun at a relatively faster speed than any other probe. As of July 23, 2010, Voyager 1 is over 17 terameters (1.7 × 1013 m, or 1.7 × 1010 km, 110 AU, 15.7 light-hours, or 10.5 billion miles) from the Sun.[46] Voyager 2 represents the first probe for 4 planets and first probe to Uranus and Neptune. Huygens is the first probe to land on Titan, while Juno is the first probe sent to Jupiter, launched on August 8, 2011.

However, these probes, while crucial to humanity's research efforts, do not exactly fit the criticality criterion. There is another class of various deep space system, tasked

---

[44] https://www.nasa.gov/mission_pagcs/mcr/ncws/mcr20110525.html.

[45] https://www.sciencedaily.com/releases/2010/04/100426113112.htm.

[46] Voyager—Mission—Interstellar Mission http://voyager.jpl.nasa.gov/.

with guarding against various threats specific to space. They watch for large asteroids and gather information on composition and trajectory so that ground teams may determine whether there is a chance for a disastrous collision course with Earth. While the kinetic extinction threat has never become an issue, there is another danger, just as dire, in the form of space weather, wherein the Sun, acting on its own cycles of activity, spews highly charged plasma particles from its surface. These solar winds have a 12-year peak of activity and are strong enough to cause auroras in areas on Earth not used to them and damage electrical equipment at ground level, let alone in orbit. Thus, such events are an existential threat to our space infrastructures, as well as our terrestrial ones.

The most important warning beacon in space is the NASA Advanced Composition Explorer (ACE) satellite which, since its launch in 1997, has been stationed at the L1 point between the Sun and Earth, approximately halfway between the two bodies. The satellite continuously provides reports of the direction and speed of solar winds as well as other particles in a state of flux for which it has the right equipment and calibration. ACE can emit a warning 15–45 min before a geomagnetic storm front reaches Earth. There are other solar observation satellites, such as the Solar and Heliospheric Observatory (SOHO) or Probe-2, which can offer a certain amount of warning beforehand, but with greater delays and less actual information. There are also radiation monitors on satellites travelling on other missions, such as METOP, GIOVE-A and B, Galileo-IOV, XMM, Integral, Herschel, Planck. While, technically, not all of them are space probes, since some are GNSS units and other types of near-Earth systems, they are, nonetheless, equipped with equipment for monitoring the space weather, to a certain degree, so that more accurate forecasts may be made.

3.5 Ground Stations

Satellites are monitored and controlled from their ground stations. One type of ground station is the control station, which monitors the health and status of the satellite, sends it commands of various kinds, and receives data sent by the satellite. The antenna that the control station uses to communicate with the satellite may be located with the station, but it need not be: to maintain constant contact with a satellite not in geostationary orbit, and which therefore moves relative to the Earth, the station needs to have antennae or autonomous stations in more than one location. Satellites may also have other types of ground stations. For example, a communication satellite's mission is to send data (voice communication, credit card authorization, video broadcast, etc.) from one user to another, and each user needs an antenna and is in effect a ground station. A satellite may therefore be communicating with many ground stations at the same time. For example, a Boeing 702 communication satellite can carry over 100 transponders.[47]

[47] http://www.apcon.aero/news/boeing702.pdf.

Military communications satellites have ground stations that range from large, permanent command headquarters to small, mobile field terminals. Ground stations are generally not highly protected from physical attack. Disabling a control station may have an immediate disruptive effect, but the disruption can be reduced by having redundant capabilities, such as alternate control centers. Computers at control centers may be vulnerable to attack and interference, especially if they are connected to the Internet. However, high value command computers will have high security, and many of the military command center computers are isolated from the Internet.

3.6 Interactions (Communication Links)

The term link refers to a path used to communicate with the satellite (and is sometimes used to refer to the communication itself):

- Uplinks transmit signals from a ground station to the satellite
- Downlinks transmit signals from the satellite to a ground station
- Crosslinks transmit signals from satellite to satellite
- **Telemetry, tracking, and command (TT&C) link** is the part of the uplink and downlink used to control a satellite's function and monitor its health. The uplinks and downlinks are vulnerable to interference since the strength of the radio signals when they reach the receiving antenna is often low, so that an interfering signal need not be strong. Links can also be interfered with by placing something impermeable to radio waves, such as a sheet of conducting material, in the path between the satellite and ground station. This would likely be done close to the receiver or transmitter, where it could achieve the greatest effect.

3.7 Administrative and Legislative Framework

Some may argue that it is premature to discuss the administrative and legislative framework of space activity since it is still in an incipient state, with an uncertain direction before it in terms of how it will evolve and how it will, ultimately, impact the exploration and exploitation of space. On the contrary, this is the best moment to not only lay out a set of priorities and example for such a framework, but also to stress its significance in shaping the future of mankind in space. For this reason, the administrative and legislative framework qualifies as a nascent critical space infrastructure, with ample precedent behind it on Earth.

The European Union has recognized from the early stages of its European Programme for Critical Infrastructure Protection that administration is an infrastructure on par in criticality with traditional ones, including Defence, Transport, and ICT among others (in the European Council Directive 2008/114/EC of 8 December 2008 on the identification and designation of European critical infrastructures and the assessment of the need to improve their protection). Various states around the world have begun assessing the risks entailed by ever complex infrastructure systems on which they are dependent and, as administration provides the means for day-to-day and emergency governance, along with coordination and distribution of resources, its underdevelopment or dysfunction poses a serious threat to the well-being of the system in its entirety.

Moving forward and under the assumption of greater development in space, a few key issues will arise, which will require an ever-developing (not necessarily expanding) administrative and legislative framework:

Given that most of human activity and, for the time being, all commercial endeavors take place in low Earth orbit, there is a growing realization of the existence of an 'orbital commons,' whose deterioration, as a result of debris, affects every space capable nation, organization and private actor.

The realization of the criticality of space infrastructures warrants investments and preparation for their protection; however, the materialization of risks in space does not hinge on nationality. As seen on Earth, critical infrastructure protection efforts relying on a system-of-systems approach must encompass the entirety of the geographical space and infrastructural substance at risk. Most of the time, these are at national levels, with a few select ones (energy, shipping and so on) for which an extra layer of coordination, at international level, is required. In space, especially in the crowded orbits, the propagation of risks defies national borders and interests, which suggests that a framework that encompasses as much of human activity in this area, regardless of nationality, is required. The alternative is CIP efforts severely lacking in efficiency and scope and the loss of opportunities for developing synergies (common standards, practices etc.).

For most of the timeline of human presence in space, the main drivers were the sovereign and often competing states, which would have eschewed a formal administrative and legislative framework as a breach of their sovereignty. Now, at least in low Earth orbit and geosynchronous orbit, private entities account for more and more of the development, signaling the need for a paradigm shift.

Private actors are used to a regulated environment on Earth, which adds needed predictability to the way they conduct their affairs. With increasing complexity and volume of activity in space, the costs of not superimposing an institutional layer, from early on, can be quite high. Frontier economics do not work in the high-risk, high-cost environment of space, and actors already stretched to the limits of their risk acceptance can be easily discouraged or eliminated from the new 'space race' by the lack of administrative amenities that they take for granted on Earth.

The International Telecommunications Union, which is a UN agency, handles the allocation of valuable geostationary orbital slots, which are important for broadcasting, especially above populous areas. From the perspective of CIP efforts in space, the ITU is an important actor, preventing unintentional jamming and administrating a scarce resource. A recent highlight of its importance was in April 2010, when the Galaxy 15 satellite in GEO orbit stopped responding to ground control signals due to interference by a nearby satellite operating on the same frequency. It began to drift away from its assigned position and was not recovered until the end of December 2010, by which time it had wandered very far away. As a result of the malfunction, other Galaxy satellites had to be repositioned to provide the original Galaxy 15 capability and other companies' satellites moved out of its way as it changed orbit (Ministry of Defence 2012). The level of communication to perform such maneuvers is a basic requirement of 'critical space infrastructure protection.' There is also an initiative for a proposed association between the major space agencies of the world. An Inter-Agency Space Debris Coordination Committee (IADC) was formed in 1993.

The existing framework, while a welcome development, is plagued by inadequacies and irrelevance to the growing private activity in space. The main current administrative components are the various treaties comprising space law and the United Nations Committee on the Peaceful Uses of Outer Space (COPUOS) which, through its various subcommittees, provides recommendations on technical and legal issues for the General Assembly of the UN. COPUOS is also responsible for the drafting and the negotiations that have led to these treaties. The treaties comprising the UN space governance framework are:

- The 1967 Treaty on Principles Governing the Activities of States in the Exploration and Use of Outer Space, including the Moon and Other Celestial Bodies
- The 1968 Agreement on the Rescue of Astronauts, the Return of Astronauts and the Return of Objects Launched into Outer Space
- The 1972 Convention on International Liability for Damage Caused by Space Objects
- The 1975 Convention on Registration of Objects Launched into Outer Space
- The 1979 Agreement Governing the Activities of States on the Moon and Other Celestial Bodies
- The 1963 Treaty Banning Nuclear Weapon Tests in the Atmosphere, in Outer Space, and Under Water.

We would also add the various national airspace regulations and regulation authorities, which are an overlooked, but important, component, of the existing framework, as they hold initial jurisdiction over space-bound flights and their content and their return phase. Moreover, countries which are active in space-related activities are also adopting their own, national regulations with regards to them. Keeping in mind that Article 9 of the Outer Space Treaty provides for state retention of authority and jurisdiction over the asset in question, the aeronautic activity regulators, also due to the implied similarities between aviation and space activity, will continue to hold sway. The United States Federal Aviation Administration, in particular, can be considered to hold great influence over space activities, since many commercial activities are American in origin and utilize American airspace. Through its requirements for allowing permission to operate, the FAA is in a position to significantly impact the development of standards, technology and practices for private actors who, through market dynamics, would influence non-Americans as well. Famously, the samples brought back from the Moon by American astronauts had to be taken through customs, and the astronauts themselves carried passports and were processed upon arriving in Hawaii. The efforts on the part Space Exploration Technologies Corporation, a leading private company in the field of rocketry, to achieve human worthiness ratings on their vehicles and previously, to obtain unmanned flight permission, are another adequate example in this regard.

While COPUOS has elaborated technical standards and recommendations for limiting the amount of space debris created during the course of human activity, the adoption and implementation of these contributions is still voluntary, meaning that, in practice, Member States get to ignore them without penalty. The following deficiencies can be highlighted with the current administrative and legislative framework:

- Fragmentation along national lines and differing standards concerning activity in an environment where a universal standard and authority would be the best option
- Limited acceptance of existing treaties, ranging from 100 nations signing and ratifying the Outer Space Treaty (with 26 more only signing) to only 13 nations signing the Moon Treaty, and none ratifying it. Most of these glaring disparities are obscured by the fact that only a limited number of countries have, historically, been present in space, however that number is set to rise with increased interest in space development and diminishing costs
- The fact that technical recommendations made by COPUOS must be adopted and implemented voluntarily, and the UN itself has virtually non-existent enforcement capabilities
- Limited ability to gauge how well and how much of the agreed measures have been implemented and punish dereliction
- An over-emphasis, in existing legislation, on interactions between states, as opposed to interactions between private actors with commercial concerns, who are the fastest growing demographic in space.

With an increasing part of the inventory of space assets being administered by private companies for commercial activities, it was inevitable that the private sector would become an active stakeholder in coordinated protection activities. An example is the initiative of the International Institute for Private Law (UNIDROIT) to formulating a Space Assets Protocol to aid in formalizing and removing the uncertainty from space entrepreneurship.[48]

The efforts and failures of UNIDROIT would serve as an adequate example to highlight the incipient struggle for establishing a legislative and administrative framework for space activity, especially with regards to commercial space development. The International Institute for Private Law (UNIDROIT) is an international organization comprising 63 Member States, based in Rome. Its remit is to "study needs and methods for modernizing, harmonizing and coordinating private and, in particular, commercial law as between States and groups of States and to formulate uniform law instruments, principles and rules to achieve those objectives". For the past several

[48]https://www.globalnegotiator.com/international-trade/dictionary/unidroit/.

years, it has engaged in formulating a Space Assets Protocol to aid in formalizing and removing the uncertainty from space entrepreneurship. This protocol began development within the context of the Cape Town Convention on International Interests in Mobile Equipment, which opened on the 16th of November 2001 to provide a generic legal framework for registration of ownership and security interests, as well as legal remedies for default. There are three specific sector Protocols:

- Aircraft equipment
- Railway rolling stock
- Space assets

Over the years, UNIDROIT has been responsible for a wide array of conventions which codified and harmonized legal practices for private parties:

- Convention relating to a Uniform Law on the International Sale of Goods (The Hague, 1964)
- Convention relating to a Uniform Law on the Formation of Contracts for the International Sale of Goods (The Hague, 1964)
- International Convention on Travel Contracts (Brussels, 1970)
- Convention providing a Uniform Law on the Form of an International Will (Washington, D.C., 1973)
- Convention on Agency in the International Sale of Goods (Geneva, 1983)
- Convention on International Financial Leasing (Ottawa, 1988)
- Convention on International Factoring (Ottawa, 1988)
- UNIDROIT Convention on Stolen or Illegally Exported Cultural Objects (Rome, 1995)
- Convention on International Interests in Mobile Equipment (Cape Town, 2001)
- Geneva Securities Convention (Geneva, 2009).

As such, it is well qualified to ascertain the need for such a framework of private law for space activity and formulate an adequate protocol. It expressed its vision as such[49]:

> *(the space assets protocol) represents the coordinated efforts of both Governments and the commercial space sector to render asset-based financing more accessible to an industry that is presently searching for innovative ways to obtain start-up capital for space-based services. Such ventures are full of risk and uncertainty and, consequently, their financing is currently still prohibitively expensive. By introducing a uniform regime to govern the creation, perfection and enforcement of international interests in space assets, notably satellites, it is envisaged that the cost of financing will be reduced as a result of the increased level of transparency and predictability for financiers, thereby making financing more widely available to a greater number of players in the commercial space sector. Such an instrument will, in particular, help bring*

[49]https://www.unidroit.org/overview-2012-space-assets.

much needed financial resources to the new space community, namely those small start-up companies that have emerged as a result of the booming commercial space sector.

UNIDROIT is very pragmatic and saw that its interests, and those of the private sector, are, for the time being, focused on commercially viable activities involving satellites. Its team of respected lawyers began codifying certain notions, such as what constitutes a space asset. In Article 1 of the Protocol the term "space asset" is defined as "any man-made uniquely identifiable asset in space or designed to be launched into space, and comprising:

- A spacecraft such as a satellite, space station, space module, space capsule, space vehicle or reusable launch vehicle
- A payload (whether telecommunications, navigation, observation, scientific or otherwise)
- A part of a spacecraft or payload such as a transponder… together with all installed, incorporated or attached accessories, parts and equipment and all data, manuals and records relating thereto."

The Space Protocol would have the following components:

- Debtor's rights and assignments of debtor's rights and rights of use
- Public service exemption from default remedies
- Insurer's salvage interest
- An international registry to record and determine priority (according to when registered) among creditor rights in space assets. It would include transponder leases, conditional sales agreements and security interests.

Unlike the other components of the Cape Town Convention, the Space Protocol had numerous revisions and, eventually, began to receive input from private parties and entities other than member states. There was significant opposition to the Protocol, especially in light of the fact that, unlike the Aircraft Protocol, which operates in a field with a low barrier to entry for a state, very few of the 63 members of UNIDROIT are space capable, yet they all would have a say in the drafting and adoption of the protocol. The other objections are of a technical nature, regarding how well the Protocol's components were designed to achieve the desired result. However, the back and forth on the issue highlighted not only growing interest in this field (with private parties being uninvolved at the beginning, in 2001, and, eventually, offering significant input by 2008), but also the fact that, even in the underdeveloped space market, an unwritten body of customs and procedures for commercial activity had already started to take shape and be developed enough to be quoted and agreed upon. Indeed, the Space Protocol is notable in the breadth of opposition mounted against it, from which it is possible to also expect input on future legal frameworks.

A UNIDROIT report on the matter highlighted a letter dated 9 December 2011, signed by more than 90 leading players in commercial satellite programmes [including satellite operators (45), spacecraft manufacturers and launch services providers

(17), insurance (11) and finance companies (11) that provide services to those in the satellite industry, and satellite and space related associations (8)], urging UNIDROIT and its Member States to reconsider the adoption of the draft Space Protocol. Opposition to the draft Space Protocol was led by the European Satellite Operator's Association (ESOA) and The US Satellite Industry Association (SIA) who have repeatedly expressed their general and specific concerns about the Protocol in letters and petitions to UNIDROIT. In the UK, the Finance and Procurement Working Group of the Space Innovation and Growth Strategy (IGS), which issued its final report in March 2010, addressed this subject and stated the concerns of the UK space sector. In September 2011, a letter was sent to the Minister of State for Universities and Science by Atrium Space Insurance Consortium representing 10 Lloyd's Syndicates, stating its serious concerns regarding the Protocol and the negative implications for the space insurance community. Almost 80% of the underwriting capacity on the huge London insurance market held the same views, which indicates just how much space activity had already entered private actors' minds.

Unlike the Protocol on Aircraft Equipment, which was adopted in 2006 and ratified by 44 members, or the Protocol on Railway Stock, which has been adopted but not yet implemented, the Space Protocol was deemed deficient. Of the 40 states and 10 international organizations present between 27 February and 7 March 2012 in Berlin to discuss adoption, only 25 signed the Final Act of the conference, though it included the biggest players in space activity and most of the medium ones. However, the Space Protocol, itself, was only signed by three countries, of the five needed to begin ratification: Burkina Faso, Zimbabwe and Saudi Arabia, of which only Saudi Arabia has any direct involvement in space-based activities.

In particular, the issue of catering to the needs of private actors, as a precondition for rapid expansion of human activity, must be addressed quickly. Their concerns mirror those of actors on Earth, where a significant body of treaties and practices governs commercial activity, with the closest approximation being not aeronautic, but maritime frameworks. For their investments to be viable and find adequate financing, private actors need to eliminate as much of the uncertainty and risk of activity in space as possible. At the very least, they need to be able to contract adequate insurance, assign adequate liability to parties in an incident which led to losses and enforce those liabilities in a competent court. Common standards aid immeasurably, by relieving private entities of the risk, uncertainty and cost of devising their own technical specifications.

Consider the following scenario—orbital collisions which are, currently, treated as an Act of God, with the existing Liability Treaty never having been invoked. If space development is to continue, this shall not be the case, and private actors will desire the adequate framework for extracting damages from the guilty party. What happens when two active satellites collide in orbit? What are the limits of the damages that can be sought and what are the mitigating circumstances? Given the added orbital debris that such a collision would generate and current tracking capacity, should the guilty party also be held liable for damages to third parties, whose assets have been damaged by collisions with debris from the initial impact? Or to fourth and fifth parties, for that matter, in the eventuality of a cascading collision event? The framework required would enable injured parties to recuperate a part of their investment, encouraging space activity, but limit the liability of the responsible party so that companies are not discouraged from entering development because of the threat of excessive litigation and award settlements. A fine balance must be struck, one which also incentivizes positive behavior and practices (like de-orbiting inactive satellites, investing in orbital tracking capacity and maneuvering capabilities and, one day, orbital clean-up activity) that has already been recommended by COPUOS, but is unlikely to be adopted or enforced unless it is done under economic duress.

Final Remarks

A CSI taxonomy can help with classification of space systems. Such classifications can then be used for rigorous study of systems, elements, and their relationships: good or bad to better understand operability of the involved systems. Five main categories are suggested—remote sensing, communication, meteorological, Global Navigation Satellite Systems, and administrative and legislation framework. Exploring the relationships in different context might address the missing elements as well as lead to the discovery of the unknown risks and vulnerabilities. Certainly, this might call for development of good taxonomic databases for CSI for studying biodiversity and the related issues. Such an approach might lead to development of different kinds of databases including those that address hierarchical classification. Moreover, such studies might lead better and accurate information management for purposes of conservation management, inventory, as well as commerce.

References

Barnhart, D. J., Vladimirova, T., & Sweeting, M. N. (2007). Very-small-satellite design for distributed space missions. *Journal of Spacecraft and Rockets, 44*(6), 1294–1306. https://doi.org/10.2514/1.28678.

Darling, D. (2002). *The Complete Book of Spaceflight: From Apollo 1 to Zero Gravity*. New York: Wiley.

Iannotta, B. (2004). Jupiter moon probe goes nuclear. *Aerospace America, 42*(3), 36–39.

Kitmacher, G. H. (2010). *Reference guide to the international space station: Assembly* (complete ed.). South Carolina: CreateSpace Independent Publishing Platform.

Liu, J., Gu, D., Ju, B., Yao, J., Duan, X., & Yi, D. (2014). Basic performance of BeiDou-2 navigation satellite system used in LEO satellites precise orbit determination. *Chinese Journal of Aeronautics, 27*(5), 1251–1258. https://doi.org/10.1016/j.cja.2014.03.006.

Maral, G., & Bousquet, M. (2002). *Satellite communications systems: Systems, techniques and technology* (4th ed.). New York: Wiley.

Pattan, B. (1993). *Satellite systems: Principles and technologies*. New York: Van Nostrand Reinhold. Retrieved from https://trove.nla.gov.au/version/43426844.

Royal Academy of Engineering. (2011). *Global navigation space systems: Reliance and vulnerabilities*. London: The Royal Academy of Engineering.

Wright, D., Grego, L., & Gronlund, L. (2005). The physics of space security: *A reference manual*. Cambridge, MA: American Academy of Arts and Sciences.

Chapter 4
Critical Space Infrastructure Interdependencies

Abstract This chapter elaborates on a key theme of critical space infrastructure: interdependency. In the context of present discussion, interdependency is described means being influenced, determined by, conditioned by, or subject to another—relationships to other systems. These relationships are examined in relation to, among other critical systems: water, energy, chemical, transportation, and food supply along with potential effects.

4.1 Interdependency in Space Systems

It is well established that hazards, risks, and threats to infrastructure systems and their missions can stem from natural phenomena and/or manmade activities. In such as case, one might consider biological attacks (e.g., smallpox), 9/11 attacks, and oil spills as made-made. On the other hand, natural events include such things as the 2004 tsunami in South Asia and Hurricane Katrina. The performance of infrastructure systems is affected by natural phenomena and/or manmade activities via intricate relationships. In attempts to making infrastructures more secure, dependable, and resilient, there is a need to understand intricate relationships among the infrastructure systems that work together as an integrated whole for the well-being of society.

Working on the assumption that the goal of maintaining and sustaining public health, economy, and security depends on the inputs and outputs of multiple well-interconnected infrastructure systems, the relationship among infrastructures is not one-to-one rather it is multidirectional (Katina and Pinto 2012; Gheorghe et al. 2018). In the multidirectional relationship concepts of risk, dependency, exposure, resiliency etc. take on a new meaning beyond their traditional formulations. The reminder of this chapter, while focusing on the concept of interdependency, examines the basis for critical space infrastructure interdependencies in relation to other critical systems.

Table 4.1 contains an interpretation of the importance of various space systems in connection with terrestrial critical infrastructures, as defined in the EU framework. What we find is that space systems, while highly influential in certain areas, have not achieved the numbers, quantities and complexity of functions required to have

© Springer Nature Switzerland AG 2019

A. Georgescu et al., *Critical Space Infrastructures*,
Topics in Safety, Risk, Reliability and Quality 36,
https://doi.org/10.1007/978-3-030-12604-9_4

Table 4.1 Relation mapping for critical space infrastructure and critical infrastructure

	Energy	ICT	Water supply	Food supply	Health	Finance	Defence, public order, national security	Administration and governance	Transport	Chemicals and nuclear energy
Remote sensing	2	2	3	3	2	1	5	5	4	3
Communication	2	6	1	2	3	5	6	5	4	3
Meteorological	3	1	4	3	2	2	4	2	3	2
GNSS	4	6	4	4	4	6	6	5	6	5
Nanosatellites	1	2	2	2	1	1	2	2	1	1
Space stations	2	2	1	1	3	1	3	3	1	2
rockets	1	1	1	1	1	1	5	2	4	1
Space probes	3	3	1	1	3	3	3	3	2	1

Where:

1. Represents the infrastructures are related at some level but the interaction is only marginally important
2. Represents the interaction is important, but only for a few components of the CI in question, leading to an indirect and diffuse dependence on CSI
3. Represents the infrastructures are strongly interacting, and, in some areas, we can say that there is a degree of direct dependency of the whole CI on CSI
4. Represents the dependency of CI on CSI is important but not an existential threat
5. Represents the CI is wholly dependent on CSI activity and its disruption would cause a potentially irredeemable deterioration and cessation of CI functions
6. Represents both infrastructures are strongly interacting, and they are mutually dependent, the failure of any one of them causing great malfunctions in the other.

a bigger impact across all significant terrestrial systems. We estimate that, if space development continues and leads to lower prices and higher accessibility, new space systems will be put in place to achieve a deeper connection to critical infrastructures. As such, the score we assigned to highlight the importance of a certain CSI to a certain CI is not always indicative of a true connection between the infrastructures in their entirety. It is also based on the importance of certain components and certain circumstances that augment the overall relationship between the infrastructures in question. For instance, we find that the prospect of natural or manmade disasters is a significant boost for the relative importance of CSI availability in the functioning of the system-of-systems. At the same time, the sixth level of critical infrastructure dependence between space and terrestrial systems has not, technically, been reached yet, due to a lack of bilateral influence. However, we have decided to award six points to those infrastructure couplets where the significant, almost existential importance of the CSI on the terrestrial CI can act as an incentive for a malicious third party to interfere with CSI functioning. Thus, terrorism, criminal activity or military actions provide the threat to CSI that enables us to consider the CSI-CI relationship as being bilateral.

Another point to consider is that critical space infrastructures, especially satellites, have evolved to be completely dependent on two unique critical infrastructures: ground stations and communication links. Without them, either the quality of service would be much lower than at the moment (errors in positioning, in coordination and in control) or inexistent, since CSI rely heavily on the transmission of various data and the reception of data and instruction from operators and other infrastructures. Ground stations, in particular, feature heavily in the CSI framework, even though they represent a terrestrial infrastructure. Moreover, their impact on CI is dependent on their contribution to CSI operations, meaning that the effect is very strong, as well as indirect. Both ground stations and communication links, which are closely integrated to a degree, are vulnerable to willful disruption of activity, serving as the prime vectors for cyber-attacks and other forms of manipulation.

Global navigation satellite systems, along with their less prominent regional counterparts, are among the most important CSI for the widest range of terrestrial infrastructures. GNSS services are invaluable for a wide range of transport options, especially air and maritime. They are also beneficial for coordination and tracking of a large number of separate actors or equipment. Such equipment, bearing GNSS capabilities, can also be used to power in situ remote sensing capabilities, such as detecting cracks in dams or monitoring water flow. The use of ground stations to reduce GNSS errors makes such services viable for even the most delicate operations. The coordination function is vital for Emergency Services and Emergency and Crisis Management, as well as various military defensive and offensive operations. Moreover, GNSS comes into play in ICT and Finance, where the accuracy of its atomic clocks for sequencing data transmission and time-stamping transactions have become essential to the proper running of these systems.

Communication satellites offer the most widely used and recognized space system critical services and have become invaluable for the proper functioning of all critical infrastructures reliant on instant and constant communication. Their impact on business continuity and quality of life is staggering, and one finds it difficult to imagine modern economic, political and social activity without mediation by these infrastructures.

Earth Observation systems, consisting of remote sensing and meteorological satellites, offer a wide range of applications to a wide range of users. The information provided is integrated in activity under normal circumstances but finds special importance under exceptionally adverse conditions. Weather satellites routinely monitor and provide early warning for extreme weather patterns, while other remote sensing satellites monitor security and environmental threats, providing instant coverage. In case of natural disasters both optic and radar satellites provide images of the affected areas. For example, when floods occur, the International Charter on Space and Major Disasters is activated and sends images of the vulnerable regions that are used for mapping. The maps are then handed out to the different authorities. In addition to monitoring city lights, these photos are a lifesaving asset in the detection and monitoring of fires. Not only do the satellites see the fires visually day and night, but the thermal and infrared scanners on board these weather satellites detect potential fire sources below the surface of the Earth where smoldering occurs. Once the fire is detected, the same weather satellites provide vital information about wind that could fan or spread the fires. These same cloud photos from space tell the firefighter when it will rain. Snowfield monitoring can be helpful to the hydrologist keeping track of available snowpack for runoff. Ice floes, packs and bergs can also be located and tracked from weather space craft. Even pollution whether it's nature-made or man-made can be pinpointed. The visual and infrared photos show effects of pollution from their respective areas over the entire earth. Aircraft and rocket pollution, as well as condensation trails, can also be spotted. The ocean current and low-level wind information gleaned from the space photos can help predict oceanic oil spill coverage and movement. Almost every summer, sand and dust from the Sahara Desert in Africa drifts across the equatorial regions of the Atlantic Ocean. GOES-EAST photos enable meteorologists to observe, track and forecast this sand cloud.

Space probes, at first glance, have little impact on terrestrial infrastructures, because, by definition, they are removed from the vicinity of Earth. However, excluding probes which are of an exploratory nature, there is a category of probes dedicated to studying threats originating from space, such as solar weather patterns. The most notable is NASA's Advanced Composition Explorer probe, stationed at the L1 Lagrange Point, between the Earth and the Sun, which can provide a warning 15–45 min before a geomagnetic storm hits Earth. The adverse effects of space weather are felt by terrestrial infrastructures as well, and have cause numerous disruptions, suggesting that CSI are particularly vulnerable to intense solar storms. Therefore, the probes' early warning capabilities directly impact user's dependent on CSI provision of critical services in transport, finance, communication, national defense and so on.

Space Launchers do not directly impact terrestrial CI, but, considering that they underpin the development of space infrastructure, the indirect effect on terrestrial systems is very high. Rockets also have an important direct effect on national security, because a very high proportion of modern weaponry depends on the presence of significant launch capabilities.

Nanosatellite constellations are in their infancy. However, their affordability and expansion of potential applications can transform them into very popular solutions for cost-conscious governments and companies. This means that, in the future, a growing proportion of CSI may be represented by nanosatellite networks, providing remote sensing, communication, orbital observation of other CSI and others.

Space stations, currently represented only by the ISS, have a role in fostering international cooperation and research. As their associated costs decrease, and other space stations are constructed, the value of a manned orbital asset becomes apparent, as it can be very flexible in terms of applications that can be conducted. The ISS is already used for remote sensing capabilities, although the fact that it is a single object in LEO limits its coverage and efficiency. It also tests and launches new equipment and might, one day, offer on-site repair of CSI.

4.1.1 Critical Energy Infrastructure

The energy infrastructure fuels the economy of the 21st century. Without a stable energy supply, health and welfare are threatened, and the world economy cannot function. More than 80% of the world's energy infrastructure is owned by the private sector, supplying fuels to the transportation industry, electricity to households and businesses, and other sources of energy that are integral to growth and production across the nation. The energy infrastructure is divided into three interrelated segments, including: electricity; petroleum; and natural gas.

The heavy reliance on pipelines to distribute products across the nation highlights the interdependencies between the Energy and Transportation Sectors. The reliance of virtually all industries on electric power and fuels means that all sectors have some dependence on the Energy Sector.

Energy critical infrastructure use of comsats is limited. Even if, in some cases we can find some direct or indirect interaction between those two infrastructures we cannot say that there is a very important relation between them. This is mainly due to the fact that energy infrastructure, especially remote location, have a proper terrestrial communications system.

Today, GNSS is an essential element of the global information infrastructure. Hundreds of applications affecting every aspect of modern life have been developed. GPS technology is now in everything from cell phones and wristwatches to bulldozers, shipping containers, and ATM's.[1]

Conventional energy sources are still cheap, but alternatives are entering the market need significant government support and technological developments because they are still not mature enough. In this context, GNSS can be a valuable asset in the challenges related to Energy, providing added-value for infrastructure, transport and resource monitoring. Interaction exist but not important—1 point.

The influence of GNSS on Energy CI is not critical but increase along with development of new technologies. For substance monitoring, in case of some harmful substances (e.g. nuclear waste) used by conventional energy sources, GNSS can do: vehicle location, geo-fencing services or fleet management.

Directive 2009/31/EC of EU requests to monitoring techniques for Carbon capture and storage (CCS)—a new an approach to mitigate the contribution of fossil fuel emissions to global warming. CCS consists in capture of CO_2 at combustion source (power, cement, gas production etc.), compression of gas to dense fluid phase, transport by pipeline (or ship) to storage site and geological storage by injection to depth over 700 m, where CO_2 remains in supercritical fluid phase. This geological storage can induce displacement of fluids, ground displacement and induced seismicity. As a complementary solution to InSAR (Interferometric synthetic aperture radar), GNSS can be used to find out vertical displacement movements with mm accuracy, locally and in real-time.

Intelligent electricity distribution such as Smart Grid concept (Katina et al. 2016) is a complex technology requiring distributed grid management, including control centres balancing electricity, advanced interoperability mechanisms and precise environmental monitoring. GNSS can help monitoring location of personnel when checking the status of the infrastructure and provides a temporal reference cheap and easy to use to synchronize different network operations.

[1]GPS Applications, http://www.gps.gov/applications.

Power companies and utilities need improved time synchronization throughout the power grid in order to enable efficient power transmission and distribution. If you place GPS-based time synchronization devices in power plants and substations is possible to trace back the exact location of a power line break by analyzing precise timing of an electrical anomaly which propagates through a grid.

Hydropower generation is very dependent of water flow. So, it is important an understanding on the available water reservoir and water flows, and the tight monitoring of dam status (cracks). In this case, GNSS is used as in situ remote sensing instruments. Multiple measurements on reference GNSS stations located along the dam (pseudolites) can provide real-time information concerning dam deformation and cracks. Also, GNSS-R can be used locally to monitor lake levels and soil moisture.

Wind is one of the most challenging renewable energy sources as it requires historical data (some years) to locate the most appropriate places where to place a new farm, and to foresee wind behavior close to surface in short, medium and long term. To respond to these requirements, GNSS can be used as broad remote sensing technique like in GNSS on board flying device, where it can help to estimate wind speed at certain altitudes. A promising approach to get near-real time wind profiles is estimation of water vapor content in the atmosphere by analyzing the tropospheric delay of GNSS signals.

With regards to the use of remote sensing and weather satellites, energy use and city growth can be monitored since both major and even minor cities, as well as highway lights, are conspicuous. This informs astronomers of light pollution. The New York City Blackout of 1977 was captured by one of the night orbiter DMSP space vehicles.

4.1.2 *Information and Communication Infrastructure*

The Information Technology (IT) Subsector is central to security, economy, and public health and safety. Businesses, governments, academia, and private citizens are increasingly dependent upon IT Sector functions. These virtual and distributed functions produce and provide hardware, software, and IT systems and services, and—in collaboration with the Communications Subsector—the Internet. The sector's complex and dynamic environment makes identifying threats and assessing vulnerabilities difficult and requires that these tasks be addressed in a collaborative and creative fashion.

The IT Sector functions are operated by a combination of entities—often owners and operators and their respective associations—that maintain and reconstitute the network, including the Internet. Although the IT infrastructure has a certain level of inherent resilience, its interdependent and interconnected structure presents challenges as well as opportunities for coordinating public and private sector preparedness and protection activities.

The Communications Sector is an integral component of the world economy underlying the operations of all businesses, public safety organizations, and government. Over 25 years, the sector has evolved from predominantly a provider of voice services into a diverse, competitive, and interconnected industry using terrestrial, satellite, and wireless transmission systems. The transmission of these services has become interconnected; satellite, wireless, and wireline providers depend on each other to carry and terminate their traffic and companies routinely share facilities and technology to ensure interoperability.

The private sector, as owners and operators of the majority of communications infrastructure, is the primary entity responsible for protecting sector infrastructure and assets. Working with the federal government, the private sector is able to predict, anticipate, and respond to sector outages and understand how they might affect the ability of the national leadership to communicate during times of crisis, impact the operations of other sectors, and affect response and recovery efforts.

The Communications Sector is closely linked to other sectors, including:

- The Energy Sector provides power to run cellular towers, central offices, and other critical communications facilities
- The Information Technology Sector provides critical control systems and services, physical architecture and Internet infrastructure
- The Finance Sector relies on telecommunications for the transmission of transactions and operations of financial markets
- The Defence, Public Order, National depends on telecommunications for directing resources, coordinating response, alerting the public, and receiving emergency calls
- The Administration Sector with its postal services uses telecommunications for its control systems, tracking shipments, and regular communications requirements.

The Information and Communication Technology infrastructure has a pretty obvious relation with the communication satellites. ICT and comsats can operate relatively independent one on other but this only in the condition of losing many key capabilities. Interaction is very important and even vital in some respects. The impact on any one of those infrastructures in the event of losing the other is significant.

Precise time is crucial to a variety of economic activities around the world. Communication systems, electrical power grids, and financial networks all rely on precision timing for synchronization and operational efficiency. 'Stability of radio communications transmission, constant digital traffic flow, time slot alignment and traditional services over next generation Ethernet based infrastructure are some of the features that good time and timing bring to communications networks' (Royal Academy of Engineering 2011).

Accuracy, stability and reliability are necessary features for time sources. Ground-based clocks, especially with atomic clocks (CSAC) are accurate enough, but the synchronization of many such clocks is problematic. Each Global Positioning System (GPS) satellite contains multiple atomic clocks that give very precise time data to the GPS signals. GPS receivers decode these signals, effectively synchronizing to the atomic clocks. Time determined by users is within 100 billionths of a second. This is obtained without the cost of owning and operating atomic clocks.

For communications applications synchronization is a necessity. For example, "wireless telephone and data networks use GPS time to keep all of their base stations in perfect synchronization. Similarly, digital broadcast radio services use GPS time to ensure that the bits from all radio stations arrive at receivers in lockstep. This allows listeners to tune between stations with a minimum of delay.[2] So, ICT CI is strongly dependent of GPS time, especially for synchronization.

4.1.3 Chemical and Nuclear Industry

The Chemicals subsector can be divided into five main segments, based on the end product produced, as follows: Basic chemicals, Specialty chemicals, Agricultural chemicals, Pharmaceuticals and Consumer products. Each of these segments has distinct characteristics, growth dynamics, markets, new developments, and issues.

The Nuclear Energy Subsector includes nuclear power plants; non-power nuclear reactors used for research, testing, and training; manufacturers of nuclear reactors or components; radioactive materials used primarily in medical, industrial, and academic settings; nuclear fuel cycle facilities; decommissioned nuclear power reactors; and the transportation, storage, and disposal of nuclear and radioactive waste.

The Chemicals and Nuclear Energy Sector is dependent on and interdependent with a wide range of other sectors, including:

- Energy as a supplier of electricity to the nation's electrical grid
- Health as a supplier of nuclear medicine, radiopharmaceuticals and in the sterilization of blood and surgical supplies
- Transport through the movement of radioactive materials
- ICT
- Defense, Public Order and National Security
- Food Supply
- Water Supply.

The interaction is important, but only for a few components of the CI in question, leading to an indirect and diffuse dependence on CSI.

4.1.4 Water Infrastructure

The Water Sector includes both drinking water and wastewater utilities. The Water Sector is vulnerable to a variety of attacks, including contamination with deadly

[2]http://www.gps.gov/applications/timing/.

agents, physical attacks such as the release of toxic gaseous chemicals and cyber-attacks.

If these attacks were realized, the result could be large numbers of illnesses or casualties and/or a denial of service that would also impact public health and economic vitality. Critical services such as healthcare (hospitals), and other dependent and interdependent sectors, such as Energy, Food Supply, and Transport, would suffer negative impacts from a denial of service in the Water Sector. Comsats may be use for improving communications with some of this infrastructure elements but no important interaction can be mentioned.

Water CI is not so dependent on GNSS. In general, GNSS is used for hydrographical surveying. Underwater surveys are required by applications such as assessment for water supply and flood-control reservoirs, construction over waterways and adjacent inland development, flood and habitat studies. The most important information collected during hydrographical surveys is the water surface elevation which can be determined from an accurate GPS elevation in real-time. In order to assure safe drinking-water it is needed to assess the water supplies. GPS is used to collect the coordinates for potable surface water intakes, and reservoir boundaries.

4.1.5 Food Supply

The Food Supply Sector has the capacity to feed and clothe people. The sector is almost entirely under private ownership and is composed of farms, restaurants, and food manufacturing, processing, and storage facilities. The Food Supply Sector has critical dependencies with many sectors, but especially with:

- Water, for clean irrigation and processed water
- Transportation, for movement of products and livestock
- Energy, to power the equipment needed for agriculture production and food processing
- Finance and Chemical.

Effects of GNSS on food CI are very limited. GNSS is used to survey fish banks in oceans. The new concept of precision agriculture was made possible by combining GPS and GIS, coupling real-time data collection with accurate position information in new applications used for farm planning, field mapping, soil sampling, crop inspection, tractor guidance, autonomous vehicle control or yield monitoring. Due to GPS, farmers can work during rain, dust, fog or darkness, when visibility field conditions are low. Precision agriculture means precise application of pesticides, herbicides, and fertilizers, and better control of the dispersion of those substances, i.e. micromanaging of fields. Use of GPS, GIS and remote sensing offers information for improving land and water, achieved by better combining of fertilizers and other soil amendments.

Accurate maps of field area, with road locations, irrigation systems, and area in crops where weeds or disease is present, can be made using location information collected buy GPS receivers. All these data can also be used by aircraft sprayers equipped with GPS. These can fly with precision over the field, applying chemicals only where needed, and minimizing chemical drift. So, the amount of chemicals needed is reduced and the environment is protected.[3]

4.1.6 Health Infrastructure

The Health System protects all sectors of the economy from hazards such as terrorism, infectious disease outbreaks, and natural disasters. Because the vast majority of the sector's assets are privately owned and operated, collaboration and information sharing between the public and private sectors is essential to increasing resilience of the Health critical infrastructure. Operating in all EU states and territories, the sector plays a significant role in response and recovery across all other sectors in the event of a natural or manmade disaster.

While healthcare tends to be delivered and managed locally, the public health component of the sector, focused primarily on population health, is managed across all levels of government: national, state, regional, local, and territorial. The Health Sector is highly dependent on fellow sectors for continuity of operations and service delivery, including ICT; Defence, Public Order and National Security; Energy; Food Supply; Transport and Waters.

GNSS effects on Health CI are rare, although the health interacts with other CI like water and national security (through rescue operations in disasters, for example). Enhanced GIS usage may help to detect air, soil and water pollution and to monitor disease propagation.

During last decades, GPS has become an integral part of modern emergency systems, guiding emergency vehicles or helping stranded drivers to find assistance. In first case, location information of police, fire, rescue vehicles or boats, provided by GPS, reduces delay in dispatch of emergency services. Today, many mobile phones incorporate GPS, so car passenger can ask for help. Also, many ground and maritime vehicles are equipped with autonomous crash sensors and GPS, so even their occupants are unable to call for help, the vehicle, through an automatic communication system is capable to do so.

Due to the widespread development of satellite infrastructure, with its spatial component providing now complete coverage of the earth's surface, a wide range of health services can be provided remotely: instant communication in case of medical emergencies, monitoring the health care of patients, diagnosing different diseases, support in search and rescue operations etc. Next, we analyze the key areas related to health care in which the satellite systems are involved.

[3]http://www.gps.gov/applications/agriculture/.

4.1.6.1 Telemedicine

Telemedicine is the delivery of health care services, where distance is a critical factor, by all health care professionals, using information and communication technologies for the exchange of valid information for diagnosis, treatment and prevention of disease and injuries, research and evaluation, and for the continuing education of health care providers, all in the interests of advancing the health of individuals and their communities (World Health Organization, WHO 2010).

As can be seen, telemedicine can simply involve two health professionals discussing a case over the telephone or be as sophisticated as using the satellite technology to broadcast a consultation between providers at facilities in two countries, using video conferencing equipment. Telemedicine is used in order to provide medical services, through various means of communication, to the population who is spread on large areas of land (especially to the population living in remote or inaccessible areas). Telemedicine helps eliminate distance barriers and can improve access to medical services that would often not be consistently available in distant rural communities. It is also used to save lives in critical care and emergency situations.

According to the World Health Organization (WHO 2010), there are four elements defining the broad area of telemedicine:

(a) Its purpose is to provide clinical support
(b) It is intended to overcome geographical barriers, connecting users who are not in the same physical location
(c) It involves the use of various types of ICT (Information and Communication Technologies)
(d) Its goal is to improve health outcomes.

One of the first published accounts of telemedicine occurred in the early 20th century, when electrocardiograph data were transmitted over telephone wires.[4] Telemedicine, in its modern form, started in the 1960s in large part driven by the military and space technology sectors, as well as a few individuals using readily available commercial equipment. Examples of early technological milestones in telemedicine include the use of television to facilitate consultations between specialists at a psychiatric institute and general practitioners at a state mental hospital (Benschoter et al. 1965). Although the paper concentrates on the advantages and the disadvantages introduced by the use of the videotape recorder during the psychotherapy sessions, the tape is later played back to the resident and his instructor for discussion. This enables the provision of expert medical advice from a major teaching hospital to an airport medical centre (Dwyer 1973) (in this paper, the author describes an interactive television system set up between Massachusetts General Hospital and a medical station in Boston. The system has proven to be feasible and acceptable to individuals and institutions in the community, providing psychiatric skills on a much wider scale, in a more accessible way, and faster than any other system. The author delineates the implications and future uses that appropriately placed IATV links could have,

[4]See notes on the paper 'Le télécardiogramme' as cites in Rivera-Ruiz et al. (2008).

such as significantly multiplying existing manpower and, more evenly, distributing psychiatric care without placing greater demands on manpower).

In 1962, the American telecommunications giant AT&T launched the world's first true communications satellite, called TELSTAR. Since then, countless communications satellites have been placed into the orbit and the technology being applied to them is forever growing in sophistication.

The first use of satellite communications for telemedicine purposes took place in 1985. The National Aeronautics and Space Administration (NASA) first used telecommunication technology to furnish disaster aid following the devastating 1985 earthquake in Mexico City. The Advanced Technology-3 (ATS-3) communications satellite provided critical voice communication support for the international rescue and relief efforts of the American Red Cross and the Pan American Health Organization. The ATS-3 communication link was vital because the earthquake disrupted all land-based forms of communication in Mexico City, except a few ham radio operations. Within 24 h of the disaster, ATS-3 gave priority to satellite communication traffic involving disaster assessment and emergency rescue operations.

For the particular case presented in this paper—the satellite communications, it should be noted that this type of communication has several features ideal for providing telemedicine services, among which:

- Global availability—communications satellites cover all land masses and there is growing capacity to serve maritime and even aeronautical markets. Customers in rural and remote regions around the world who cannot obtain high speed Internet access from a terrestrial provider are increasingly relying on satellite communications
- Reliability—satellite communications can operate independently from terrestrial infrastructure. When terrestrial outages occur from man-made and natural events, satellite connections remain operational
- Instant access—satellite services can be instantly accessed through dedicated ground—based equipment (VSAT—Very Small Aperture Terminal, widespread in recent years). Satellite has proven its value as a provider of 'instant infrastructure' for commercial, government and emergency relief communications
- Broadcast/multicast capabilities
- Cost effectiveness—cost of satellite capacity does not increase with the number of users/receive sites, or with the distance between communication points. Whether crossing continents or staying local, satellite connection cost is distance insensitive.

The way medical services are provided via satellite communication is illustrated in the following diagram (Fig. 4.1).

The implementation of satellite communications in the field of 'Telemedicine' can be done in three ways, depending on the type of satellite links used: point to point, multipoint to single point, and multipoint to multipoint.

Point to point satellite link: represents the connecting method, via satellite, of no more than two nodes (facilities), as shown in Fig. 4.2.

Although there are many practical examples of point-to-point satellite connections, the implementation of telemedicine services based on this concept presents

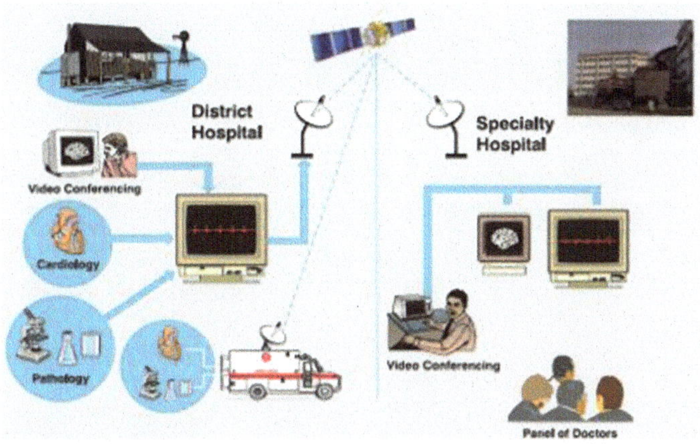

Fig. 4.1 Illustrating the concept of telemedicine via satellite communications, adapted from WHO (2010)

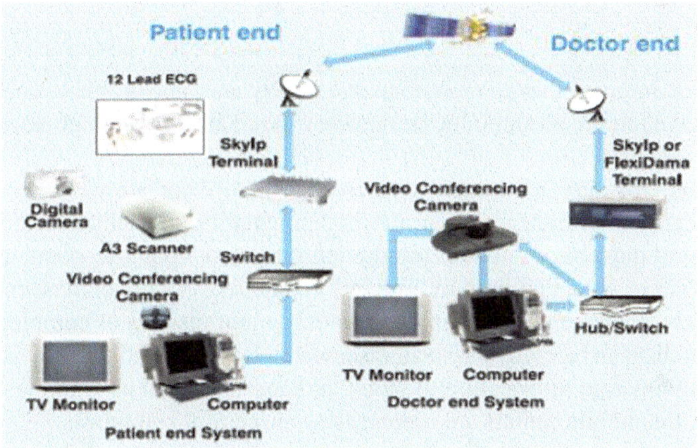

Fig. 4.2 A point-to-point satellite connection between patient and doctor/hospital, adapted from WHO (2010)

some significant disadvantages. Although the use of two dedicated channels for connecting patient to doctor and, obviously, doctor to patient will translate into a very good performance of the communication (which is continuous), the performance is reflected in the cost of the implementation, namely satellite terminal equipment costs and the channel transmission capacity cost. These costs are unjustified because building a single medical point for patients (which include medical equipment and satellite infrastructure technology), is not a viable solution, although this medical point could be a fixed one or a mobile one. The solution is not viable because telemedicine solutions implemented by use of satellite communications addresses mainly patients from large areas, patients that can hardly benefit from these services. In other words,

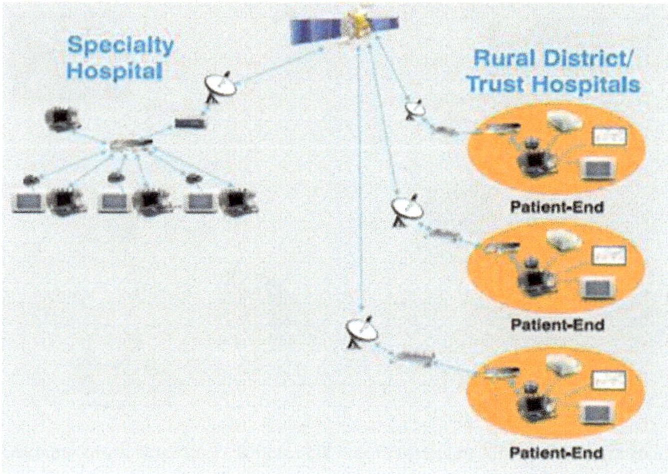

Fig. 4.3 Multipoint-to-point connection between many patients and one doctor/hospital

this solution imposes that no more than one patient at a time is able to communicate with the counterpart hospital/doctor, which is not a flexible enough solution to be implemented.

However, for very hard accessible areas, where there are no other means of communication and transportation, the implementation of mobile medical centers (regardless of the type of satellite communications used—point-to-point, multipoint-to-multipoint or single point to multipoint) is required. Although these mobile centers, already implemented in many countries, cannot dispose of complete medical equipment that can be used to treat all diseases, in terms of routine medical examinations, disseminating information to target audiences, as well as organizing existing resources, the mobile centers are representing successful solutions.

Multipoint-to-point satellite link: in addition to the more well-known point-to-multipoint communication type (used in broadcast transmissions—a single transmitter, many receivers), multipoint-to-point links are two—directional communication links, i.e. the transmitter may receive data (in this case, it is called 'hub') and the receivers have the capability of transmitting data. Figure 4.3 provides a model for such a satellite connection.

This type of connection is financially more efficient than the point-to-point communication-type, as costs are reduced twofold: firstly—the uplink channel capacity (the patients channel) is divided by the number of the patients (and hence costs) and, secondly, the complexity of the satellite equipment necessary to patients is reduced. This translates into lower total implementation costs (per patient) than the point-to-point scenario (it is true that in this case the complexity of the hub increases, but overall equipment costs decrease system-wide).

Due to the similar characteristics to the 'broadcast' communications type, this type of connection is suitable to the following areas of telemedicine, providing a

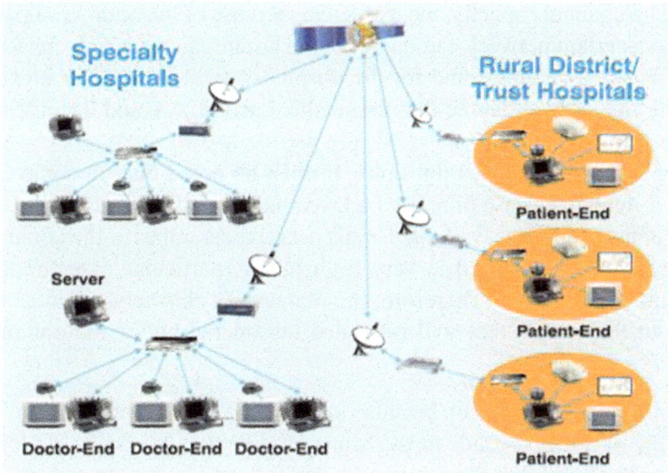

Fig. 4.4 Multipoint-to-multipoint satellite connection among many patients and many doctors/hospitals

very high level of performance: tele-training and tele-education. Accessing these services may be made by a large number of patients—type medical centers, the only requirements that must be met are:

- To have adequate satellite equipment
- To be in the area of coverage of the satellite
- To have a valid satellite connection.

Multipoint-to-multipoint satellite link: connections are used to connect a large number of nodes (facilities) via satellite, shown in Fig. 4.4. In terms of the subject studied herein—telemedicine—this solution offers maximum performance per cost, because connecting the patients to a large number of hospitals increases the likelihood that, regardless of the medical problems encountered by patients, they could be successfully resolved.

From the figure above, one can better see the advantage of implementing "multipoint to multipoint" satellite links: creating networks of patients who can connect through various communication media to a single satellite terminal, who is connected, at the other side of the link, to a network of specialty hospitals combines the advantage of a lower implementation cost to the wide flexibility in the design of the two networks (the hospitals and the patients).

However, it must be noted that this solution has, like all the other solutions, some drawbacks are articulated:

- Sharing the satellite resources between the members of each network. In other words, to get in touch with a particular specialty hospital, each patient must share the available resources with the other patients—and the same apply to the other end of the link. Although this is reflected in lower operating cost of satellite

resources—channel capacity, the excessive increase of the number of patients that belong to a certain network can decrease the communication link capacity for the satellite-hospital channels and for the hospital-satellite channels below a certain threshold, in a certain degree that the medical activities could become inefficient for the patients.
- Viability of the communication links. It must be noted that, in some cases (like satellite failure, or in case of really bad weather, or in case of any other factor that can make the satellite unavailable for his users) the number of the communication links that will be disrupted is very high in this particular case (multipoint-to-multipoint connections). Therefore, the planners of this network must take great care about the factors that will influence the operativity of the communication links and of the satellite.

As a conclusion to this point, because in medicine can be found special situations when a delay of a few seconds in the transferred data may lead to life losses, it all comes down to ensure that the optimum balance between the insured resources and the implementation cost is met.

The applications included in the broad area of 'Telemedicine fall into two categories, depending on the mode of data transmission between patients and healthcare professionals (or possibly just between medical staff): there are applications that require real-time transmission of data and applications that do not require real-time transmission of such data.

By 'real-time transmission of data' it is understood that the delay between the transmission of the data from the emitter and the arriving of the same data to the receiver does not exceeds a certain threshold, specific from application to application, but often limited to fractions of a second. Examples of applications that require real-time transmission of data are: telephone conversation and video conferencing. It is important to note that, for the particular case studied herein, of satellite communications, the uplink and the downlink communications channels introduce significant delays, dependent of the orbit of the satellites used—for example, for Low Earth Orbit satellites, the total delay introduced is less than 40 ms (round trip), for Medium Earth Orbit satellites the delay is approximately 125 ms, while for geostationary orbit satellites it still takes a radio signal about 250 ms, or about a quarter of a second, to travel to the satellite and back to the ground.

The first category—real-time transmission of data—includes services such as 'Remote Monitoring' and 'Real-time interactive medical services.' Remote Monitoring allows remote monitoring of certain parameters of interest from the patient by a healthcare professional (this application is used to monitor the progress of certain diseases like heart diseases, diabetes, asthma).

Remote monitoring may include video-conferencing, messaging reminder and surveillance questioning, and/or one or more sensors—such as electrocardiogram, pulse oximetry, vital signs, weight, glucose, and movement or position detectors (Stachura and Khasanshina 2007).

One of the first papers—and one of the best in the field—that created the bases for remote health monitoring is the 1995 article: 'Remote monitoring of health status of the elderly at home. A multidisciplinary project on aging at the University of New South Wales' (Celler et al. 1995). The paper discussed the design and implementation of a multidisciplinary research project and associated field trials to test the hypothesis that functional health status amongst the elderly can be accurately determined remotely by continuously monitoring relatively simple parameters that measure the interaction between participants and their environment. One of the primary goals of the project was to automatically prompt appropriate, timely and cost-effective intervention of medical and community-based services to help reduce morbidity and maintain an independent high quality of life for the elderly. Also, targeted intervention could diminish the demand for high cost medical services.

'Real-time interactive medical services' provides real-time interaction (such as telephone conversations or online communications) between patient and doctor. These services allow the achievement of certain medical activities that can be normally undertaken only by visiting specialized medical practices (physical examination, psychiatric evaluation, eye tests, access to medical history of the patients etc.).

The second category—applications that do not require real-time transmission data—includes services such as 'record and forward.' The 'Record and forward' is the component of telemedicine involving medical data acquisition, recording and subsequent transmission of the data to a doctor or a medical specialist, for analysis. This type of service does not involve the simultaneous presence of the two parties at the ends of the link.

From a medical point of view, the subdomains of the 'Telemedicine' field, as they are presented in 'Telemedicine, healing touch through space,' are the following (WHO 2010):

I. **Tele-consultations**—represent the exchange of pictures, text or data between health care professionals, in the purpose of collaboration/consultation on a particular case. Tele-consultations can be divided into real-time consultations and 'record and forward' consultations. Real-time consultations represent the modern methods of consultation between doctors, using technologies such as video-conferencing, which allows simultaneous interactions. 'Record and forward' consultations-type consist in distributing static images or audio/video clips, which in turn can be used for analysis at a later time. This method is more efficient than the tele-consultation method, especially when the participants are in different time zones. Also, this method is less practical in case of limited bandwidth.
II. **Tele-diagnosis**—the diagnosis is carried out from a remote location to the patient's disposal place and is based on the evaluation of the data submitted by instruments that monitor the patient health. The data is transmitted via communication links to the Diagnostic Center, where the evaluation of the data is performed.

An interesting new feature in the field of tele-diagnosis is presented in the paper "Design and Development of Communication and Control Platform for Medical Telediagnosis Robot (MTR)". This paper, appeared in 2013, presents a communication

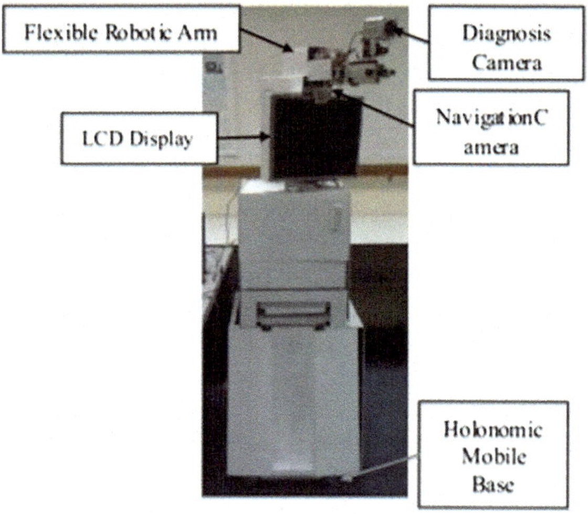

Fig. 4.5 The components of an MTR: medical tele-diagnosis robot

and control platform for Medical Tele-diagnosis Robot. The Medical Tele-Diagnosis Robot is a system which allows physicians to communicate with the patients, virtually present to give direction in a medical procedure, to diagnose patients and to round at the patients' ward. The Medical Tele-Diagnosis Robot (see Fig. 4.5) is equipped with basic medical instruments to allow specialists or doctors diagnose patients remotely, like a non-interventional flexible robotic arm to position and orientate the diagnosis camera (a high-resolution video camera) (Mariappan and Khoo 2013).

The communication and control platform for the Medical Tele-diagnosis Robot consists of a peer-to-peer based network, an audio-video communication module, a robot-computer interfacing device, a distributed medical information storage and retrieval system and a remote robot control interface. The MTR P2P overlay is intended to provide three types of functions, communication between Control Station and Robot, sharing of patients' medical data and multicasting of the tele-diagnosis session to the intended audiences. The MTR's P2P overlay is developed on Microsoft's Peer Infrastructure. The audio-visual system plays a very important role in MTR because it handles the core functionality of the system—visual (video) and audio (voice) communication for doctors, patients and nurses in (1) diagnosis, (2) discussion and exchanging of ideas between medical personnel and (3) coordinating a medical procedure. MTR requires two different cameras to operate at the same time during a tele-presence (tele-diagnosis) session: a high-resolution camera that is mounted on a multi-degree-of-freedom robotic arm—which gives the viewer at the Control Station a clear real-time image on the body part, and a lower resolution camera—used to provide the doctor at the Control Station eye contact with the person he or she is communicating with. The MTR's audiovisual system is developed based on Microsoft Media Foundation platform that is shipped together with Windows operating system since Windows Vista. Medical Information Management handles the storage, retrieval, presentation, creation, indexing, sharing and transmission of patient medical data using the P2P functionality supplied by the network engine. This component is shared by all the client entities in the network including robot, control station, local server and regular client (Mariappan and Khoo 2013).

Some of the critical components of the MTR System were tested for different criteria, depending on their functions and purposes. The MTR Network, the audiovisual system, the Robot Control Interface and the Medical Information Management were tested and simulated and they all were proven feasible and efficient.

III. **Tele-treatment**—represents the bi-directional link between the patient and the physician and is composed of data relating to the patient's vital signs and the communication to medical professionals on treatment methods appropriate to the patient's medical issues. Implementation tests have been made in this field. Some of these tests are presented in 'tele-monitoring and tele-treatment applications' section, which is based on communication via Internet between patient and doctor.
IV. **Tele-education** is the way to ensure professional education services at a distance. Satellite-based tele-education is not a new concept—for many years satellite equipment is used for the distribution of TV shows with educational content. However, only in recent years, along with the spread of IP technology and the decrease of the satellite equipment costs, there have been opened a lot of new opportunities in terms of broadcast and multicast distribution of educational content.

Internet-based medical education can take a variety of forms: online readings, online examinations, comprehensive educational programs, the use of electronic libraries, the use of specialized database fields. All these examples are found in satellite-communication based tele-education.

One of the first satellite contributions to telemedicine (more specifically, in tele education) is presented in the paper, 'Satellite contributions to telemedicine: Canadian CME experiences' (Chouinard 1983). In the article 3 experiments are described, all of them using the Hermes Satellite, a satellite launched in 1976 by Canada and USA. The first experiment consisted in providing medical services to remote locations and the second experiment consisted in production and broadcasting of native health information programs.

The third experiment was done at Memorial University of Newfoundland in St. John's, which has a long and varied experience in outreach education, particularly in sending videotaped programs to students in remote areas. After 5 years of planning, a 3-month experiment was performed in 1977 at the Health Sciences Centre of Memorial University, which was linked to four peripheral hospitals via one-way video and two-way audio communications. This was the first attempt to provide continuing medical education (CME) from a Canadian medical school to remote areas via satellite. Although funding was a big problem, the experiments were enthusiastic, and several important spin-offs ensued. The 150 h of programming demonstrated clearly that one-way television and two-way voice provided an ideal interaction for CME programs. However, live video in real time was too expensive and the satellite link was abandoned and replaced by a ground-based teleconferencing system (Chouinard 1983).

V. **Tele-training**—represents the professional training of medical and paramedical staff. In recent years a number of tests have been conducted by different countries on the effectiveness of these methods in the medical field. In some of these tests satellite links were used to transport the information. One of the last pilot programs developed, which includes an important component of tele-training, is the SAHEL program (African Satellite eHealth validation).[5] This program is currently implemented in Kenya and Senegal through Indra's telemedicine system and aims at achieving and further expanding the telemedicine services that are financially viable in Africa, by using satellite technologies. The tele-training component of the SAHEL program is provided in France by the Université Numérique Francophone des Sciences de la Santé et du Sport (UNF3S) and consists of eLearning services provision to the staff located in remote healthcare centers.

In recent years, many projects providing medical tele-training services through the use of satellite communication links have emerged, one of the most important of all being the MIST (Medical Interagency Satellite Training) program, funded by the Veterans Health Administration and DoD Joint Incentive Fund Program. The goal of the program is to significantly improve the amount and quality of training available to the healthcare community, without the students having to leave their houses. Programming offered on the MIST comes from a variety of sources and covers a wide range of topics, like immunization, pan flu planning, emergency response or the latest medical techniques.

The network that provides these services is managed by the ATN (Air Technology Network) Program Management Office, which coordinates the broadcast schedule for all users and oversees the technical operation and development of the network.

VI. Tele-monitoring—represents the communication of data concerning a patient's vital signs to some specialized medical personnel. Numerous tests have been made in this area on the effectiveness of this method, through the use of videoconference systems, whose main quality is given by the high degree of applicability and the possibilities for re-use (Broens et al. 2007).

VII. **Tele-support**—represents the assistance of the management operations in emergency situations. A good example of the support offered through satellite communications took place in 2008 when Zambia was hit by massive floods, that affected nearly 400,000 people, of which 36,000 have lost their homes. ITU (International Telecommunication Union) offered help by deploying 25 satellite terminals that were used to restore vital communication links and to coordinate the search and rescue operations of the victims.

Below are examples of telemedicine projects already implemented. In India, ISRO (Indian Space Research Organisation) has managed to connect hospitals and health

[5] https://business.esa.int/projects/sahel.

centers in remote rural areas with specialty hospitals in cities via INSAT satellites since 2001. That year started the 'Telemedicine Pilot Project' that connected the Apollo Hospital in the city of Chennai with the Aragonda—village hospital. The project continued to evolve and, in 2005, over 20 specialty hospitals in urban areas of India were connected with about 78 hospitals/medical centers in rural or difficult to access areas.

Many States have come forward to introduce Telemedicine on a regular operational basis and have planned to equip all the District Hospitals with Telemedicine facility both for ambulatory and Intensive care treatment. Karnataka, Chattisgarh, Kerala and Jarkand are some of the states which have initiated the establishment of Satellite Based Telemedicine Facility for all their district hospitals and a few trust hospitals.[6]

Within the same project, a mobile unit of intervention was established. This mobile unit of intervention consisted of a vehicle equipped with medical equipment and satellite communications. The role of the mobile unit was to create a telemedicine center wherever there was a need for it.

As a result of ISRO's Telemedicine endeavor, remote areas like Kargil and Leh in the North, offshore islands of Andaman and Nicobar and Lakshwadeep, as well as some of the interior parts of Orissa, Karnataka, Kerala, Chattisgarh, J&K, North-eastern states of India and some tribal districts in certain other states have access to specialty healthcare from some of the major specialty hospitals in the country today. Also, a part of the ISRO project, Continuing Medical Education efforts provide doctors at rural healthcare centres a chance to upgrade their medical knowledge and skills through interactions with experts at the specialty hospitals through satellite based tele-link. Such interactions indirectly result in significantly enhancing the quality of healthcare available to rural patients. The Continuing Medical Education programme has been integrated with the tele-education programme by linking some of the Medical Institutions with the Specialty Hospitals and Research Centres.[7]

The ISRO's Telemedicine facilities at GB Pant Hospital, INHS Dhanvantari Hospital at Port Blair and Bishop Richardson Hospital at Car Nicobar were used during the post Tsunami disaster relief work that took place after the country was hit by a tsunami caused by an earthquake on 26 December 2004 (although the impact of the tsunami on Thailand was far more considerable). The remote population of Andaman and Nicobar Islands was helped as part of this mission.

Encouraged by the steady growth of its Telemedicine programme, ISRO launched in 2011 a next generation communications satellite, called GSAT-12. GSAT-12 is a geosynchronous satellite equipped with 12 Extended C-band transponders (each with 36 MHz usable bandwidth) of which two transponders have been offered to Telemedicine activities. The two transponders have the capacity to absorb more than 500 telemedicine nodes. Currently, ISRO has installed around 400 nodes, enabling around 330 remote hospitals, 60 super-specialty hospitals and 14 mobile units to render services as teleconsultation, tele-follow-up and tele-education.

---

[6]http://www.televital.com/downloads/ISRO-Telemedicine-Initiative.pdf.

[7]http://www.televital.com/downloads/ISRO-Telemedicine-Initiative.pdf.

The GSAT-12 communication satellite replaces the Edusat satellite which went out of operation at the beginning of 2011 (the telemedicine services had been seriously affected following the disruption in connectivity linkages and have been down for several months). GSAT-12 has resolved the problem of lack of extended C-band transponders, which support tele-education and tele-medicine services, along with disaster management support.

In the U.S., the first mobile telemedicine unit—that belonged to the Army—was released in 1993 and included a ruggedized video conference system connected through a modem to an Inmarsat A-type terminal. This unit has been used by the United Nations in the Macedonian operations (February 1994) and in Haiti, but its bulky size and its massive weight (151 lb) made it suitable only for locations where a move at short notice was unlikely. That led to its subsequent use only within the fixed centers (Navein et al. 1998).

The second U.S. telemedicine mobile unit that belonged to the Army included a PC connected to an Inmarsat B-type terminal through a modem. A series of medical sensors, a video camera and one camera for capturing movies and images were connected to the PC, that could send the information forward, by e-mails (the 'record and forward' category of the telemedicine). The mobile unit provided video-conferencing services at speeds up to 64 kbps, that could be used from anywhere on Earth (except certain areas in the polar regions), and, unlike its predecessor, weighted only 60 kg. This unit has been used successfully in Bosnia, Macedonia, Saudi Arabia, Thailand, Panama and Belize.

There were a lot of telemedicine projects developed in the U.S. that were centered on the satellite communications links. Below are some of the associated projects.

In 1992, Mayo Foundation (Rochester—Minnesota, Jacksonville—Florida, and Scottsdale—Arizona), the National Aeronautics and Space Administration, and the Defense Advanced Research Projects Agency collaborated to create a complex network of fiberoptic landlines, video recording systems, satellite terminals, and specially developed data translators linking Mayo sites with other locations in the continental United States on an on-demand basis. The purpose was to transmit data via the asynchronous transfer mode (ATM) digital communications protocol over the Advanced Communications Technology Satellite. The links were intended to provide a conduit for transmission of data for patient-specific consultations between physicians, evaluation of medical imagery, and medical education for clinical staffs at remote sites. Low-data-rate (LDR) experiments went live late in 1993. Mayo Clinic Rochester successfully provided medical consultation and services to 2 small regional medical facilities. High-data-rate (HDR) experiments included studies of remote digital echocardiography, store-and-forward telemedicine, cardiac catheterization, and teleconsultation for congenital heart disease. These studies combined landline data transmission with use of the satellite. The complexity of the routing paths and network components, immaturity of available software, and inexperience with existing telecommunications caused significant study delays (Gilbert et al. 1999).

These experiments demonstrated that next-generation satellite technology can provide batch and real-time imagery for telemedicine. The first-generation of the ATM and satellite network technology used in these experiments created several technical problems and inconveniences that should be overcome as the network infrastructure matures.

In 1993, the Telemedicine Spacebridge Demonstration Project (a joint U.S and Russian program) emerged. The program's purpose was to further the application of telemedicine both internationally, domestically, and in space. The system has been set up to use a Russian satellite over the Atlantic Ocean and a U.S. domestic satellite to allow physicians a two-way video and audio link between various sites of medical centers in the United States and the Central Hospital in Moscow, Russia. The U.S. and Russian governments have entered into agreements and committed to the Telemedicine Spacebridge Demonstration Project which began regular operation on September 10, 1993 and continued its operational phase till May 20, 1994 (Zuzek et al. 1994). The project provided a two-way video link between medical specialists at participating medical institutions in the US and their counterparts in the Russian Republic. The system consisted of two earth stations located at NASA's Lewis Research Centre in Cleveland, Ohio (one is an US Ku band earth station and the other is a Russian earth station), coupled with video converters and crypto equipment. The Russian government made available to the demonstration project, satellite time on the WSDRN (Western Satellite Data Relay Network) satellite. The WSDRN or 'Loutch' satellite was the key segment of the overall Spacabridge system. This satellite was designed and launched by the Russian government to support the Russian space program. The satellite's features and its position at 16° West longitude allowed the configuration of a link between NASA Lewis Research Center in Cleveland, Ohio and Moscow, Russia for the Telemedicine Project.

The Telemedicine Spacebridge Demonstration Project has performed well, despite numerous technical and bureaucratic problems.

In Sweden, the first project of this type was the HECTOR (Health Emergency Which Through Telematics Operational Resources) project, a maritime medical consultation system. The system was designed that, in the case of an emergency aboard a military ship, to provide a direct connection to the Emergency Department of a certain hospital. The system could transmit data, voice, images, videos and also could guarantee videoconferencing communications between the ship and the hospital. The system was composed of: videoconference terminal, telemedicine terminal, medical sensors and photo/video cameras.

In a 1999 article, 'Telemedicine in Sweden—A diffusion study', a survey of telemedicine use in Sweden was carried out. The results are amazing: the response rate by hospital management was high at almost 90%. Sixty per cent of hospitals were involved in some kind of telemedicine activity in Sweden, with a further 15% planning telemedicine implementation. In rural areas of north Sweden good links exist between primary and secondary care. In the more populated southern region good links exist between the county hospitals and specialist university hospitals. Fifty-four per cent of telemedicine applications were used for specialist consultations, 13% for consultation between paramedics and hospitals and 10% for rounds. Teleradiology was the most frequent service (Sjögren et al. 1999).

In Romania, the first telemedicine project implemented was "The Telemedicine Demonstrative Pilot with Diagnostic, Clinic and Educational Applications", a part of the Aerospatial program, that was implemented starting with 2001 by the Fundeni Clinical Institute (as coordinator)—its Department of General Surgery and Liver Transplantation and its Department of Medical Imaging were the departments involved in this project, and by the Institute for Space Science and the InterNET SRL PRIVATE Company (as cooperators).[8]

The description of the project is best presented in the next picture (Fig. 4.6) and covered the Department of General Surgery and Liver Transplantation (DLT), as well as the Department of Medical Imaging (DMI) of the Fundeni Clinical Institute:

In the Pilot phase of the project, the Fundeni Telemedicine Pilot was connected with standing, accessible medical sites in Romania and abroad. Communications were made using the links as shown in Fig. 4.7.

The goals of the project were:

- To demonstrate the competence and infrastructural capabilities to set-up and use a true telemedicine system
- To increase the public and governmental awareness for the benefits of telemedicine
- To perform specific and public medical education
- To develop national wide partnership and international partnership
- To offer the pillar for the development of a National Telemedicine System
- To perform lobby for the telemedicine concept, system and funding (public and private)
- To integrate the medical system, space application system, and private initiative efforts.

At the end of the pilot project, the conclusions were that all the goals have been met and sometimes exceeded (the development of international cooperation allowed

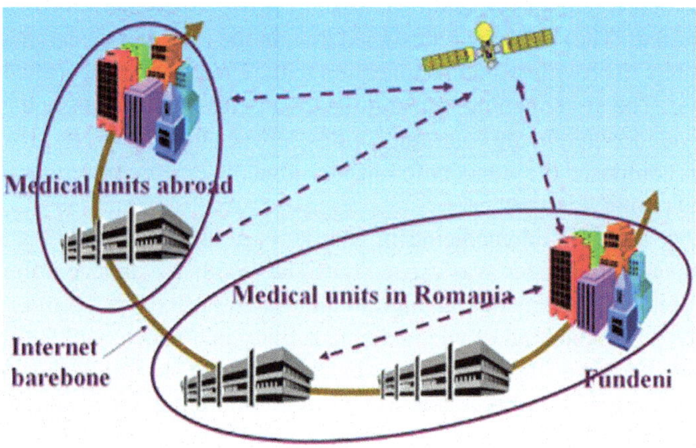

Fig. 4.6 The Romanian telemedicine demonstrative pilot with diagnostic, clinic and educational applications (Fundeni Telemedicine Pilot)

[8]http://www.e-medicina.ro/fundeni_telemed/english/pilot/pilot.html.

some pioneering, circumstantial projects, funded on international cooperation or sponsorship basis) and that the project should be continued. Due to the lack of necessary funding, the project was ended.

Another Romanian telemedicine project (actually one that remained only at the feasibility study level) was the Romanian Pilot centre for Telemedicine, an initiative of the Norwegian centre for Telemedicine (Tromsø, Norway), financed by Innovation Norway and by the project partners (Norwegian centre for Telemedicine and National Communications and Research Institute—Bucuresti, Romania). The project was thought to take place between June 2008 and March 2011, but the feasibility study only took place between June 2008 and Dec 2008.

Through this project, a pilot centre for e-health and telemedicine in a rural area of Romania must have been established. The Romanian partner, National Communications and Research Institute (INSCC) had already made research in the area around Pitesti, had established the framework for a centre and had initiated the dialogue with the local health personnel. Based on the feasibility study, an application was submitted to Innovation Norway, in December 2008, for a full project development.

One of the latest telemedicine projects implemented in Romania involves Infoworld, a Romanian company that provides integrated solutions for the health sector, who is compatible with the requirements of any type of health care provider. Infoworld's product portfolio includes systems for hospitals (HIS) and medical centers, radiology information systems (RIS/PACS), clinical laboratory software (LIS), family doctors, blood banks (CTS) and home/funds from public or private health insurance.

Infoworld has a vast amount of experience in the telemedicine area, being already involved in a telemedicine project—the Health Optimum project (HEALTH-care delivery OPTIMisation throUgh teleMedicine), along with Alcatel. Also, Infoworld has supplied and installed a data system for emergency telemedicine during the first emergency telemedicine project in Romania, operating at the emergency

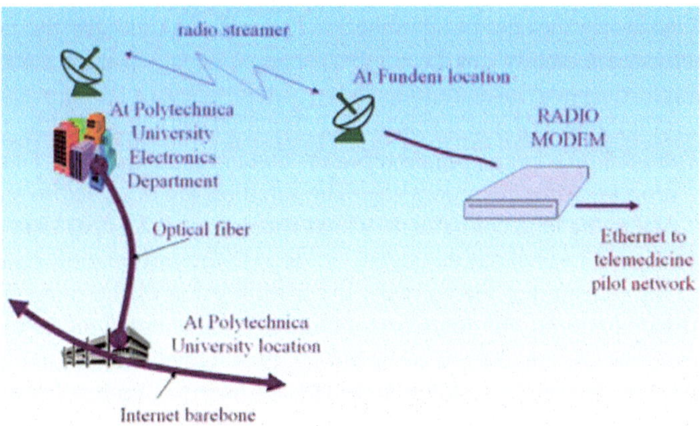

Fig. 4.7 The Pilot permanent communication link

unit of the Emergency County Hospital Targu-Mures, later extended in two regional centers in Bucharest and Iasi.

The project that is currently under development by Infoworld and his partners is the first Romanian telemedicine project implemented for medical emergencies and it involves an interconnecting system of audio-video and data through which 95 hospitals (municipal, city and county hospitals) are connected to regional centers (the Emergency County Hospital Targu Mures, the Floreasca Emergency Hospital and the County Hospital St. Spyridon) and to SMURD (Urgency and Reanimation Mobile Service). In this manner, these smaller hospitals can get help and medical advice for critically ill patients.[9]

The regional coordination centers are able of also monitoring the patient's vital signs—pulse, heart rate, respiratory rate, 12-lead ECG, pulse oximetry, CO_2, blood pressure. The images collected by the regional coordination centers can be simultaneously distributed to other several hospitals. One of the key parameters of the communications between those hospitals is that the confidentiality of medical care is provided by security mechanisms like authentication and authorization.

The monitoring devices used are all compatible with the HL7 standard and the application and data management services are also built in accordance with the HL7 Resource Location and Update Service standard. The database used to store the telemedicine sessions and the patients' data is Microsoft SQL Server 2008.

4.1.6.2 Search and Rescue Operations

Search and rescue represent a process of searching and helping the people who are in distress or imminent danger (and not only). Search and rescue operations using satellite connections as a means of communication can be described by the following scenario: an airplane, a boat or a climber, having an emergency radio transmitter on board (which is called Emergency Position-Indicating Radio Beacon—in the case of a vessel, Emergency Locator Transmitter—in the case of an aircraft or Personal Locator Beacon—in the case of a climber) is in danger. In this case, the radio transmitter is activated manually (for the climber) or automatically (for the aircraft/vessel, if, for example, the plane loses altitude with a certain speed, or the ship is submerged below a certain threshold) and it transmits a signal which is received by the closest satellite. The satellite contains signal repeaters that bring the signal to a certain level and signal processors for its processing. The satellites are designed to forward the received signal to the land component, which is composed of Local User Terminals (they are designed to analyze the signal, to decode the information related to the location of the transmitter and to make the identification of the transmitter based on an existing database), Mission Control Centres (which, based on the information received from the LUT and from other MCCs, is designed to store, to classify and to transmit these information away to Rescue Coordination Centres/Search and Rescue Points of Contact) and Rescue Coordination Centres/Search and Rescue Points

[9]http://www.infoworld.ro/en_telemedicina.html.

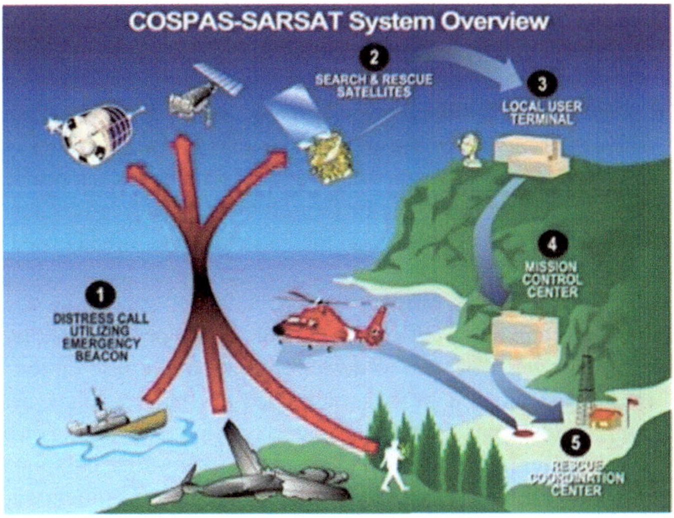

Fig. 4.8 An overview of Cospas—Sarsat system

of Contact (they are designed to coordinate the search and rescue activity of the aircraft/ship/climber).

One of the first search and rescue programs using satellite links as a means of communication was the Cospas—Sarsat program[10] (see Fig. 4.8), which was developed by Canada, France, USA and Russia in 1979. The services provided by this system are the ones presented in the previous paragraph: emergency transmitters location and detection—the transmitters activated by aircrafts, boats or people in distress. Many other countries have joined the project over the years, ensuring either ground segment components or joining as mere users of the system. In 2011, 26 countries, including Argentina, Australia, Brazil, China, India, Italy, Japan, Saudi Arabia, Spain, Turkey and the UK were suppliers of the ground segment, while other 11 states, including Denmark, Finland Germany, the Netherlands, Poland, Serbia, Sweden and Tunisia were simple users.

The satellitar segment of the system consists of 5 satellites in geosynchronous orbits—called GEOSAR, and 6 satellites in low polar orbits (low-earth polar orbit)—called LEOSAR. The GEOSAR satellites have signal repeater equipment on board, while the LEOSAR satellites have repeaters and signal processing equipment on board.[11]

The four founding countries (Canada, France, United States of America and Russia) have achieved the development of the Emergency Position-Indicating Radio Beacon (EPIRB) equipment on the 406 MHz frequency, so that it could be detected by the system. EPIRB has been seen as a key element in the search and rescue

[10]https://cospas-sarsat.int/en/.

[11]https://www.cospas-sarsat.int/en/system-overview/cospas-sarsat-system.

operations that occurred in dangerous environments. Before the founding of the Cospas-Sarsat System, the civil aviation used the 121, 5 MHz frequency for signaling in case of distress, while the air force used the 243 MHz frequency as the main signaling frequency in case of danger, with the 121, 5 MHz frequency used as an alternative. Thus, the Emergency Locator Transmitter (ELT) equipment-type for the civil aviation were built to transmit on the 121, 5 MHz frequency, a frequency already monitored by civil aircrafts, while the ELTs used by the air force was manufactured to transmit on the 243 MHz frequency, in the band that was already used by the air force.

At the beginning of its development, the COSPAS-SARSAT system was designed to detect the transmitters that used any of the 406, 121.5 or 243 MHz frequencies. Recently, the system was designed to detect only the alerts sent on the 406 MHz frequency. This allows the system to be optimized for this frequency, avoiding the traditional problems (such as false alarms) that the equipment working on both 121.5 and 243 MHz frequencies had. However, nowadays, many ELT equipment include both a transmitter on 406 MHz (for satellite tracking) and a transmitter on 121.5 MHz (that can be received by local search teams, using localization equipment).[12]

ELT equipment design has evolved significantly since 1982. The newest radio beacons on the 406 MHz frequency include today GPS receivers. Such equipment can provide, almost instantly, extremely accurate positions in case of distress, to the ground search and rescue centers, via GEOSAR satellites.

The SARSAT (Search and Rescue Satellite Aided Tracking) system, operated by United States Department of Commerce National Oceanic and Atmospheric Administration, is an integral part of worldwide search and rescue operations and is used to detect and locate mariners, aviators, and recreational enthusiasts in distress almost anywhere in the world at any time and in almost any conditions. The SARSAT system uses NOAA satellites in low-earth and geostationary orbits. The satellites relay distress signals from emergency beacons to a network of ground stations and ultimately to the U.S. Mission Control Center (USMCC) in Suitland, Maryland. The USMCC processes the distress signal and alerts the appropriate search and rescue authorities to who is in distress and, more importantly, where they are located. NOAA-SARSAT is a part of the international Cospas-Sarsat Program to which 41 nations and two independent SAR organizations belong.[13] From its starting point, in 2001, the system helped the following number of people in getting rescued (Table 4.2).

Galileo Support to Search and Rescue Service (SAR) represents the contribution of Europe to the international COSPAS-SARSAT co-operative effort on humanitarian Search and Rescue activities. Galileo is to play an important part of the Medium Earth Orbit Search and Rescue system (MEOSAR). Galileo satellites will be able to pick up signals from emergency beacons carried on ships, planes or persons and ultimately send these backs to national rescue centres. From this, a rescue centre can know the precise location of an accident. At least one Galileo satellite will be in view of any point on Earth so near real-time distress alert is possible.

---

[12] https://www.cospas-sarsat.int/en/system-overview/cospas-sarsat-system.

[13] http://www.sarsat.noaa.gov/.

Table 4.2 Number of people rescued since 2011

Year	People rescued
2016	307
2015	250
2014	240
2013	261
2012	263
2011	207
2010	293
2009	195
2008	282
2007	353
2006	272
2005	222
2004	260
2003	224
2002	171
2001	166

Galileo is a global navigation satellite system (GNSS) currently being built by the European Union (EU) and European Space Agency (ESA). On 21 October 2011 the first two of four operational satellites were launched to validate the system. The next two followed on 12 October 2012. The system comprises 30 in-orbit spacecrafts (including 3 spares) located at 23, 22 km orbital altitude. The satellites have 675 kg and the following dimensions: 2.7 m × 1.2 m × 1.1 m. Their lifetime is more than 12 years. The Galileo support to the Search and Rescue service, herein called SAR/Galileo, represents the contribution of Europe to the international COSPAS-SARSAT cooperative effort on humanitarian Search and Rescue activities. SAR/Galileo shall[14]:

(a) Fulfil the requirements and regulations of the International Maritime Organization (IMO) via the detection of Emergency Position Indicating Radio Beacons (EPIRBs) of the Global Maritime Distress Security Service and of the International Civil Aviation Organisation (ICAO) via the detection of Emergency Location Terminals (ELTs)
(b) Be backward compatible with the COSPAS-SARSAT system to efficiently contribute to this international Search and Rescue effort.

SAR/Galileo will allow for important improvements of the existing COSPAS-SARSAT system:

[14]http://www.navipedia.net/index.php/Galileo_Search_and_Rescue_Service#cite_note-GALHLD-3.

Table 4.3 The service performances requirements for the Galileo search and rescue service

Galileo support to search and rescue service (SAR/Galileo)	
Capacity	Each satellite shall relay signals from up to 150 simultaneous active beacons
Forward system latency time	The communication from beacons to SAR ground stations shall allow for the detection and location of a distress transmission in less than 10 min. The latency time goes from beacon first activation to distress location determination
Quality of service	Bit error rate <10e-5 for communication link: beacon to SAR ground station
Acknowledgment data rate	6 messages of 100 bits each, per minute
Availability	>99.8%

(a) Near real-time reception of distress messages transmitted from anywhere on Earth (the average waiting time is currently one hour)
(b) Precise location of alerts (a few meters for EPIRBs and ELTs equipped with Galileo receivers, while the current specification for location accuracy is 5 km)
(c) Multiple satellite detection to avoid terrain blockage in severe conditions
(d) Increased availability of the space segment (27 Medium Earth Orbit satellites on top of the four Low Earth Orbit satellites and the three Geostationary satellites in the current system).

In addition, SAR/Galileo will introduce a new SAR function, namely the return link from the SAR operator to the distress emitting beacon, thereby facilitating the rescue operations and helping to identify and reject the false alerts. The service is being defined in cooperation with COSPAS-SARSAT, and its characteristics and operations are regulated under the auspices of IMO and ICAO. The Search and Rescue Transponder on Galileo satellites detects the distress alert from any COSPAS-SARSAT beacon emitting an alert in the 406–406.1 MHz band, and broadcasts this information to dedicated ground stations in the "L6" band. COSPAS-SARSAT Mission Control Centres (MCC) carry out the position determination of the distress alert emitting beacons, once they have been detected by the dedicated ground segment. Table 4.3 articulates the service performances requirements for the Galileo Search and Rescue Service.

The Galileo Search and Rescue (SAR) Service begun to be tested after the second pair of the Initial Operation Capability (IOC) satellites were put into orbit, since these are the first satellites equipped with SAR search and rescue repeaters. The first successful test was conducted on 17 January 2013 and basically consisted of sending an UHF signal from ESA's Redu Centre in Belgium and reception of the reply in the L-Band antenna of the ground segment. After that it was signed two contracts for the deployment of the Galileo SAR Ground Stations: the first one on March 2013 with INTA (Spain) for the deployment of the Maspalomas (Spain) Ground Station in, and the second one with KSAT (Norway) on 17 April 2013 in order to develop the Spitsbergen (Svalbard), Norway station.[15]

4.1.6.3 Monitoring Disease Propagation

Tele-epidemiology is the science that consists in monitoring and studying the propagation of human and animal diseases (water, air and vector borne diseases) which are closely linked to climate and environmental changes, based on space technology.[16]

Vector borne diseases create the greatest burden in the world, because they are deadly and cover large area. World statistics indicates that vector diseases affect ½ of the world population in 140 countries. Every year 300–500 million clinical cases occur with 2–3 million fatalities. Children, old people and pregnant women are the most vulnerable targets. Africa is the most affected continent, which contributes 60% of the global cases and 80% of death. Climate and ecosystems determine distribution of vector and weather affects timing, duration, area and intensity of outbreaks. In general, warm and wet weather stimulated mosquitoes hatching, activity and the rate of diseases transmission to people. Such weather parameters as precipitation, temperature and relative humidity serve as the indicators of diseases and their development. However, weather station network is not dense enough—especially in Africa—for effective diseases monitoring. Therefore, satellite data have been used in recent years for global watch of vector diseases. New techniques have been developed and applied successfully for early detection and monitoring diseases from the operational environmental satellite.[17]

New Vegetation Health (VH), techniques have been developed and applied successfully for early detection and monitoring diseases from the operational environmental satellite. VH was developed from reflectance/emission measured by the Advanced Very High-Resolution Radiometer (AVHRR) flown on NOAA polar-orbiting satellites since 1981. The calibrated measurements in the visible and near infrared wavelength were converted to the Normalized Difference Vegetation Index (NDVI) and in the infrared wavelength to brightness temperature (BT), which were expressed as a deviation from 30-year climatology. Three indices characterizing moisture (VCI), thermal (TCI) and vegetation health (VHI) conditions were produced and calibrated against in situ data. They were applied to identify vector disease early enough to mitigate their consequences.[18] Depending on the values of these indices, complex formulas can be applied to obtain maps showing (e.g., see Fig. 4.9) the risk degree in terms of disease emergence.[19]

---

[15]http://www.navipedia.net/index.php/Galileo_Search_and_Rescue_Service#cite_note-GALHLD-3.

[16]http://www.isprs.org/proceedings/XXXVIII/8-C23/pdf/Lafaye.pdf.

[17]https://ams.confex.com/ams/93Annual/webprogram/Paper223622.html.

[18]For example, see https://ams.confex.com/ams/93Annual/webprogram/Paper223622.html.

[19]http://www.isprs.org/proceedings/XXXVIII/8-C23/pdf/Kogan.pdf.

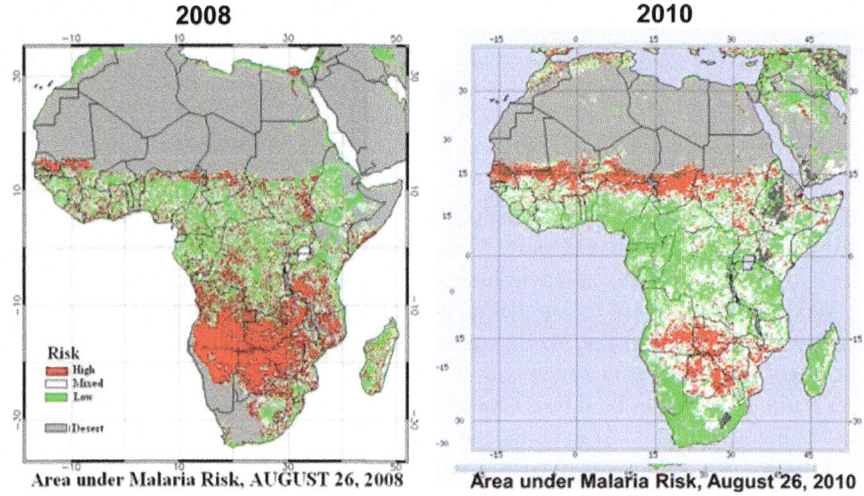

Fig. 4.9 Map of Africa with the risk degree in terms of malaria, for two arbitrary time points

Other projects regarding the monitoring and the studying of the way diseases are transmitted by carriers accomplished for many years by other entities, such as CNES ("Centre National d'Etudes Spatiales") in France, but the lack of adequate software tools to study the information that is transmitted by means of satellites is the main factor that slows down the development of this science. However, positive results were obtained and dedicated software applications for this science are constantly developed. For example, in 1998, CNES, in association with the Pasteur Institute, the French Institute for Agronomy Research (INRA) and the Veterinary Science School in Lyon (partnership called S2E), initiated a tele-epidemiology program using environmental data to better understand how the epidemics spread. The MEDES space clinic[20]—a biomedical experimentation centre and a CNES subsidiary—manages the operative part of this program. From 2002 till 2008, the S2E partnership has been involved in a number of projects, such as: EMERCASE (Rift Valley fever in Senegal and Mauritania), Guyana S2E (dengue fever in Guyana), S2E Argos (tropical diseases in Niger and Burkina Faso) and MATE (dengue fever in Argentina).

A first experiment took place in Senegal in 2001 and consisted in the deployment of portable electronic systems for the purpose of collecting health status data. This experiment took place in order to monitor the Rift Valley fever—a fatal disease to cattle, for which there was no cure at the time. This type of fever can be also transmitted to humans and is reported to be fatal in 5% of cases. The experiment consisted not only in the deployment of the portable electronic systems (for collecting the health status data), but also in the bio-mathematical shaping of the epidemics dynamics, through the use of remote-sensing satellites. Also, measurements of the epidemic factors (such as vegetation, hydrology, weather data or population) were made with

[20]http://www.medes.fr/en/the-space-clinic.html.

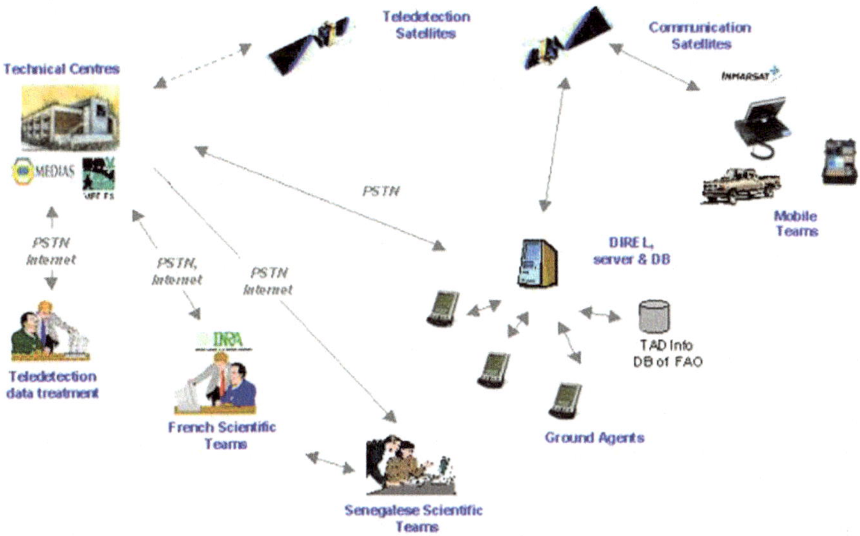

Fig. 4.10 Implementation scheme of a tele-epidemiology project by MEDES

the help of Meteosat, SPOT, Poseidon and Envisat satellites. Communication satellites were used for the rapid transmission of the collected data to the analysis centers, as can be seen in Fig. 4.10.

In July 2003, the system detected a high epidemical risk, after a period of intense hailstorms. A vaccination campaign was immediately initiated to try and stop the epidemic. Because the experiment has been successful, the Senegalese Animal Culture Board has decided to extend the pilot project at country level.

4.1.7 The Financial Infrastructure

The finance sector represents a vital component of any nation's critical infrastructure. Large-scale power outages, recent natural disasters, and an increase in the number and sophistication of cyber-attacks demonstrate the wide range of potential risks facing the sector. Financial institutions provide a broad array of products from the largest institutions with assets greater than one trillion dollars to the smallest community banks and credit unions. Whether an individual savings account, financial derivatives, credit extended to a large organization, or investments made to a foreign country, these products allow customers to: deposit funds and make payments to other parties, provide credit and liquidity to customers, invest funds for both long and short periods and transfer financial risks between customers. Financial institutions are organized and regulated based on services provided by institutions.

The financial sector and its component systems and infrastructures are critical to the economic performances needed to maintain and improve the level of prosperity of participant actors, from governments and companies, up to individual citizens. The financial system provides critical products and services for a number of human economic activities[21]:

- Price discovery process, which ensures the allocation of limited resources in a productive manner
- Mobilizing capital to the most productive uses
- Reducing the information asymmetry between different economic actors
- Instruments to reduce transaction costs and facilitate economic exchanges between different economic actors under conditions of uncertainty
- Instruments for reconciling preferences and price of supply and demand for capital
- Instruments for storing and providing liquidity value properties for immediate recovery.

The symbolic economy present from the beginnings of banking activities, has taken a particular scale in recent decades, exceeding in value the product of the real economy. The sophistication and diversity of financial products allowed a disconnection between directly productive economic activities and entities benefiting from their surpluses, leaving financial markets to intermediate the distribution of profits, transfer of ownership and translate information from daily market prices. This revolution has allowed a considerable economic expansion at the cost of vulnerability to threats that affect the functioning of markets and expectations of participants in them. As the barriers to participation in the financial sector have been reduced, the number of participants and their exposure to these risks has increased considerably.

Today, it is possible that extensive crisis occurs without any basis in the real economy, with only a financial dimension to a real effect on the wealth and welfare of society. These effects are enhanced by pessimism or irrational exuberance, as well as network links debt and accounting value that determines the welfare of modern business and how it is estimated the economic success. The fact that a large part of the governance in the financial sector addresses the area of financial risk where there has been most destructive 'innovation' in recent years, increases the potential for economic damage to real effects, as demonstrated by the 2008 financial crisis and the European sovereign debt crisis that followed.

Financial infrastructures consist of the following components and entities: banks, insurance companies, finance companies, national treasury and financial departments of non-financial companies, markets (capital, shares, futures, commodity), trading systems, regulatory frameworks, regulators, all of which are present in large numbers at local, national, regional and global level, forming networks of extensive linkages that ensure a coherent system.

Financial infrastructure developed over the last decades has become an irreversible global scale, possible only through the existence and development of digital technologies and support provided by space infrastructures. In tandem with their development,

---

[21] http://ideas.repec.org/p/wop/pennin/95-08.html.

financial sector infrastructure became connected with the ITC, the most internationalized and less dependent on the limitations provided by the physical space and geography. This allows not only greater economic efficiency by allowing investments in distant markets, access to foreign capital and realization of international trade, but also becomes a source of fragility, the potential cascading outages proper functioning of the various systems of the financial infrastructure. The complexity of financial relations between different entities served, in 2008, to the obscuration of the true risk propagation of financial shocks. With this experience in hand, it becomes clear how dangerous it can be any disruption to the critical space infrastructure on which the financial sector hardly depends.

The dependence of financial critical infrastructures to space systems may be indirect or direct. The former depends on the degree of dependence on space systems of other systems of critical infrastructures which are necessary for proper functioning. According to the scientific framework of critical infrastructure, the degree of interdependence of various infrastructures is high and rising as human activity becomes more alert and more complex. An example of this is the energy without current financial activity would be impossible. It should be noted that the financial system is composed not only of current transactions, but an entire continuity of transactional relationships that determine the positions and responsibilities of the participants in the market in the past, present and future. Thus, an interruption of power supply of the technical systems would not only hinder the development of international financial activity, but also locally, where financial players, brought face to face, would be deprived of the information necessary for any activity. And proper functioning of the energy sector is dependent control functions, control and coordination of critical infrastructure space provided. Moreover, all financial systems are to enable better governance risks to energy infrastructure, be it industrial accidents and slowdowns of transport, or the potential danger of cosmic storms, able to affect power grids and generating unit's electricity. Face dependence of the proper functioning of financial markets to the real economy and linking them more than the proper functioning of critical space through extensive use of them, now and in the future, more and more economic sectors, from agriculture and health to transport, industry and services.

The direct dependency on space systems is an area that has seen many advances in recent years, as space capabilities were successfully adopted the business model of financial actors. Of course, given the global nature of the financial infrastructure, communications satellites are very important to the daily functioning of the financial system. As intensive in terms of information and a horizon of activity increasingly shorter (high frequency trade, carried out by specialized algorithms can involve closing and opening positions in less than a millisecond, based on constantly updated information), financial sector shows a critical dependence of continuous and reliable communication capabilities and high capacity. Substitutability to terrestrial infrastructure is limited.

Another spatial critical infrastructure are the global navigation systems, not only for their positioning ability, but for their utility in synchronization with the help of atomic clocks on-board on each satellite of the positioning constellations. Using them is much cheaper than buying and maintaining a proper atomic clock by each actor and at the same time, it is more reliable than using synchronization data transmissions over the Internet.[22] As such, American network systems such as Global Positioning System (GPS) is used to record the exact date of completion of a transaction, providing an excellent way to keep records available and verifiable. Given the volume and value of daily transactions in international markets, correct timing of the performance of an act specifically in the market can make the difference between loss and gain. For this reason, financial markets such as New York and Tokyo are perfectly synchronized. Major banks are using atomic clocks placed in space to synchronize operational activities of the various divisions worldwide, coordination being imperative in the era of high frequency trading algorithms. Non-financial companies, both large and small, use GNSS for time stamping data transactions or billing their customers.

Disruption of these levels of synchronization would cause significant losses in terms of the operational capacity of the financial infrastructure. A good example of this is the 'Flash Crash' of May 6, 2010, when the Dow Jones indicator recorded the second biggest daily loss in history to more than 1000 points, which were then regained in a short period of time. The probable cause was a faulty trading, but the result was a cascading disruption of the functioning of other automated systems, as detected its deregulation of the market and its closed position to withdraw from a situation increasingly volatile. As sell orders flooded the market and the share price jumped from a few cents to a few thousand dollars in just a few seconds, the market's liquidity decreased, and so more risk averse automatic algorithms chose to leave the market. The vicious circle continued several hours until sufficient human actors-imposed order for pricing to regain credibility (Kirilenko et al. 2017). An error in timing information and controls would create as much chaos as high risk of cascading disruptions and extend this situation would lead to the withdrawal of human actors in the market, causing losses hardly recoverable. There is potential for criminal elements to jam the GNSS signal or replace it, not for causing the closure of markets, but to induce a delay of a few milliseconds, enough to make arbitrage between the de-sync markets and the guaranteed earnings (Humphreys et al. 2008).

Other 'critical space infrastructures' are the Earth observation systems, which provide important information on the financial decisions of the actors in the system and, finally, for the proper functioning of markets (Higgins 2012). Thus, insurers are using space systems to collect data on the risks they subject their customers, and to include these factors in calculating premiums they require.[23] For example, a farmer who wants to ensure their crop will be faced with historical data, current models about the quality of the soil, the severity of atmospheric phenomena such as rainfall or groundwater, which will influence the estimated crop and its risks and therefore will affect the amount insured. These data are obtained at a far less cost using the space systems that can be the only option in areas without the network of various measuring

[22]Official U.S. Government information about the Global Positioning System (GPS) and related topics, http://www.gps.gov/applications/timing/.

[23]See for example, https://www.esa.int/Our_Activities/Observing_the_Earth/Earth_observation_moves_closer_to_the_risk_business.

factors on the ground. Moreover, investment and development banks,[24] which have to channel tens of billions of dollars per year to projects in certain regions using the information provided by the 'Earth Observation' system to plan their strategy for the dissemination of funds. Such a project is assisting the World Bank, the European Space Agency to collect and interpret data to influence funding decisions. Thus, these systems become indispensable instruments for identifying opportunities to financial institutions for the judicious allocation of capital to the area's most 'profitable' in terms of economic or social, depending on the institutional profile.

4.1.8 Transport Infrastructure

The transportation system quickly, safely, and securely moves people and goods through the country and overseas. Transport infrastructure is crucial to the mobility of resources, assets and people in a world increasingly economically globalized. As a result of the internationalization process and then globalization, global transport infrastructures have become increasingly complex and diverse, having to solve problems of coordination and control that were not so pressing when the focus of economic activity was still locally and based on short trips. As a result, the transport sector was among the promoters to integrate the capabilities offered by spatial critical infrastructures to enhance the efficiency of infrastructure. Whether it is commercial, military or public transport, the penetration of space technologies is growing, new applications being developed by a leading actor and adopted by the rest of the market, part of a competitive process. The main space applications used are those related to positioning and navigation and monitoring weather conditions, but applications such as the management of system—of—systems through telecommunication satellites (coordination packages, just-in-time supply, etc.) are becoming increasingly prevalent.

The Transportation Systems Sector consists of six key subsectors, or modes:

- Aviation includes aircraft, air traffic control systems, commercial airports, heliports, and landing strips
- Highway Infrastructure and Motor Carrier encompasses roadways, bridges, and tunnels. Vehicles include automobiles, motorcycles, trucks carrying hazardous materials, other commercial freight vehicles, motorcoaches, and school buses
- Maritime Transportation System consists of coastline, ports, waterways, Exclusive Economic Zone, and intermodal landside connections, which allow the various modes of transportation to move people and goods to, from, and on the water
- Mass Transit and Passenger Rail includes service by buses, rail transit (commuter rail, heavy rail—also known as subways or metros—and light rail, including trolleys and streetcars), long-distance rail and less common types of service (cable cars, inclined planes, funiculars, and automated guideway systems)

---

[24]See for example, http://www.esa.int/Our_Activities/Observing_the_Earth/World_s_major_development_banks_look_closer_at_Earth_observation.

- Freight Rail consists of major carriers, hundreds of smaller railroads, active railroad, freight cars, and locomotives.

Most transport elements have a proper terrestrial communications system (including terrestrial radio communications). Some of the transport CI, for example air and maritime transport are strongly interacting with comsats and at some level we can identify a dependency on comsats.

Arguably, transport critical infrastructure is the most important critical infrastructure that depends on GNSS. For today, aviation or maritime navigation presence of GNSS is critical and became critical for road transportation and railways.

GNSS services have become vital to the smooth running of business within the trade routes.[25] Raising trade due to globalization of production systems has created the need for capabilities to manage trade flows safely. An example of this is the rise of commercial Southeast Asia as a manufacturing center and commercial transport, where 50% of the world's containers are passing through several straits increasingly crowded. Also, in shipping, port infrastructure development cannot keep up with the dynamic trade relations. Increased port activity occurs both in the number of vessels and their increasing size, leading to problems with the safe maneuvers inside the port while unloading or loading their goods. Without the coordination provided by the use of spatial critical infrastructure, normal operations might not carry out in the same conditions and with the same rapidity and safety, reducing the effective operational capacity of the transport infrastructure. Also, GNSS services are used to manage the next stage of the transport, such as coordinating the distribution of containers for takeover by other links of transport infrastructure.[26]

A similar situation is encountered in air transport, where increasing numbers of passengers and the development of certain airports as major transport hubs generated congestion problems which can be solved only by space capabilities. Aviation in itself is highly dependent on global positioning, especially as it automates the process of piloting ever more.

In aviation, most commercial aircraft now use GNSS for en-route navigation. Three-dimensional position determination for all phases of flight from departure, en route, and arrival, to airport surface navigation is possible from space-based position. GNSS was licensed by several states for initial approach and non-precision approach to specified airfields, even at remote locations where traditional ground-based services are unavailable. GNSS with improved signals is used for landing during poor visibility conditions. Automatic Dependent Surveillance—Broadcast (ADS-B) is increasingly used in areas of the world where there is no radar coverage; this involves aircraft calculating their position using GNSS and other sources and broadcasting it to other aircraft (Royal Academy of Engineering 2011). The new Area Navigation concept means a greater role for GPS because aircraft can fly user-preferred routes from waypoint to waypoint, where waypoints do not depend on ground infrastructure. GNSS also serves as an essential component for the Enhanced

[25] https://www.transportation.gov/regulations-fr/notices/2018-05864.

[26] http://insidegnss.com/karen-van-dyke-re-engineering-the-airways/.

Ground Proximity Warning System (EGPWS) that has proven successful in reducing the risk of Controlled Flight into Terrain, a major cause of many aircraft accidents.[27]

Meanwhile, agglomeration of air corridors would generate unacceptable risks in the absence of continuous updating of aircraft position, which may come to pass up to 500 m from each other during a normal flight. Air collisions would be more frequent or the air transport more rarefied if there was not the GNSS services. For road and rail transport, space applications have become increasingly used, however without the acquired critical aspect that the sea and air have. The development of individual transport comfort due to positioning applications is also a benefit, but without the critical connotations that the global supply GNSS system has acquired. Other benefits of using GNSS are able to reduce fuel consumption as a result of identifying a route and a walk rate that minimizes fuel consumption—be it on land or the air traffic, judicious application of spatial information from sensors may help determine an optimal course. Since fuel is an important component of the price of operating the transport sector, the ability to achieve savings is important in the broader context of the overall global efficiency.

GNSS provides the fastest and most accurate method for mariners to navigate, measure speed, and determine location, one method to increase level of safety and efficiency for all mariners. For ocean and inshore navigation, accurate position, speed, and heading are needed to ensure the vessel reaches its destination safely, economically and timely. Need for accurate position information becomes even more critical for port approaches, harbor entrance and docking by reason of vessel traffic and other waterway hazards making maneuvering more difficult and increasing the risk of accidents. GNSS data are used for underwater surveying, buoy placement, navigational hazard location and mapping. Navigation to optimum fishing locations and tracking of fish migrations, ensuring compliance with regulations need also GNSS data. Together with geographic information system (GIS), GNSS is playing an increasingly important role in the management of maritime port facilities such as management and operation of automated container placement, facilitating the automation of the pick-up, transfer, and placement process of containers by tracking them from port entry to exit. GNSS information is embedded within Automatic Identification System (AIS) transmission. The AIS, which is endorsed by the International Maritime Organization, is used for vessel traffic control around busy seaways, providing governments with greater situational awareness of commercial vessels and their cargo.[28]

At present, road transport applications are the majority users of GNSS signals, for in-car navigation, commercial fleet management, taxi services, public transport monitoring and passenger information, and emergency vehicle location, dispatch and navigation. Today GNSS is used to monitor vehicle location, making possible effective strategies that can keep transit vehicles on schedule and inform passengers of precise arrival times. Other important application is surveillance of road and highway

[27] http://www.gps.gov/applications/aviation.

[28] http://www.gps.gov/applications/marine/.

networks. GNSS help to identify the location of features on, near, or adjacent to the road networks.[29]

Rail systems throughout the world use GNSS to track the movement of locomotives, rail cars, maintenance vehicles, and wayside equipment in real time, reducing accidents, delay and operating costs, and increasing track capacity, customer satisfaction, and cost effectiveness. Positive Train Control (PTC) systems, a combination of real-time location information with sophisticated command and control systems are used to monitor and control train movements. These systems can automatically vary train speeds, re-route traffic, and safely direct maintenance crews onto and off tracks. For rail systems without PTC, GNSS gives dispatchers and passengers more accurate information on train arrivals.[30]

The complexity of the transport infrastructure is growing and the need for operational efficiency determines the implementation of new solutions offered by spatial infrastructures. Using satellites to monitor weather conditions (especially extreme weather events) is not new and may have benefits especially in maritime and air transport, but new capabilities allow Earth monitoring of ocean currents, sandbanks and other phenomena that may affect shipping. Sandbanks' problem is particularly critical in some areas, such as the Suez Canal and other areas with deceptive depth variations.[31]

Meanwhile, a combination of all-the-above space capabilities is useful not only for daily operation and emergency transport sector, but also for planning the future development of transport infrastructures. The best routes can be selected for the construction of new roads and railways, the busiest ports can be enlarged, and the analysis of urban, national and international flow of people and goods can constitute the foundation of future infrastructure projects that provide an increase utility due to the incorporation of this information in decision making. Furthermore, observation of the orbit technologies can provide an early warning about security threats or serve the daily management of the risks of such dynamic infrastructures. In addition, maintenance activities, needs repair detection and diversion of traffic flow becomes lighter when using the spatial critical infrastructure services (McCord et al. 2003).

The transport sector registers a myriad of indirect dependency linkages to the spatial critical infrastructure. The Theory of Critical Infrastructure Protection advances the idea of extensive interconnections between the various critical infrastructures that provide a propagation vector, cascading, of the effects of a critical system malfunction. As such, the ICS' influence is also felt by the close relationship between the transport and energy sectors, or between transport and the food and health, particularly in epidemics and their spread in the geographical space.

[29] http://www.gps.gov/applications/roads.

[30] http://www.gps.gov/applications/rail.

[31] https://weather.msfc.nasa.gov/land/ncrst/dot.html.

4.1.9 Defence, Public Order and National Security

The defense sector is the worldwide industrial complex that enables research and development, as well as design, production, delivery, and maintenance of military weapons systems, subsystems, and components or parts. The Sector provides products and services that are essential to mobilize, deploy, and sustain military operations. A system of prevention, preparedness, response, and recovery elements, the Emergency Services represents any nation's first line of defense in the prevention and mitigation of risk from terrorist attacks, man-made incidents, and natural disasters. It encompasses Law Enforcement, Fire and Emergency Services, Emergency Management, Emergency Medical Services and Public Works.

Remote sensing—Remote sensing makes it possible to collect data on dangerous or inaccessible areas. Military collection during the Cold War made use of standoff collection of data about dangerous border areas. In principle, these satellites provide imaging data to anyone who will pay for them. In practice, a country could exercise 'shutter control,' as the United States did during the beginning of the war in Afghanistan by buying the exclusive rights to the images in some parts of the Ikonos orbit.

The used of comsats for national security purposes is extended, especially in remote operations. Comsats provide the means to conduct important operations in almost any territory. The use of comsats as part of communications infrastructure is also important for national security.

For National Security CI GNSS/GPS remains critical, its applications being integrated into every part of military operations. Nearly all new military assets, from vehicles to munitions, are equipped with GPS. GPS is also used for tracking of vehicles and valuable cargoes and tracking of suspects.

GNSS has an important role for any successful rescue operation. The time of rescue is reduced based on knowledge of the precise location of landmarks, streets, buildings, emergency service resources, and disaster relief sites. As example, GPS has played a vital role in aid efforts for disasters such as the tsunami that struck in the Indian Ocean region in 2004, Hurricanes Katrina in 2005, and the Pakistan-India earthquake in 2005. In all these interventions, search and rescue teams used GPS, GIS, and remote sensing technology to create maps of the disaster areas for rescue and aid operations, as well as to assess damage. GPS is used also in the management of wildfires. Aircrafts combine GPS data with infrared scanners data in order to design fire maps that identify fire boundaries and 'hot spots' (Vamanu et al. 2016). In case of storm tracking and flood prediction, meteorologist can assess water vapor content based on transmissions of GPS data through the atmosphere.[32]

[32]http://www.gps.gov/applications/safety/.

4.1.10 Administration Infrastructure

The Government Facilities Sector includes a wide variety of buildings that are owned or leased by federal, state, and local governments. Many government facilities are open to the public for business activities, commercial transactions, or recreational activities while others that are not open to the public contain highly sensitive information, materials, processes, and equipment. These facilities include general-use office buildings and special-use military installations, embassies, courthouses, national laboratories, and structures that may house critical equipment, systems, networks, and functions.

In addition to physical structures, the sector includes cyber elements that contribute to the protection of sector assets (e.g., access control systems and closed-circuit television systems) as well as individuals who perform essential functions or possess tactical, operational, or strategic knowledge.

The Postal and Shipping Subsector moves million messages, products, and financial transactions each day. Postal and shipping activity is differentiated from general cargo operations by its focus on letter or flat mail, publications, or small- and medium-size packages and by service from millions of senders to nearly 152 million destinations. The services include automated processing facilities; local delivery units; many and varied collection, acceptance, and retail operations; transport vehicles including vans, trucks, tractor trailers, and aircraft; and information and communications networks.

The Education Facilities Subsector covers pre-kindergarten through 12th grade schools, institutions of higher education, and business and trade schools. The subsector includes facilities that are owned by both government and private sector entities and, therefore, faces unique challenges.

Every sector of the economy depends on the service providers in the Postal and Shipping Sector to deliver time-sensitive letters, packages, and other shipments. In particular, the Finance, and Health Supply Sectors rely heavily on the Postal and Shipping Sector for the shipment and delivery of critical documents and packages. Major interdependencies with other sectors include those with the ICT, Energy, and Transport Sectors.

In case of natural disasters both optic and radar satellites provide images of the affected areas. For example, when floods occur, the International Charter on Space and Major Disasters activates and sends images of the regions that are used for mapping. The maps are then handed out to the different authorities. In addition to monitoring city lights, these photos are a lifesaving asset in the detection and monitoring of fires. Not only do the satellites see the fires visually day and night, but the thermal and infrared scanners on board these weather satellites detect potential fire sources below the surface of the Earth where smoldering occurs. Once the fire is detected, the same weather satellites provide vital information about wind that could fan or spread the fires. These same cloud photos from space tell the fire-fighter when it will rain. Some of the most dramatic photos showed the 600 Kuwaiti oil fires that the fleeing Army of Iraq started on February 23, 1991. The night photos showed huge flashes, far outstripping the glow of large populated areas. Snowfield monitoring can be helpful to the hydrologist keeping track of available snowpack for runoff. Ice floes, packs and bergs can also be located and tracked from weather space craft. Even pollution whether it's nature-made or man-made can be pinpointed. The visual and infrared photos show effects of pollution from their respective areas over the entire earth. Aircraft and rocket pollution, as well as condensation trails, can also be spotted. The ocean current and low-level wind information gleaned from the space photos can help predict oceanic oil spill coverage and movement. Almost every summer, sand and dust from the Sahara Desert in Africa drifts across the equatorial regions of the Atlantic Ocean. GOES-EAST photos enable meteorologists to observe, track and forecast this sand cloud. In addition to reducing visibilities and causing respiratory problems, sand clouds suppress hurricane formation by modifying the solar radiation balance of the tropics. Other dust storms in Asia and mainland China are common and easy to spot and monitor, with recent examples of dust moving across the Pacific ocean and reaching North America.

In remote areas of the world with few local observers, fires could rage out of control for days or even weeks and consume millions of acres before authorities are alerted. Weather satellites can be a tremendous asset in such situations. Nighttime photos also show the burn-off in gas and oil fields. Atmospheric temperature and moisture profiles have been taken by weather satellites since 1969 (Cook 1969). Having established and defined relationship between CSI and other Cis, the following section attempts to establish quantitative and qualitative criteria for defining the 'criticality' of infrastructure systems.

4.2 Quantitative and Qualitative Parameters for Infrastructure Criticality

The criticality of space infrastructures can, with some exceptions, be measured by the same methods used to assess the criticality of terrestrial critical infrastructures, which benefit from a more substantial stock of research and other literature. While new criteria and parameters for space infrastructure criticality will be defined in this section and highlighted as such, the main purpose of this paper is to demonstrate that space systems and assets should be an integral part of the Critical Infrastructure Protection (CIP) paradigm, which has, until now, been confined to Earth-based systems. Neither should 'critical space infrastructure' protection take place "in a vacuum", focusing excessively on endogenous and specific threats such as space debris or space weather to the detriment of the powerful effects of other factors like technical standards, business considerations or international diplomacy which are driven by Earth-based systems. This diminishes the very synergies that arise from a complex, integral system-of-systems approach that lies at the core of CIP. Future studies may even come to the conclusion that, from a strategic protection perspective, individual Critical space infrastructure (CSI) systems have more in common with their terrestrial counterparts, with which they are interconnected, than with other CSI, with which they share the orbital medium and a number of specific threats such as debris and space weather.

At the same time, the very notion of criticality precludes the idea that all infrastructure systems are going to be critical. This is also true in the case of space systems, however the percentage of total space assets which can be identified is critical will, most certainly, be larger than in the case of their terrestrial counterparts. Some of the reasons why will be expounded in a later section of this paper, but the main reason is that the high economic and physical barriers to developing space systems, as well as the inherent risks of the incredibly hostile environment, will preclude the development of assets with low marginal benefits or with trivial uses. This may change with the anticipated lowering of the costs to access space but, for now, most space systems not engaged in pure scientific research are heavily relied upon by their users and operate with razor thin margins of error and redundancy. This is why the vast majority of space systems, excluding probes, are located in the various Earth orbits, with low-Earth orbits along the most useful corridors and geosynchronous orbits being the most crowded. The best candidates for critical status are the various systems which provide GNSS, communications and Earth observation capabilities, while launch systems and ground stations can be considered critical in the sense of maintaining the system-of-systems (for instance, by providing altitude control, or by ferrying additional satellite units to expand networks). Again, anticipated future development will probably increase reliance on space systems, through cost factor or diversification of application, while decreasing the criticality of any one system. However, as we shall see further, overdevelopment breeds new danger in the most valuable and crowded orbital bands.

The following is a list of quantitative and qualitative criteria and parameters that can define the criticality of a generic space infrastructure. The list is neither exhaustive, nor is it meant to be firmly grounded in the reality of current space assets. The first five parameters correspond to a vision of how the limited breadth of interactions between space systems, as well as their specific interactions, compel us to develop new indicators. The rest of the indicators are a generic list of parameters, also applicable to the terrestrial infrastructures for which they were developed and are best suited (Theoharidou et al. 2009). By adopting a more general and, implicitly, speculative view of the nature of space infrastructures, the relevance can be assured for the near future as we speculate that the complexity and size of the space infrastructure systems grows.

4.2.1 Scalability

We define scalability is the ability of a space-based infrastructure to expand its applications and size while maintaining or improving initial functions and performance. In general, this would apply to satellite constellations and other systems which require numerous investments and interventions to reach desired functionality, but possess, from an earlier stage, limited application and the capacity for incremental improvement. The most obvious examples are the global navigation systems: they have limited accuracy and coverage from the moment the constellation consists of three satellites, but every added unit increases the system's reliability and usefulness.

Therefore, scalability can be taken as a function of added capacity, added functions and time to completion as indicated in the following formula:

$$\mathrm{S} = \mathrm{f}\left(\text{added capacity, added or improved functions, } \frac{1}{\text{time to completion}}\right) \quad (1)$$

There are insufficient examples to establish a baseline for assessing how scalable an infrastructure really is. Given development times, launch programming and examples from previous extensive constellations, such as GPS and GLONAS, a rate of a few units per year seems to be an adequate target. The more scalable an infrastructure is, the more critical it may become, as the services it renders become available to more users at better quality, and extra demand is more easily absorbed. Scalability also means greater security, as the loss of a unit can be quickly rectified, and the size of the constellation minimizes the loss of service capacity and quality. This can be explained as establishing redundancy.

4.2.2 Redundancy

We define redundancy as the ability of a CSI to immediately compensate for the loss, due to malfunction or destruction, of some of its functionality. Either a component or a unit of a larger system has become erratic or inoperative, which would, under normal circumstances, lead to a lower provision of critical services and, in the future, goods. However, systems with redundancies in place take into account just such an eventuality and possess untapped capacity or other fail safes to ensure that disruptions are minimal or non-existent. An example can be of a global navigation system, which when at full strength, maintain a number of other satellites in reserve for any unforeseen disruptions. Redundancy's relationship to criticality is an inverse, as an CSI with back-ups in place is less likely to cause major damage to quality of life and business continuity, unless the disruptive event is of such magnitude, that it would exceed all redundancies put in place. This leads to the following redundancy formulation:

$$\begin{aligned} \mathrm{R} &= \mathrm{f}\left(\text{reserve capacity, } \frac{1}{\text{time to access}}\right) \\ \mathrm{C} &= \mathrm{f}\left(\frac{1}{R}\right) \end{aligned} \quad (2)$$

Such an event would be a significantly more powerful solar flare than has been registered so far, one that would exceed the shielding capacity of the CSI. Given advances in space development capacity, an increase in the scalability of space-based infrastructures is likely to diminish the perceived need for extensive redundancies. This should be guarded against, since it implies that the purpose of redundancies is, mainly, to guard against the total loss of the investment in the infrastructure, not the effects that the interruption of critical services will have on users and beyond. Either legislation or a system of economic incentive will have to be put in place to encourage, even with rising scalability, the maintenance of extensive redundancy as a strategic imperative.

4.2.3 Substitution

Unlike redundancies, where a defective component is replaced by an identical one or similar functionality is provided, the substitution (S_u) parameter expands the likely pool of systems to supplement damaged functionality to include infrastructures that have other main functions, owners, or operating parameters. Sufficient compatibility between systems must exist for substitution to be possible. This compatibility is, currently, an issue, as many space infrastructures are of a unique design which is, by economic necessity, confined to their own, narrow, mission. With time, global standards for compatibility may be adopted, and economies of scale finally be implemented in space-oriented industries. In a possible future, where many companies buy mass-produced assets such as satellites, probes, and station modules, with minimal differentiation, substitution beyond one's own limited system or mission type becomes not only possible, but also desirable as a form of low-cost insurance. One can even imagine a possible future where the owners of competing or compatible networks of nano or micro satellites agree to pool their redundancies together for the benefit of a hypothetical distressed constellation, until adequate repairs or replacements may be performed.

The capacity for substitution is not just dependent on technical considerations, but also those relating to property rights, inter-governmental or inter-company agreements, or national or international protocols for quick and temporary filling of a capacity gap in the provision of a critical service. Substitutes may be found on Earth, where local infrastructure can make up for the loss of capacity in its area of operations. Another likely example is in manned space activity where, in the near future, an experiment that can no longer be performed on the ISS may still take place in a space station build by private actors such as Bigelow Aerospace for a client. Or, in the event of a vehicle malfunction, astronauts may take refuge on one of these stations and use another vehicle for their safe return.

$$S_u = \mathrm{f}\left(\text{assured extra capacity}, \frac{1}{(\text{total required capacity for normal operations})}\right) \tag{3}$$

Faced with the threat of cybernetic attacks against its existing remote sensing assets in space, the United States is developing its capacity to substitute for the unexpected loss of one of its space imaging systems. That way, a surprise first strike is unlikely to cripple its intelligence potential and information feed to ground agents. For example, unmanned aerial vehicles (UAVs) can be used to augment satellite reconnaissance. In a conflict where the area of interest is confined to a theatre of operations, the reduced field of view available at these lower altitudes may be compensated for by using multiple UAVs (DeBlois et al. 2004). An airplane-based radar system, such as the U.S. Air Force synthetic aperture radar system JSTARS,[33] could also be used for tactical reconnaissance. Using UAVs and airplane-based radars is viable only in a region where the party has air superiority.

Another example was given in the first part of this paper, when discussing space probes and their role in monitoring space weather. It also mentioned satellites such as METOP, GIOVE-A and B, Galileo-IOV, XMM, Integral, Herschel, Planck, which also contain equipment for monitoring space weather, despite having altogether different mission. For instance, the two GIOVE satellites are GNSS prototypes for the Galileo network. The fact that they have these other capabilities as well is a very good thing in terms of monitoring space weather, but also a vindication of the idea that wholly different assets may have sufficient functional compatibility so that one may supplement the other's activity, especially in case of malfunction.

4.2.4 Dependency Between CSI and CI

Measures the criticality of a space infrastructure system by its importance in the normal functioning of a larger, terrestrial infrastructure system-of-systems. The degree of interdependence and interconnectivity are so high, that such systems-of-systems have worldwide scope, if not risk governance. A SI can be integral to the whole system's proper operation, or it may not yet possess the size or level of use which would make it a truly vital link in the system. Measurements are devised on a case by case basis, due to the varying nature of the purpose of certain systems.

An example within the global communication system would be to measure the flow of information through the world's or a certain group of communication satellites, and then compare it to global data flows, and judge their importance accordingly. Distinctions can be made between local, national, regional, and global data flows, with local and national being the least dependent (although still fairly so) on satellites, due to availability of terrestrial lines of communication. One can also have a situation, such as the global positioning systems, where the space-based infrastructure virtually represents the whole system-of-systems, with the ground control units confined to a secondary role, at least in normal operations. In either of the extremes, where

[33]For more information, see the U.S. Air Force Fact Sheet on E-8C Joint Stars, http://www.af.mil/factsheets/factsheet.asp?fsID=100 and the Air Force Technology.com's web page on JSTARS, http://www.airforce-technology.com/projects/jstars/.

the space infrastructures or the ground ones have limited functionality in the larger system, the dependency is close to zero and rises with certain circumstances.

In the future, we can expect cross-dependency between CSIs, where solar power plants are built in advantageous locations and send their energy through microwaves to a space-based recipient such as a moon base that desires full solar coverage. There is also a distinct reliance on the part of planetary probes, such as NASA's Curiosity, on a satellite to relay information back to Earth, whereas initial probes had to have the capacity to transmit data on their own. Criticality is also set to increase when economic efficiency will become an important factor, and different space infrastructures come to rely on the same provider of services, such as communication and power. To this end the following formulation is suggested:

$$\mathrm{f(CI, CSI)} = \mathrm{f}\left(\frac{\mathit{CSI\ total\ capacity}}{\mathit{CI\ capacity}}, \frac{\mathit{CSI\ capacity\ critical\ to\ CI}}{\mathit{CI\ capacity}}, \frac{\mathit{CSI\ capacity\ critical\ to\ CI}}{\mathit{CSI\ total\ capacity}}\right) \quad (4)$$

4.2.5 Effect of CSI on Other Systems

This is a parameter that identifies the basic degree of connection between one critical infrastructure system and another. The whole basis of Critical Infrastructure Protection is that physically distinct infrastructures impact one another through their role in each other's operations. The previous point was made for infrastructures belonging to the same class of function and their weight within the whole. CSI effects can be direct or indirect, the first implying criticality in relation with the normal operation of target infrastructure and the other a less critical relationship. For instance, weather and global navigation satellites are vital to air transports, but can, more or less, be disregarded in road transportation with a minor loss in comfort. Conversely, they have only limited impact in the health and education systems, which are also recognized as critical infrastructures.

The following parameters for assessing the criticality of a space-based infrastructure are based on a similar framework designed specifically to describe the criticality of terrestrial, or traditional, critical infrastructures. A wholesale adoption, followed by adaptation of the general parameters used in discussing the importance of CI is necessary if we are to create establish the concept of CSI not as a stand-alone concern, but as a highly important and timely addition to the general framework of recognized critical infrastructures on which modern activity depends. As mentioned before, this allows proper integration of CSI into the risk governance and disruption mitigation processes, while keeping in mind their unique properties and challenges.

Impact factors relating to scope do not measure the depth of damage done, merely its breadth. It is one of the most important categories of parameters when assessing terrestrial infrastructures, which are numerous, highly variable and highly individualized, but, given the nature of CSI, one can already postulate that the scope of CSI disruption, especially those defined as α type in the previous chapter, is, at the very least, regional or continental in scale.

4.2.6 Population Affected

Using a four-tiered approach to defining thresholds for impact, populations affected would range from 100 individuals to over 10,000. This would be patently untrue for most Critical space infrastructures, since their expense and challenge of creation means that they are a collaborative effort, servicing a wide array of end-users. Otherwise, their creation would not be justified. We would propose a scale ranging from the high millions, to the low billions, with the caveat that, while certain disruption effects might be diffuse and low intensity, others, within the same negative event, would be compounded by other negative circumstances. This means that factors such as population affected should be judged on a case by case basis.

For instance, the disruption of weather satellite connections over an entire region would severely impact prognosis ability. For most individuals, as well as businesses, the inconvenience would be minor. However, in a certain geographic area, prone to extreme weather patterns like hurricanes, weather satellites are invaluable to observing and tracking a coming storm. Without early warning, there are few opportunities to mount an effective evacuation or economic damage mitigation efforts. Therefore, the scope of a weather satellite breakdown is in the tens and hundreds of millions, but the most severe outcomes are registered in a certain area, where only a few million people might live. It would be in the interest of future CIP efforts if scope were also contingent on severity, so that the quoted affected population would be amended to highlight the plight of those few millions living in areas prone to experiencing natural disasters.

Another example can be a terror strike, which might be directed at a certain area or target but, given the inclusiveness of CSI services, would register many other targets as collateral damage. In this case, the following formulae is suggested:

$$\mathrm{P} = \mathrm{sum}\left\{\frac{(\mathrm{S_i} * \mathrm{P_i})}{i} = 1 \ldots \mathrm{n}\right\} \quad (6)$$

where,

p is abstract concept of infrastructure impact on affected population
$P_{1 \ldots n}$ is the population segments
$S_{1 \ldots n}$ is the degree of severity of consequences; incorporates factors other than disruption

4.2.7 Concentration

Again, four tiers can be used to address the issue of population concentration per sq. km in which numbers go from 0 to 250, 250 to 500, 500 to 750 and over 750. This scale remains applicable for CSI, when discussing disruptions of critical services and goods provision at local or regional level. Otherwise, population density is irrelevant as a measure of criticality in itself. Where it becomes apparent that density is very important is when we take into consideration other factors, such as economic impact. Places with high population density are often the richest in their respective region, regardless of whether we are discussing a rich country or one in the on-going process of development, which amplifies the economic damage associated with the loss of CSI services. Places with high population density also feature the highest concentration of terrestrial infrastructures, which raises the risks associated with cascading failure of infrastructure operations.

Rich countries, in particular, are the recipients of most of the consequences associated with CSI malfunction or destruction, as they are the most heavily invested in the use of such services. Low density areas in such places are especially vulnerable to the effects of CSI service disruptions, as they may lack viable substitutes (such as landlines for communications) and may have developed to depend heavily on CSI-dependent contact with high density regions for their quality of life. A good example would be inhospitable regions, such as Alaska in the United States, or the northern areas of Scandinavia, which, despite their remoteness, are, nonetheless, entitled to the same level of services as the heavily populated, more temperate areas. Political commitment to not letting quality of life suffer in such regions, as Finland's broadband initiative for rural area proves, it not only costly, but also raises dependence on CSI.

$$\mathrm{C} = \mathrm{sum}\left\{ \frac{\left(\frac{S_i * P_i}{A_i}\right)}{i} = 1 \ldots \mathrm{n} \right\} \tag{7}$$

$P_{1 \ldots n}$ population segments
$S_{1 \ldots n}$ degree of severity of consequences; incorporates factors other than disruption
$A_{1 \ldots n}$ geographic area corresponding to P_i.

4.2.8 *Range*

In this case, range can be described in terms of local, national, regional, or international. This is essentially, the same for CSI. However, as was highlighted before, the low volume and high efficiency of CSI in supplying certain services, means that the main areas of effect are at regional and international level, which corresponds to the coverage provided by a certain CSI at any given moment. National and local ranges are less about technical standards and operating efficiency, and more about the political nature of a certain CSI.

For instance, new global navigation systems, such as the BeiDou network under construction by China and the one proposed by India, would initially cover only their respective national territories. It should be noted that most of these would invariably plan to expand their coverage in order to be commercially viable, so they would amplify their range beyond national level. A national level political focus could remain, by maintaining the right to unilaterally terminate the provision of service to third parties in case of war or other national crisis, but, again, this is not the result of a technical element of design, but a political one. The US Global Positioning System, which is the premier provider of such services worldwide, has maintained its military backdrop and insists (and has even conducted exercises in this regard) that it has the right to terminate service even to allies in a vaguely defined case of national emergency. The only emerging network not to have this provision is the European Galileo GNSS.

Returning to the description of range, it is quite possible that advances in transportation, functionality and cost of satellites could lead to economically viable networks (incipient CSIs) whose focus would remain at local or national level. Such satellites would be nanosatellites, or their owner and operator would be various important subnational entities, such as rich municipalities and the like. This is not so much of a stretch, when considering that the megacities of the world fund their own research institutes and initiatives for cooperation, within the confines of their national legal framework, of course.

The next group of impact factors describe the severity of the effects of the disruption of CSI services and, one day, the provision of goods.

4.2.9 Economic Impact

The disruption of the services provided by critical space infrastructures would lead to significant economic losses, at multiple levels, highlighting the significant interdependences between infrastructures that constitute the framework underlying economic life in advanced societies. It is in this framework that critical space systems have integrated themselves and acquired true importance, especially as information providers and mediators. The richer a country is, the more substantial its business and individual dependence on services provided by existing CSI and the more likely it is for that dependence to grow in the future.

The first level to consider is the cost of CSI destruction or malfunction itself, entailing the need to realize expenditures for repair and replacement and registering a loss if the asset was not fully amortized. While the entity that has to swallow these costs would not agree, this is, by far, the smallest component of the damages—up to the low tens of millions of euros. It will, likely, get even smaller as the cost of the systems themselves and their placement in orbit are reduced steadily by economies of scale and new technologies.

The next level is the loss of revenue for the entity, if any, that lack of system functionality will entail which is again, if not negligible, at least not systemically important. Next, we have the economic loss of direct users relying on proper infrastructure operation. Either they suffered a direct and avoidable financial loss, or they lost an economic opportunity, or they suffered a loss of utility that can be translated in economic terms. These figures, in the case of today's heavy concentration of use on a handful of space infrastructures, are liable to be astronomical in certain cases, such as communications, and very difficult to ascertain. Then, we have the secondary losses, of entities affected by the losses of primary users. Given heavy interconnections, these secondary victims are also primary ones, losing on two fronts. The second rank losses consist mostly of lost economic opportunity in relation to the primary victims of the disruption.

By now, they are almost impossible to predict, though the first rank losses should be larger than the secondary ones. Finally, we have an eminently measurable loss, at least in the short-term—loss of confidence in the affected companies, the affected states, the responsible parties and the general economic climate, leading to rising interest rates and a lowering of valuations on the entire stock market, especially of those most affected by the CSI disruption (the owner and the primary business users). While the wealth lost this way may be recovered in a subsequent recovery of the market, its existence is not illusory, since stock valuations, even when not liquidated, are the basis of monetization through financial instruments, such as collateral for contracting debt. The daily variations, including the violent swings occasioned by disasters such as critical infrastructure disruptions, are very important to the economic well-being of many actors, as proven by the on-going world economic crisis, initially triggered by the lowering market value of over-invested assets. The economic losses suffered at this stage will also be described in the next section, on public confidence, but mainly in the next chapter, where market expectations will be further addressed. As mentioned, the low point of economic impact starts at a few million euros, and is liable to fall with time, and the high points are beyond prediction, but measure in the billions of euros at least.

Growing dependence on a limited set of infrastructures, both of existing users and new users, will also lead to an increase of these amounts, however, further development of space systems may lead to adequate redundancies and substitutes so as to mitigate the worst of the disruptions and, consequently, losses. The criticality of distinct assets will, itself, fall, to the point where the economic impact is limited to the cost of the asset itself. As of yet, the pool of assets for important operations, such as communications and coordination of activity, is still very small and, so, the potential for significant losses associated with a disruption, very high. Even if the systems are multiplied, they are still vulnerable, en masse, to disruptions by space weather and debris fields, which would negate the advantage in numbers. At the very least, their replacement would be faster and cheaper.

$$E = E_{CSI} + E_{user} + E_{opportunity} + E_{secondary} + E_{market\ confidence} \qquad (8)$$

Traditionally, economic scale impact ranges from 1 million to over 100 million dollars and it should be safe to say that, as things stand, the low estimates for CSI malfunction impact should be in the tens of millions of euros and high estimates should begin in the billions. Even at cost levels fewer than one billion euros, the criticality of that infrastructure should be called in question and, as space development matures (and our reliance grows deeper), a catastrophic and cascading CSI disruption will inflict far worse damages than our most pessimistic scenarios. As of yet, there has not been an episode of serious disruption of CSI service, for a baseline of resulting costs to be established.

4.2.10 Public Confidence and Perception

Public perception counts for a great deal in the operation of politics and markets, today. Even if the results of a significant negative event are contained to a degree which experts would find satisfying, public confidence and perception, independent as it is sometimes from facts and reality, will sanction public and private actors according to its own internal calculation. At its worst, confidence in markets and institutions is eroded, leading to self-fulfilling and reinforcing predictions of negative occurrences.

To a certain extent, critical infrastructures are unaffected by public perception, since they are technical constructs. However, the enlargement of the palette of systems considered critical and integrated into the framework of Critical Infrastructure Protection has led to the introduction of human oriented systems, such as administration and governance. The further integration of CSI into the larger CIP framework is due to the realization that a disruption of CSI activity, since it is very visible for the individual, will, in the end, impact public confidence and perception. Citizens and businesses expect governments to provide a minimal safety net against the worst negative occurrences, and an interruption in communications, digital services, positioning and other such elements of daily life would elicit a powerful response from the public. How a government or another authoritative entity handled previous disruptions, in the past, will impact present perceptions.

There is an ample body of evidence to suggest that dissatisfaction and lack of confidence in modern day public institutions, even in democratic societies, are widespread. Private institutions, such as private companies, the mass media and others, are similarly affected by a lack of public esteem. An erosion of confidence in the major institutions of society, especially those of representative democracy, is a far more serious threat to democracy and well-being than a loss of trust in other citizens or politicians. For that reason, the fundamental technical backdrop to activity in advanced societies, to which we have added CSI, must be adequately safeguarded, both in reality and in the public eye. Public perception, as well as its confidence in authority, hinges on the estimation of the risk of a negative event occurring, along with the estimation of the capacity for adequate resolution of such events. In the case of complex events, such as CSI malfunctions, the public may have no reasonable perception of risk, making the appearance of control even more important. The stakes are societal and political stability, the legitimacy of current political models, as well as the well-being of perception based economic life, which is a large component of the wealth of any advanced nation.

$$P = f\left(\text{governance capacity}, \frac{1}{\text{chance of risk}}\right)$$

$$f(P) = (1 - f(Risk)) * f(Control) \tag{10}$$

4.2.11 International Relations

space based activity, with its expense and risks, is naturally conducive to cooperation between states themselves, as well as other parties, but also leads to a tangled web of responsibility for the failure or delay of a common project or, even more serious, the disruption of critical shared infrastructures. The development of certain applications, such as military surveillance, can be interpreted as a direct threat, and so can the development and testing of anti-satellite weaponry, with the unspoken message of capacity to rapidly disrupt critical services. In the case of an actual risk materializing, whether natural or man-made, it can be unclear if there is a responsible party and who that party may be. Cyber-attacks are, invariably, anonymous or tend to be organized as such. Proof of guilt may be lacking and, if more than one party is affected, as is the case with CSI disruptions, it can lead to an atmosphere of mutual recrimination, especially since geography is an important coordinate of CSI activity and the affected parties are, invariably, neighbors and may already have a long history of enmity.

Moreover, not just states have the capacity to disrupt CSI, but also non-state entities, such as terrorist groups and other political or criminal organizations, and they are more likely to do it, since they are less sensitive to threats of retaliation in kind. Moreover, deterioration in International Relations can also take place when the disrupted space systems are used as an instrument of verifying adherence to certain agreements. Remote sensing capabilities can witness efforts at stifling illegal immigration or trafficking in various items across borders or can subject a certain government to scrutiny regarding its activity, whether political, nuclear, environmental and so on. The disruption of those capabilities can, at worst, be interpreted as an evasion of oversight, or it may simply be an unfortunate occurrence that, nonetheless, makes compliance with international demands difficult to ascertain.

Finally, there is also the possibility of a willful disruption of services which is clearly attributed to a certain party with power over the CSI. The American GPS system, administered by the military, and with full authorization to unilaterally withdraw support of all civilian application at a moment's notice and without explanation, is just such an example. It would be a terrible precedent if such a protocol were to be enacted, especially now, with limited substitution capability; the IR scandal would be of intense magnitude, pitting the US against its own allies. And such a denial of service wouldn't have to be the result of malicious intent towards the affected parties, just a prioritization of one's own narrow interest in a certain emergency situation, but the damage it would incur would render such an absence of malice moot.

$$\mathrm{IR} = \mathrm{f}\left(\mathrm{diplomacy}\ \&\ cooperation, \frac{1}{\mathrm{threat}\ \&\ uncertainty}\right) \tag{11}$$

4.2.12 Public Order

History as well as recent events indicate that civil disturbances are inevitable in all types of society, with the trigger as a variable. Hence, the need for Public Order, P_0. It is possible to rely on documented cases of civil disturbances caused by an interruption in critical services and products provisioning to conclude that critical infrastructures, through the underlying importance to the well-being of citizens and the normal operation of society, are a potential source for civil tensions. Strikes and riots caused by the economic failure of an industrial plant which is critical for the local economy are well-known, as are the extensive public disturbances registered when electricity grid failure leaves areas sweltering in heat or freezing in the cold (with the United States providing the most recent examples in the case of advanced nations). While their areas of direct effect on the citizenry are not very large, CSI have become such a central pillar of support for modern lifestyles, that the disruption of their operations could provide a similar impetus for unrest.

Unlike terrestrial infrastructures, the wide-reaching nature of CSI means that if conditions for unrest are to appear, they will manifest themselves on a larger scale and in multiple places, eschewing localism and limited reaction. This is, often, in direct relation to the issue of public confidence, addressed above. The rising importance that CSI have in relation to the system-of-systems means that CSI can indirectly become the trigger for a disruption, such as in energy supply, that incites unrest, leading to complicated situations. This is certainly the case when it comes to breakdown in communication. A break breakdown can prove immensely disruptive to the daily routine of individuals, providing a reason for misbehaving or enabling the manifestation of a past grievance. Similarly, in a digitally monitored world, a perceived lack of temporary supervision can elicit antisocial behavior. Not least, improper handling of Critical Infrastructure Protection can lead to periodic disruptions in vital services, which then foment legitimate protests against an ineffectual government.

Already, the rise of transmission capacity means that cloud-based services have divorced users from the location of their data, with CSI as efficient intermediaries. We could imagine a time, in the near future, when most payments, even small ones, shall be digital and cash is unavailable. Then, a cyber-attack similar to the one that hit Estonia in 2007 (for good reason, since Estonia is one of the most digitally-integrated societies in the world, with banking, voting, taxing and government services conducted as frequently online as possible) hits a digitally-integrated state, only far larger than Estonia ever was. Or it could just be a breakdown due to structural weaknesses in the infrastructures used. Suddenly, the citizenry cannot conduct its daily affairs and cannot even supply itself with goods—in the absence of a strong response by the state and a guarantee of quick resumption of services, chaos is bound to break loose, with looting, violence, and disruption of business continuity. Even at the lower end of the scale, it may lead to hoarding behavior, which tends to compound supply issues. In this case, offer P_0 as:

$$P_o = \text{f(final critical service provision level, confidence in government,}\ \frac{1}{\text{time to relief}}, \frac{1}{\text{incurred damages}}, \frac{1}{\text{time to resumption}}) \quad (12)$$

A cyber-attack could also be directed at disrupting a political process, or compromising government databases, like the hacker that recently stole identity data for two thirds of Greece's population. It may even take place to elicit severe unrest against the proper authorities and sabotage them politically. Space systems provide a natural conduit for such disruptions, willful or otherwise, of services on which our dependence, in advanced societies, is growing strongly. We would see, in order of severity: minor protests and vandalism, large-scale demonstrations and lobbying, widespread action (disrupting industry, violent protests) and, finally, a direct threat to the internal stability of a nation. Severe unrest can be self-reinforcing, as intensive action can lead to further disruptions of other services, or interruption of relief efforts, amplifying the anger of those affected and eliciting unrulier behavior.

4.2.13 Policy and Operations of Public Services

While there is a tradition in criticizing public services in most countries as being inefficient and backwards, significant strides are being made in harnessing the latest technologies for reducing their associated administrative costs and improving their quality and efficiency. This, invariably, means a greater reliance on digital communications, automation and centralization of databases and processing. This drive for efficiency and modernity leads to a dependence on certain space systems, especially communication arrays, which is set to deepen since the demands of the public are accompanied by less resources with which to ensure the provision of public services. Already, trends for the foreseeable future include e-government accompanied by disbandment of physical location for interaction with authorities, driverless public transportation, digitally administered and unified public utilities, with an ever-greater emphasis on private-public partnerships. For most, that would mean that the private actors will compensate for the deficiencies of the public, but, in fact, such projects suffer from the security discrepancies of the weakest link, which, in the digital age, almost certainly means the public half of the administration of the project.

A high performing public service system will be dependent, to a significant extent, on capacities afforded to them by the use of space systems, whose disruption is likely to cause significant operational losses in areas where they are prevalent. The country which is safest from these threats is also one which has not achieved significant public efficiency in the first place. The model for a country prone to cyber disruption is Estonia, which was attacked in 2007 by hackers using space systems as gateways, and whose significant e-government apparatus ground to a halt under the strain of the attack, including voting, transfers, and other vital operations. The more the government involves itself in the economic and social life of the country, through public services and state companies such as public utilities, the more responsibility it carries for ensuring against a disastrous interruption of service due to an CSI disruption. Otherwise, the legitimacy of its involvement in the economy, as well as its capacity for properly administering the well-being of its citizens, will be called into question.

$$\text{Public Services Deficit} = \text{Public Services} - \{\text{sum}(I_i * P_i)/i = 1 \ldots n\} \tag{13}$$

$P_{1\ldots n}$ public service P_i
$I_{1\ldots n}$ impact on public service P_i of CSI malfunction.

Turning back the clock on technological advances in government is not an option, but rather for the government to take an active role and set an example on how to safeguard against the threat of disruption, through substitution, backup services, redundancies and contingency plans. Otherwise, the effects of CSI disruption could range from impeding the management or operation of a public-sector entity to the indefinite cessation of its activity. Given the importance of CSI in systems which have integrated them, the result of disruptions will generally lean towards the latter.

4.2.14 Safety

CSI is a major contributor to the safety of human activity and, conversely, its malfunction poses unique threats to safety. This vulnerability is registered at multiple levels.

Firstly, there is the reliance of emergency and crisis management efforts on CSI for invaluable services, such as localization, coordination, communication and remote sensing. The absence of these significantly impacts the effectiveness of ECM efforts, rendering them impossible under certain conditions. The potential loss of life and permanent damage to communities, businesses and the environment that would occur if a natural disaster or other type of emergency were to take place and ECM efforts were hobbled by the absence of CSI defies conventional measurements. What is certain is that CSI importance in the mitigation of the impact of an extreme event grows with the severity of this event.

A second contribution of CSI is in providing security to the daily operations of society and, thereby, preventing the kind of disasters that are brought on by a lack of oversight and environmental awareness. This is doubly true in those fields whose safety is compromised by increasing activity in that particular sector. GNSS is important not just in arriving at a destination, but in performing delicate maneuvers within narrow harbors in the great commercial ports of the world and in avoiding mid-air collisions between planes on the busiest routes (where they often come to within 500 m of each other). Monitoring the weather to regulate activity, using remote sensors to detect sand-bars and other hazards and simply enabling communications between public service providers are all applications of CSI capabilities. In the case of terrestrial infrastructures and the impact of their disruption on safety, the following categories are possible: minor injuries, severe injuries and chronic illnesses, then potential casualties and, finally, widespread loss of lives.

It is also important to remember that the positive CSI impact on the safety of human activity also enables the augmentation of said activity without a corresponding lowering of safety, as long as the CSI remains operational. If there were no GNSS capabilities, port administrators would not risk allowing so much traffic within their vicinity, air plane density would not be so high and, consequently, neither would the associated economic activity experience such growth or reduced costs of business.

$$\mathrm{f(safety)} = \mathrm{f}\left(\text{mitigation capacity}, \frac{1}{\text{threat} * \text{casualty}}\right)$$

$$\mathrm{S} = 1 - \{\mathrm{sum}(\mathrm{T_i} * \mathrm{C_i})/\mathrm{i} = 1\ldots\mathrm{n}\} \tag{14}$$

$T_{1...n}$ chance for casualty C_i to take place

$C_{1...n}$ casualty group, meaning injuries, illnesses, deaths and other safety related concepts.

4.2.15 Impact on Issues of Defence

In this case 'impact on issues of defense' means defense against threats deriving from human action, whether a foreign power, a terrorist group or a lone individual who has been contracted to sow chaos. The final scenario is a real possibility, given increased opportunity for disruptions through low-cost cybernetic attacks which are very efficient, in spite of very expensive countermeasures. While there is no actual way of computing a score to measure how defensible or how secure a country or a group of allies are from enemy actions, it is safe to say that dependence on critical infrastructures, including space systems, cuts both ways—they have enabled new capabilities for defense, new efficiencies and new synergies when acting in concert with others. This takes place in long established groups (such as NATO) where cooperation mechanisms are, presumably, already in place, but also in new groupings, alliances, or ad hoc and temporary cooperation, where the communication and coordination capabilities afforded by CSI makes it easier for two unfamiliar allies to establish a working relationship with regards to defense. More and more anti-terrorism operations take place cross-border, with the sharing of information and concerted strikes. While international cooperation has advanced rapidly in this field, there are still noticeable institutional deficiencies which are aggravated when cooperation must happen at a moment's notice.

In reverse, this means that the disruption of CSI represents a tremendous blow to the defensive capabilities, both internally, within a country's security apparatus, and among allies. This is enforced by an exclusive reliance on the capabilities enabled by CSI, as evidenced by the NATO war games where the United States switched off GPS support for participating naval forces, which were then left at a significant disadvantage, due to the lack of training and equipment for non-CSI navigation.

Three broad levels of impact on defense capabilities are suggested: minor damage to the security of a nation, grave damage to the security of a nation, grave damage to the security of allied nations. Two 'other' impacts can be added, given the indiscriminate nature of reliance on most CSI today: minor damage to the security of multiple nations, and grave damage to the security of multiple nations, with the concept of allies irrelevant in both cases. This, in fact, is a positive trait, since reliance on the well-functioning of a, virtually, irreplaceable CSI, is an effective common ground to bring together, in cooperation, nations that were neither allies, nor particularly friendly to one another.

$$\text{Defence deficit} = \mathrm{f}\left(\text{defence capabilities, sum}\frac{\left(\frac{1}{\mathrm{D_i} * (1-\mathrm{I_i})}\right)}{i} = 1 \ldots \mathrm{n}\right) \tag{15}$$

$D_{1\ldots n}$ defence capability D_i
$I_{1\ldots n}$ impact on defence capability D_i of CSI malfunction.

Final Remarks

It has been said our energy system are increasingly vulnerable to disruptions from interconnected systems and therefore susceptible to failure in other systems (e.g., information services). Now might be the time to see of this statement holds true for system and the related systems. Exploring this issue might start with the consideration of the notions of 'interdependency' and a closely related terms of dependence and explore. This notion of interdependencies affirms that cross-national borders and the jurisdiction of the competent national authorities ensure that security for the whole is skewed by the lowest common dominator, regardless of the security outcomes for individually managed components in the system-of-systems. This might serve as a key point of interest of these with the most advanced and space-realized systems and actors (they might have more to lose). Therein lies a need to know types of interdependencies, their implications, and well as development of scenarios exploring what-if-analysis with tangible methods and tools. Certainly, this analysis will need to address failures in the context of space systems along the three dimensions of failures as previously described (cascading, escalating and common cause).

References

Benschoter, R. A., Eaton, M. T., & Smith, P. (1965). Use of videotape to provide individual instruction in techniques of psychotherapy. *Journal of Medical Education, 40*(12), 1159–1161.

Broens, T., Van Halteren, A., Van Sinderen, M., & Wac, K. (2007). Towards an application framework for context-aware m-health applications. *International Journal of Internet Protocol Technology, 2*(2), 109–116. https://doi.org/10.1504/IJIPT.2007.012374.

Celler, B. G., Earnshaw, W., Ilsar, E. D., Betbeder-Matibet, L., Harris, M. F., Clark, R. et al. (1995). Remote monitoring of health status of the elderly at home. A multidisciplinary project on aging at the University of New South Wales. *International Journal of Bio-Medical Computing, 40*(2), 147–155.

Chouinard, J. (1983). Satellite contributions to telemedicine: Canadian CME experiences. *Canadian Medical Association Journal, 128*(7), 850–855.

Cook, A. K. (1969). *The breakthrough team*. Rockville, Maryland: ESSA World. Retrieved from http://docs.lib.noaa.gov/rescue/journals/essa_world/QC851U461969jul.pdf.

DeBlois, B. M., Garwin, R. L., Kemp, R. S., & Marwell, J. C. (2004). Space weapons: Crossing the U.S. Rubicon. *International Security, 29*(2), 50–84. https://doi.org/10.1162/0162288042879922.

Dwyer, T. F. (1973). Telepsychiatry: Psychiatric consultation by interactive television. *The American Journal of Psychiatry, 130*(8), 865–869. https://doi.org/10.1176/ajp.130.8.865.

Gheorghe, A. V., Vamanu, D. V., Katina, P. F., & Pulfer, R. (2018). *Critical infrastructures, key resources, and key assets*. Cham, Switzerland: Springer International Publishing. https://doi.org/10.1007/978-3-319-69224-1_1.

Gilbert, B. K., Mitchell, M. P., Bengali, A. R., & Khandheria, B. K. (1999). NASA/DARPA advanced communications technology satellite project for evaluation of telemedicine outreach using next-generation communications satellite technology: Mayo Foundation participation. *Mayo Clinic Proceedings, 74*(8), 753–757. https://doi.org/10.4065/74.8.753.

Higgins, P. (2012). *Earth observations, science, and services for the 21st century*. Washington D.C.: The American Meteorological Society.

Humphreys, T. E., Ledvina, B. M., Psiaki, M. L., O'Hanlon, B. W., & Kintner, P. M. (2008). Assessing the spoofing threat: Development of a portable GPS civilian spoofer. In *Proceedings of the Institute of Navigation GNSS (ION GNSS, 2008)*, Savannah, GA.

Katina, P. F., & Pinto, C. A. (2012). On critical infrastructure interdependency. *Annual International Conference of the American Society for Engineering Management 2012 (ASEM 2012): Agile Management: Embracing Change and Uncertainty in Engineering Management* (pp. 29–38). Virginia Beach, VA: Curran Associates Inc.

Katina, P. F., Keating, C. B., Zio, E., & Gheorghe, A. V. (2016). A criticality-based approach for the analysis of Smart Grids. *Technology and Economics of Smart Grids and Sustainable Energy, 1*(1), 14. https://doi.org/10.1007/s40866-016-0013-2.

Kirilenko, A., Kyle, A. S., Samadi, M., & Tuzun, T. (2017). The flash crash: High-frequency trading in an electronic market. *The Journal of Finance, 72*(3), 967–998. https://doi.org/10.1111/jofi.12498.

Mariappan, M., & Khoo, B. (2013). Design and development of communication and control platform for medical tele-diagnosis robot (MTR). *International Journal of Networks and Communications, 3*(1), 12–20.

McCord, M. R., Hickman, M., Bronzini, M. S., Goel, P. K., Gomez, R. B., Merry, C. J. et al. (2003). *Remote sensing for transportation—Products and results: Foundations for the future. Session 3: Transportation planning: Overview of remote sensing products and results*. Washington D.C.: Transportation Research Board. Retrieved from https://trid.trb.org/view/665541.

Navein, J., Fisher, A., Geiling, J., Richards, D., Roller, J., & Hagmann, J. (1998). Portable satellite telemedicine in practice. *Journal of Telemedicine and Telecare, 4*(Suppl 1), 25–28. https://doi.org/10.1258/1357633981931326.

Rivera-Ruiz, M., Cajavilca, C., & Varon, J. (2008). Einthoven's string galvanometer: The first electrocardiograph. *Texas Heart Institute Journal, 35*(2), 174.

Royal Academy of Engineering. (2011). *Global navigation space systems: Reliance and vulnerabilities*. London: The Royal Academy of Engineering.

Sjögren, L. H., Sandberg, C., & Törnqvist, H. (1999). Telemedicine in Sweden–a diffusion study. *Journal of Telemedicine and Telecare, 5 Suppl 1*, S63–65.

Stachura, M. E., & Khasanshina, E. V. (2007). *Telehomecare and remote monitoring: An outcomes overview*. Washington D.C.: The Advanced Medical Technology Association.

Retrieved from http://archive1.telecareaware.com/telehomecare-andremote-monitoring-an-outcomes-overview/.
Theoharidou, M., Kotzanikolaou, P., & Gritzalis, D. (2009). Risk based critical analysis. In C.C. Palmer & S. Shenoi (Eds.), *Critical Infrastructure Protection III—Third Annual IFIP WG 11.10 International Conference on Critical Infrastructure Protection*. IFIP Advances in Information and Communication Technology Series (311). Hanover, New Hampshire, USA: Springer. ISBN 978-3-642-04797-8.
Vamanu, B. I., Gheorghe, A. V., & Katina, P. F. (2016). *Critical infrastructures: Risk and vulnerability assessment in transportation of dangerous goods—Transportation by road and rail* (Vol. 31). Cham, Switzerland: Springer International Publishing. Retrieved from http://link.springer.com/10.1007/978-3-319-30931-6.
Wold Health Organization. (2010). *Telemedicine: Opportunities and developments in member states*. Geneva: WHO. Retrieved from http://www.who.int/goe/publications/ehealth_series_vol2/en/.
Zuzek, J. E., Cauley, M. A., & Hollansworth, J. E. (1994). The telemedicine spacebridge project: A joint US/Russian venture in long distance medicine via satellite. In *Presented at the 15th International Communications Satellite Systems Conference*, San Diego, CA, US. Retrieved from https://ntrs.nasa.gov/search.jsp?R=19940022924.

Chapter 5
National and Global Effects—Malfunction and Destruction of CSI

Abstract Critical space infrastructures, similar to other systems, can fail. These failures can have national and global effects. The effects of critical space infrastructures malfunctions and destructions are the basis for this chapter. Specifically, effects are seen through global clientele and significant percentage of critical services output per unit. These are examined through a number of characteristics.

5.1 Characterization of CSI Malfunctions

Critical space infrastructures, through their global clientele and significant percentage of critical services output per unit, are also liable, when destroyed or disrupted, to have the following characteristics.

5.1.1 Cascading Disruptions Phenomena

At the foundation of any ambitions for development, one finds a complex system of critical infrastructures which ensure that the society and the economy have the material basis on which to exist and to develop. Modern societies are highly dependent on the continuous operation of critical infrastructures that ensure the supply of essential goods and services. The more advanced a nation is, the more infrastructure it has, the more infrastructure it needs to build, and the more infrastructures become critical to its very survival. These critical infrastructures range from pipelines and other industrial assets, to healthcare, water supply, food supply and, finally, education, communication and financial capabilities. Removing any one of these from the equation makes it impossible for a society to maintain its standard of life and develop economically.

The notion of critical infrastructure protection (CIP) addresses the numerous vulnerabilities that such a complex system will generate (Gheorghe et al. 2018). These vulnerabilities can be triggered either intentionally or through simple attrition, either

© Springer Nature Switzerland AG 2019

A. Georgescu et al., *Critical Space Infrastructures*,
Topics in Safety, Risk, Reliability and Quality 36,
https://doi.org/10.1007/978-3-030-12604-9_5

alone or in a chained event, and can have devastating consequences. At the very least, they impact investor and citizen confidence and the uncertainty of safety makes it harder to plan for the future.

Cascading phenomena occur when the disruption of normal operations for one critical infrastructure asset or system leads to similar, rapid failures across the entire system-of-systems to which it belongs. The conceptual framework behind Critical Infrastructure Protection has evolved to consider that almost all infrastructures, critical or otherwise, have a degree of connection with one another, meaning that the system-of-systems is an all-encompassing concept. Therefore, cascading failures can propagate throughout the entire geographic, economic, social and political planes occupied by the system-of-systems, as well as through its individual components and layers of infrastructures. Moreover, disruptions in one infrastructure system can have self-reinforcing effects after impacting another system with which there is a bilateral interdependence. Keeping this in mind, vulnerabilities observed in one component or another are heightened, risks are multiplied, and the prospect of their materialization becomes dire. Recovery from a cascading failure event can only take place when it has run its course and whatever is left of the system, especially its soft components (with humans as main units), can reset its operations.

The easiest cascading disruption chain to visualize is the one with its initial point in the energy sector. A blown-out power transformer can leave millions of people in the dark, especially in bottlenecked grids. This can curtail natural gas extraction and transportation, amplifying the potential for further energy disruption. In the meantime, aside from the individual consumers left in the dark, businesses in many fields, including industrial ones, have been unable to operate. The administrative infrastructure is running without underlying digital support, banking and finance are effectively shut down, as are higher functionalities within the health system (lab analysis, high-tech diagnostics, surgeries). Depending on the local environment and climate, further damage to sensitive equipment, such as servers, as well as to population health, can be expected. Public order may be impacted if the resumption of service is more difficult than anticipated and the personal inconveniences greater than normal (hot weather, spoiled food due to lack of refrigeration). This has grave economic costs, depending on the location where it happens. The greatest economic costs are registered in locations belonging to advanced economies, where the greatest value is created. They are also the most likely to absorb the damages and recover quickly, as well as have mitigation measures in place. Poorer countries face even longer delays in achieving normal functioning of infrastructures, and lower economic impact in absolute terms belie the greater human tragedies that may take place. Since energy is at the foundation of almost all human activity, the cascading disruption will reverberate throughout the whole economy and of the affected area. Nothing will remain untouched, unless it has enough substitution capacity to wait for normal service to resume.

The potential for cascading failures builds up naturally, alongside the development of a local, national, regional or international economy and its underlying infrastructures (Pugliese et al. 2016). Economic synergies arise from the sort of interconnections and efficiencies between systems that help propagate future cascade events.

Therefore, in parallel with development, one also must take into account the possibility that new interdependences and vulnerabilities will arise which need to be identified and then mitigated. For this reason, CIP concerns itself with anticipating future developments, both in existing infrastructures, but also new ones, such as space systems. It must then integrate its findings into a framework that can provide adequate risk governance and damage control, for the purpose of ensuring quality of life and business continuity in adverse conditions. CIP is not only reactive, it also strives to be proactive, and cascading phenomena, while impossible to prevent entirely without costly substitution, redundancy and system insulation, can be effectively targeted by planning in advance for the minimization of vulnerabilities in new infrastructures, thus generating resilience for the whole system-of-systems.

Critical space infrastructures have the potential to generate wide-ranging cascading phenomena, due to their global field of operations and the volume of services they render. Their central importance to communications, coordination and so on have made us wholly dependent on the operational capacity and resilience of a limited pool of assets in the most hostile environment known to men. It is so hostile, that it features two unique threats to infrastructure operation—space debris and space weather, the latter of which can also impact terrestrial systems. The example given above, with the blown-out power transformer, could easily have been the result of particularly intense activity in the Sun, which spews out plasma particles at high speeds and can impact terrestrial systems, let alone orbital ones. And space systems are central to studying and anticipating space weather activity.

Due to low levels of development of space infrastructures, in applications and in numbers, interconnection between space systems is limited. Likewise, being designed for continuous operation in a hostile environment with difficult resupply capabilities, they are removed from the maintenance rituals that accompany most terrestrial infrastructures. As such, they rarely have the potential to be a link in a cascading infrastructure failure; rather, they are more adept at being the starting point of the disruption, especially given the role of the most important space infrastructures as bottlenecks in information transmission and intermediaries between controller and controlled asset. Coordination capacity among many different entities, such as airplanes and cargo ships maneuvering in limited spaces, would be lost, leading to accidents or a slowing down of the coordinated system. In economies that practice 'just-in-time' resupplying to cut costs in acquiring and storing inventory, such coordination is vital, and its interruption reverberates throughout entire supply chains and industries.

Being so close to the top of the potential cascading failure means that it makes sense for critical space infrastructure systems to be as rugged and resilient as possible. This means not just physical hardiness, but also inclusion into a wider strategic vision for ensuring the provision of vital goods and services, which would advocate for separate redundancies, extra capacity in the system and potential for substitution of capacity at a moment's notice. Critical space infrastructures have become the globalizing element which provides an intimate link between systems-of-systems whose main bodies are located at national or regional level. Thus, CSI malfunctions or destructions can trigger multiple cascading events through otherwise mostly separate infrastructure chains. Of course, systems dependent on CSI should also become more rugged and resilient, especially in relation to the CSI itself, since this dependence is likely to grow in the future. Again, we find proper motivation for the inclusion of CSI understanding and protection within a wider strategic framework.

5.1.2 Effects on Market Expectations and Actor Behavior

The financial activity underlying the world economy at all levels has suffered a paradigm shift within the last decades, driven by instant communications and rapid economic internationalization, globalization and liberalization. In parallel, there has been unprecedented innovation in the creation and widespread use of ever more complex financial instruments. This has led to a financial topography where risk is widely spread but opaque, in which financial positions are assumed and discarded within a few seconds and, even more disturbing, the wealth of the world, by extension, the financial health of corporations and governments depends on the day to day valuation of financial positions. These, in turn, are intimately linked to consumer and investor sentiment, which are highly volatile and prone to exuberant optimism or pessimism. This leads to patterns of self-reinforcing market downturns which, as the current sovereign debt crisis shows, are difficult to stop and need little factual basis to set in.

The link between critical infrastructure disruptions and a pessimistic outlook on behalf of economic actors, persisting well after the disruption has been eliminated, is well documented and attested to. Unless the disruption is severe, the direct damages it has caused are far exceeded by the lost opportunities or dwindling economic activity engendered by frightful economic actors of all stripes. Nowhere is this more apparent than on the financial markets, where large downturns are possible in an instant. It has become almost a given that bad weather, an industrial accident, a trivial piece of news or even a positive economic report less in line with market expectations will trigger a 'dip'.

By arguing that space systems also satisfy the requirements for being critical infrastructures, we have to also imply sufficient importance so that their malfunction or destruction will impact global markets and have lasting effects on the investor, consumer and decision maker psyche. This will, invariably, be proven true. Far from simply affecting the bottom lines of the owners and the business continuity and quality of life of the users, CSI disruptions, more so than their terrestrial counterparts, which are still fragmented on a regional basis, will plant doubts into the minds of global market participants. They will engage less in the market and will price their transactions more pessimistically. For stocks, that means lower valuations and loss of wealth for owners. For debt, that means higher interest rates, lower liquidity and worsening outlooks for the debtors. This is the exact same pattern as registered in any downturn and economic crisis, regardless of its origin. Historically, regional blackouts, disruptions in services, natural disasters, have all led to drops in markets even outside the affected area, as interdependences abound, and many investors are active at an international level.

We would be hard pressed to predict a likely scenario for the damage done by market reactions to a partial communications blackout, on account of heavy solar activity. Likewise, a very public disaster such as, in the past, the explosion of one of the space shuttles, will also register in emotive markets. In the same category, we may find that the failure of a high-profile attempt at adding to the critical infrastructure pool located in space—a defective satellite, a failed launch for a new vehicle, disappointing functionality from new, space-based instruments—impacts more than just the entity responsible but will also impact the outlook with regards to space based activities.

The criticality of existing space infrastructures suggests that further development is advisable, as it can have a positive impact on users' welfare and the security of the provision of critical space-based services and, in the future, goods. However, this development will depend on access to financial resources on favorable terms, which has not been very forthcoming in the past for companies, which is why the various states have led the way in financing new development, ensuring adequate demand for business viability and, in many cases, ensuring the provision itself, in the absence of private appetite for such an endeavor.

It is vitally important to recognize the potential for depressing the markets and the minds of economic agents that disruptions of different CSIs may possess in varying degrees. The staggering reliance of our economic systems on what, even in accounting, is euphemistically termed "goodwill", which may account for a substantial amount of the world's wealth, means that the cost of CSI disruptions can be too high to afford not to award them the attention they deserve.

5.1.3 Chilling Effects in Various Fields

The chilling effect is one of the most insidious and easily overlooked consequences of a disruption in services rendered by critical space infrastructures, either due to malfunction or destruction. However, we would count it amongst the most influential in its long-term effects, precisely because it defies normal quantitative measurements or actual diagnosis.

Chilling effects are registered when a certain event, invariably a negative one, subtly influences the decisions matrix of actors it has affected, whether economic, social or political. The easiest example to understand is when a very well frequented commercial centre is hit with a crisis that renders it inoperative, causing great discomfort to those who rely on it; when its functionality is fully restored, it may observe that some of its clients have found substitutes for it, even though it was the most convenient venue. While it might have eliminated the cause of its disturbance entirely, the risk perception of its clients, or users, has been invariably affected to the point in which one can state that a chilling effect has taken place in commercial operations. Other examples include tourism in places affected by social unrest or one of a kind terrorist attacks, and the effects persist for a while after the cause has been removed.

In social matters, rampant litigation and vague libel laws can have a chilling effect on free speech, as can any other situation in which one actor feels discouraged by the vague and uncertain prospect of a negative occurrence to the extent to which he gives up on his intended actions. Laws are most often cited as sources of chilling effects; however, critical infrastructure disruptions can have the same result with regard to consumer and investor sentiment, within a certain geographic area.

Critical space infrastructures are even more predisposed towards generating high intensity chilling effects, as they perform global functions, are highly visible and have easily verifiable and publicized disruptions. Moreover, as this paper argues, they perform vital services for regional and planetary levels of populations within a very limited pool of fragile and hard to replace assets. CSI malfunctions and destruction have, at the very least, a national level impact and even higher, if they are to be identified as critical. Therefore, since the distribution of consequences, loss and suffering would be very wide, it stands to reason that chilling effects would also be widely distributed.

Potential chilling effects include:

- Delay or outright cancellation of planned orbital activities, in the case of incidents within the transportation infrastructure between ground and space levels or, in the future, between two points in space. Further cancellations can be attributed to economic actors unwilling to expose themselves to newly perceived risks or to market judgement of those risks, which would entail rising insurance premiums, negative publicity or unfavorable stock market evolutions. Ex: less satellite launches due to multiple orbital debris incidents.
- Temporary or even permanent drives towards substitution of functions performed by CSI with terrestrial equivalents which are perceived to be safer, but are also less capable providers, leading to a loss in utility for the end-users. Since space is such a harsh environment, with still considerable exploitation costs, it is likely that those engaging in the construction, maintenance and use of CSI have exhausted all possible terrestrial variants for achieving the same results, performing the same functions and working at the same scale that orbital solutions provide. Turning to inferior Earth-based solutions, while tempting in the short-term, would raise costs and have diminishing returns, even in terms of CI security. Ex: replacement of an extra weather satellite with a network of high-altitude platforms and automated ground monitoring points, especially as drone technology and costs continue making large strides; also applicable to surveillance and communications.
- A slowdown of scientific exploration and other activities situated at the technological frontier of space exploration and exploitation, with long-term repercussions from the space development perspective, due to a reassessment of priorities for governmental decision makers. Examples include near-permanent abandonment of avenues of research, such as nuclear propulsion, or the pursuit of new milestones, such as manned exploration of the solar system. Historical examples include, for example, the Challenger and Columbia disasters. These are examples of disasters that, for a while, grounded the US Space Shuttle fleet (for years, in fact) and

delayed the construction of the ISS and affected other missions, as well as wasting shuttle lifetime usability.

- A technological and philosophical divergence, beginning at the point in which the CSI related malfunction took place, which would lead to the favoring of one technology or method of space development over another, with no rational underlying basis. Historical example: the preference for robotic probes to the ultimate exclusion of manned space exploration, leading to great advancements in robotics to the detriment of human transportation systems, further skewering the preference matrix of decision makers.

The given historical examples combine one or more of the scenarios described above and, given time and further quantitative and qualitative development of human activity in space, more such examples will follow, along with further enlargement of the palette of chilling effects.

5.1.4 Loss of Synergies and Delayed Error Accumulation

The purpose of most Earth-centric space systems is to provide raw information and communication and coordination capabilities to users. Earth-based infrastructure chains have grown around to take advantage of these capacities, as well as be dependent on them. Therefore, once a disruption of these capacities is underway, the terrestrial users they were servicing begin operating out of sync.

One might argue that the resulting damage stops here: services are resumed quickly, or the system continues to shed its efficiencies and operate in an increasingly erratic manner in relation to its environment. Thus, the period before the next phase of a cascading failure sets is beset by operational degradation, to the point where operating at a semblance of previous efficiency becomes impossible. Delays mount in tightly controlled supply systems, transportation slows to avoid catastrophic accidents and electricity grids suffer malfunctions as system safeguards fail to adequately respond to changes in flows. At any point before utter breakdown, the process is still reversible, even though cascading inertia means that effects are already being felt in the next infrastructure node.

This discussion is might to suggest the importance of '*protection of critical infrastructure*'[1] and more importantly question current efforts to mitigate damage at micro level, to reduce the potency of cascading effects and to hasten the recovery of full operational capacity once the CSI interruption of activity is, hopefully, resolved. As we expand our demands from critical infrastructures, especially those located in space, without a correspondingly rapid increase in capacity or security, we should expect not just a devastating systemic failure, but also frequent minor disruptions

[1]For example, RAND addresses homeland security and critical infrastructure needs through objective research that assists national, state, and local agencies in preventing and mitigating terrorist activities, https://www.rand.org/topics/critical-infrastructure-protection.html. *Could similar approached be undertaken for 'space critical infrastructure'?*

that can serve as warning for greater dangers. Taken together, the damages resulting from these theoretical disruptions can be substantial, but also remedied by continuous marginal improvements in CIP efforts. Protection of CSI is discussed in the following chapter.

Final Remarks

There exist chilling effects associated with disruption in services rendered by critical space infrastructures. Insidious and easily overlooked, as they rarely happen, at times they might sounds as something from a sci-fi movie. Suffice to say that leading actors have no idea when the next Carrington event or Chelyabinsk impact will occur. NASA's Charles Bolden, when addressing the US House of Representatives Science Committee on 19th March 2013 said: "From the information we have, we don't know of an asteroid that will threaten the population of the United States…But if it's coming in three weeks, pray. The reason I can't do anything in the next three weeks is because for decades we have put it off".[2] Now that we have a good reason to consult and "involve other nations to the extent appropriate", the time is ripe for trans-world leaders to emerge and take the lead in addressing space threats and develop protective measures from knowledge gain in the past 30 years (National Research Council 2009, p. 10). A starting point might as well be addressing the low-hanging fruits:

- Delay or outright cancellation of planned orbital activities. For example, increasing multiple orbital debris incidents might lead to less satellite launches.
- Temporary or even permanent drives towards substitution of functions performed by space systems with terrestrial equivalents which are perceived to be safer, but are also less capable providers, leading to a loss in utility for the end-users. For example, replacement of an extra weather satellite with a network of high-altitude platforms and automated ground monitoring points, especially as drone technology and costs continue making large strides; also applicable to surveillance and communications.
- A slowdown of scientific exploration and other activities situated at the technological frontier of space exploration and exploitation, with long-term repercussions from the space development perspective, due to a reassessment of priorities for governmental decision makers.

It is not be surprise that these efforts can only be undertaken if there is a recognition that critical space systems, can and will fail, similar to other man-made systems. These failures can have national and/or global implications. Charactering system failures provides a great impetus for CSI failure exploratory research. However, this exploratory research should lead to development actionable 'solutions' against space system failures and consequences.

[2]David Cornish, "Nasa: 'prayer' currently best defence against unidentified nearby asteroids," *Wired*. http://www.wired.co.uk/news/archive/2013-03/20/nasa-on-space-threats.

References

Gheorghe, A. V., Vamanu, D. V., Katina, P. F., & Pulfer, R. (2018). *Critical infrastructures, key resources, and key assets*. Cham, Switzerland: Springer International Publishing. https://doi.org/10.1007/978-3-319-69224-1_1.

National Research Council. (2009). *Severe space weather events—Understanding societal and economic impacts*. A workshop report, Washington, DC: National Academies Press.

Pugliese, A., Salado, A., & Nilchiani, R. (2016). Effects of fractionation on system security in space systems. In *AIAA SPACE 2016* (pp. 1–10). Long Beach, California: American Institute of Aeronautics and Astronautics. https://arc.aiaa.org/doi/10.2514/6.2016-5292.

Chapter 6
Critical Space Infrastructure Protection

Abstract The purpose of this chapter is to provide a forward-looking summary to protection of 'critical space infrastructures'. Critical space infrastructure protection is necessary due to increasing potential failures as well as impact on space systems as well as interconnected systems on surface and underground systems. Emphasis is placed on need for protection measures against special phenomena including space weather, space debris, electromagnetic pulse, and anti-satellite attacks.

6.1 CSI Protection

Modern society already registers a profound dependence on the capabilities of critical space systems at an economic, political and social level. These systems are, at the same time, operating in the most hostile environment known to man and under harsh economic barriers. The current stock of space assets is, thus, limited in extent, and fragile in its constitution. It is subject to numerous specific threats, such as the risk of collision with space debris and the full extent of the energies unleashed by "space weather phenomena". They are also an easy target for certain kinds of attacks, some of which are their exact opposite in cost and sophistication. These realities imply the need to protect especially those systems which can be classified as critical, a task made even more difficult by the international nature of space infrastructure dependency, ownership and sovereignty. Critical space infrastructure Protection (CSIP) will not truly be achieved through national efforts, even those of the premier space powers, but through consensus-building, international cooperation and the creation of an institutional, legislative and administrative framework to govern human activity and, in particular, security-related activity in space. Most daunting of all is the fact that current trends in space industry development suggests that the next few years will bring rapid changes to the infrastructure landscape of space and to the underlying economic conditions governing the development of space. While the prospect of further growth in infrastructure capacity in space is a welcome development for decreasing the reliance on a handful of units and systems, security actors cannot ignore the potential for compounding the issue of orbital crowding or of vulnerability to attacks.

© Springer Nature Switzerland AG 2019

A. Georgescu et al., *Critical Space Infrastructures*,
Topics in Safety, Risk, Reliability and Quality 36,
https://doi.org/10.1007/978-3-030-12604-9_6

One of the great drivers of CSIP is military interest both in ensuring the security of its own systems of communication and surveillance, and in ensuring the provision of critical services to their respective nations. The military is also a key driver and stakeholder in the development of critical infrastructure protection for the systems such as pipelines, transportation, health and energy. NATO, in its bid to reinvent itself, has identified energy security for its members as a strategic goal to be pursued and enforced wherever the global energy system-of-systems registers a threat. Historically, the impetus of the development of the United States interstate highway systems in the 1950s, the largest in the world, was also the military need for adequate infrastructure in case of a US invasion at the height of the Cold War. While the military is now joined by other proactive stakeholders, such as private companies (who, in Western countries, own and operate 60–90% of all technical assets described as critical), civilian authorities and institutions in applying CIP at terrestrial level, the incipient nature of CSIP and the breadth of military risks it faces has mandated a greater military involvement. At the same time, the military is a heavy user of civilian space systems, and so is directly interested in its resilience (Ministry of Defence 2012). For instance, in the United States, the US Strategic Command and, before 2002, the US Space Command, identified the 'protection of space assets' as a 'crucial war fighting and peacetime national objective because space products and services are integral to joint war fighting capability and an increasingly important part of national politics, economics, and culture'.[1] While the protection of space assets was the most important priority, the Space Command identified three other priorities related to it in order to achieve control of the space environment—surveillance, prevention and negation. There are more advanced paradigms for CSIP now, involving multiple stakeholder models and international cooperation in addition to military support, however the basic requirements for a successful program remain the same.

Other requirements are also interspersed in the list below:

- Identifying and understanding threats
- Identifying interdependencies
- Identifying third party dependencies, which carry significant and often poorly understood risks
- Ensuring adequate sensor capacity and sensitivity
- Ensuring adequate defensive information operation (anti-jamming, backup communication links)
- Training personnel
- Modelling and simulation
- Hardening and shielding system components
- Ensuring mobility of systems
- Developing or maintaining robust replacement capacity for offline systems (Ministry of Defence 2012)—in the case of the military, it can also involve maintaining obsolete systems as emergency replacements (for instance, aerial and maritime navigation aids in case the GNSS signal is down).

[1] See http://www.wslfweb.org/docs/dstp2000/jwstppdf/16-psba.pdf.

6.2 Space Specific Risks to Critical Space Infrastructures

Currently, one of the main foci when it comes to risk prioritization for space system protection is digital protection, which is understandable. Cybernetic systems permeate throughout the entire system-of-systems and, as we have established, the main role of 'critical space infrastructure' within the infrastructure system-of-systems is to facilitate information flows for command, communication, coordination and control capabilities. Therefore, it stands to reason that one of the main risks affecting space systems are cyber-attacks which are a low cost and geographically independent means of attack. The insidious fact that such attacks may not be launched solely to disrupt the functioning of space systems, but also to corrupt it, such as hackers modifying the GNSS clocks that timestamp transactions to desynchronize financial markets for the purpose of fraud, is another justification for increasing vigilance. However, there are a number of unique threats facing space systems, most of which are related to the complexities and hardships of their unique environment, as well as the level of exposure to particular threats which the nature of orbital dynamics subjects them to. The most well-known are the risks associated with "space weather" and with the increasing number of natural and artificial debris that crowd their orbital paths. These have been described in more detail in the sections that follow. We also mention a series of other risks and challenges which deserve closer scrutiny.

6.2.1 Space Weather

Space weather encompasses the conditions on the Sun and in the space environment that can influence the performance and reliability of satellites and ground-based systems, as well as endanger life or health. Along the years there have been different space weather events, out of which the Carrington event in 1859, distinguished itself as the most intense. Other events took place in 1956, 1989 and 2003 and affected areas close to the polar regions and not only, as is the case of South America. Along with the evolution of technology came the vulnerabilities, as societies are more dependent on small scale electronics, space systems and other critical infrastructure. Between 2006 and 2010 there was the lowest level of space weather activity for nearly 100 years (Hapgood 2010), while another solar maximum takes place between 2012 and 2015. The risk that space weather poses is magnified by the interdependencies that exist between critical infrastructure, thus leading to cascading failures.

Space weather comprises a variety of space phenomena influenced by a 11-year Solar cycle. These phenomena are the coronal mass ejections (CMEs), solar flares, solar radiation storms, solar radio bursts, galactic cosmic rays and high-speed solar wind streams.

Usually the sun emits a low-density plasma (ionized matter) that do not affect the Earth due to its magnetosphere. However, during the CMEs, which is denser plasma ejected from the Sun at a very high speed, contain strong south pointing magnetic fields—opposite to Earth's magnetic field, overcome the magnetospheric shielding, reaching the Earth. This causes a change in Earth's magnetic field, affecting the power grids, pipelines, railway signaling, magnetic surveying and oil and gas drilling. Moreover, the high-density plasma heats the upper atmosphere, affects its density and disrupts radio communications. Sometimes associated with the CMEs, the solar flares are explosions on the Sun's surface. Their impact is limited to radio effects, such as the X-ray flash. This can produce an atmospheric layer that absorbs the UF radiation and extra layers of ionized material that slow down radio signal from GNSS satellites, thus rendering an error in positioning.

The solar radiation storms are charged particles with very high energies released by the Sun. They pose a major threat to spacecraft, as they can damage electronics and power systems. If these particles enter the Earth's atmosphere, they can rise the radiation levels, and can impact aircraft digital systems, aircrew and passengers. They can also produce an atmospheric layer that absorbs high frequency (HF) radio waves across polar regions.

The solar radio bursts are natural radio emissions that interact with radio signals on Earth and have the potential to interfere with satellite navigation, wireless internet and mobile telephones. Table 6.1 are based on NOAA's tables describing space weather intensities and effects.

The galactic cosmic rays are energetic charged particles produced by supernovae and pervading the interstellar space, sometimes reaching our Solar Systems, and even the Earth. Here they can damage spacecraft similar to the radiation storms. The number of cosmic rays that affects the Earth is influenced by the solar wind and has a maximum at Solar cycle minimum.

The solar wind emitted from the Sun's polar regions is much faster than the wind from the Equator, usually not reaching the Earth. The high-speed solar wind streams take place in the declining phase of the solar cycle, when the magnetic field streams migrate from the polar regions towards the Equator, therefore enabling the polar wind to interfere with our planet, increasing the fluxes of energetic electrons in the outer radiation belt. These can penetrate spacecraft, generate electric discharges and deposit electric charge inside insulating material. These streams can affect both geostationary satellites at 36,000 km altitude, and GNSS satellites at 20,000 km altitude. Table 6.2 provides categories of geomagnetic storms as well as descriptions of possible effects.

'Space weather' is primarily generated by high-speed ejections of plasma from the sun, which experiences periodic solar flares, but also incorporates other sources of radiation and charged particles coming from deep space. Unlike extreme weather on Earth, which disproportionately impacts the populations of poor countries, space weather impacts rich countries above all others, since they are the biggest consumers of space services and they derive the greatest economic value from their integration with their respective economies. There has never been a truly destructive solar-flare event, because only recently has humanity become vulnerable to them through industrialization, the advent of the knowledge economy and the widespread adoption of extremely vulnerable technological systems, which are dependent on the continuous provision of electricity or operate based on vulnerable electronic circuitry. One

Table 6.1 Descriptions of space weather intensity and effects

Category		Effect	Physical measure	Average frequency (1 cycle = 11 years)
Scale	Descriptor	Duration of event will influence severity of effects		
Solar radiation storms			Flux level of $\geq$10 meV particles (ions)	Number of events when flux level was met (number of storm days)
S 5	Extreme	**Biological**: unavoidable high radiation hazard to astronauts on EVA (extra-vehicular activity); passengers and crew in high-flying aircraft at high latitudes may be exposed to radiation risk **Satellite operations**: satellites may be rendered useless, memory impacts can cause loss of control, may cause serious noise in image data, star-trackers may be unable to locate sources; permanent damage to solar panels possible **Other systems**: complete blackout of HF (high frequency) communications possible through the polar regions, and position errors make navigation operations extremely difficult	10^5	Fewer than 1 per cycle
S 4	Severe	**Biological**: unavoidable radiation hazard to astronauts on EVA; passengers and crew in high-flying aircraft at high latitudes may be exposed to radiation risk **Satellite operations**: may experience memory device problems and noise on imaging systems; star-tracker problems may cause orientation problems, and solar panel efficiency can be degraded **Other systems**: blackout of HF radio communications through the polar regions and increased navigation errors over several days are likely	10^4	3 per cycle

(continued)

Table 6.1 (continued)

Category		Effect	Physical measure	Average frequency (1 cycle = 11 years)
Scale	Descriptor	Duration of event will influence severity of effects		
S 3	Strong	**Biological**: radiation hazard avoidance recommended for astronauts on EVA; passengers and crew in high-flying aircraft at high latitudes may be exposed to radiation risk **Satellite operations**: single-event upsets, noise in imaging systems, and slight reduction of efficiency in solar panel are likely **Other systems**: degraded HF radio propagation through the polar regions and navigation position errors likely	10^3	10 per cycle
S 2	Moderate	**Biological**: passengers and crew in high-flying aircraft at high latitudes may be exposed to elevated radiation risk **Satellite operations**: infrequent single-event upsets possible **Other systems**: small effects on HF propagation through the polar regions and navigation at polar cap locations possibly affected	10^2	25 per cycle
S 1	Minor	**Biological**: none **Satellite operations**: none **Other systems**: minor impacts on HF radio in the polar regions	10	50 per cycle

example of the potential severity of the cyclical highs of solar activity which can generate powerful space weather phenomena is of the Carrington event—an extremely powerful solar storm named for its principal observer, British astronomer Richard Carrington, which, in 1859, led to auroras manifesting at the Equator, measurement devices becoming erratic and world telegraph networks being heavily damaged. At the time, humanity was not critically dependent on systems vulnerable to these threats, but a reiteration of the Carrington event today would have dire consequences. Since then, the vulnerability of developed societies, despite never being severely tested, has grown with each passing year, and one can safely say that it has become an existential threat. One of the largest modern instances of a solar storm, on 13–14 September 1989, led to a loss of contact with numerous space assets for over a week and left 6 million inhabitants of the Canadian province of Quebec without electricity for several hours, and many planes grounded or rerouted (Cannon 2013). Solar weather can also lead to disruptions of services and significant damage at terrestrial

Table 6.2 Categories and effects of geomagnetic storms

Category		Effect	Physical measure	Average frequency (1 cycle = 11 years)
Scale	Descriptor	Duration of event will influence severity of effects		
Geomagnetic storms			Kp values	Number of storm events when Kp level was met; (number of storm days)
G 5	Extreme	**Power systems**: widespread voltage control problems and protective system problems can occur; some grid systems may experience complete collapse or blackouts. Transformers may experience damage **Spacecraft operations**: may experience extensive surface charging, problems with orientation, uplink/downlink and tracking satellites **Other systems**: pipeline currents can reach hundreds of amps, HF (high frequency) radio propagation may be impossible in many areas for one to two days, satellite navigation may be degraded for days, low-frequency radio navigation can be out for hours, and aurora has been seen as low as Florida and southern Texas (typically 40° geomagnetic lat.)	Kp = 9	4 per cycle (4 days per cycle)
G 4	Severe	**Power systems**: possible widespread voltage control problems and some protective systems will mistakenly trip out key assets from the grid **Spacecraft operations**: may experience surface charging and tracking problems, corrections may be needed for orientation problems **Other systems**: induced pipeline currents affect preventive measures, HF radio propagation sporadic, satellite navigation degraded for hours, low-frequency radio navigation disrupted, and aurora has been seen as low as Alabama and northern California (typically 45° geomagnetic lat.)	Kp = 8, including a 9	100 per cycle (60 days per cycle)

(continued)

Table 6.2 (continued)

Category		Effect	Physical measure	Average frequency (1 cycle = 11 years)
Scale	Descriptor	Duration of event will influence severity of effects		
G 3	Strong	**Power systems**: voltage corrections may be required, false alarms triggered on some protection devices **Spacecraft operations**: surface charging may occur on satellite components, drag may increase on low-Earth-orbit satellites, and corrections may be needed for orientation problems **Other systems**: intermittent satellite navigation and low-frequency radio navigation problems may occur, HF radio may be intermittent, and aurora has been seen as low as Illinois and Oregon (typically 50° geomagnetic lat.)	Kp = 7	200 per cycle (130 days per cycle)
G 2	Moderate	**Power systems**: high-latitude power systems may experience voltage alarms, long-duration storms may cause transformer damage **Spacecraft operations**: corrective actions to orientation may be required by ground control; possible changes in drag affect orbit predictions **Other systems**: HF radio propagation can fade at higher latitudes, and aurora has been seen as low as New York and Idaho (typically 55° geomagnetic lat.)	Kp = 6	600 per cycle (360 days per cycle)
G 1	Minor	**Power systems**: weak power grid fluctuations can occur **Spacecraft operations**: minor impact on satellite operations possible **Other systems**: migratory animals are affected at this and higher levels; aurora is commonly visible at high latitudes (northern Michigan and Maine)	Kp = 5	1700 per cycle (900 days per cycle)

level. The 2003 'Halloween storm' was another peak of solar activity[2]—alongside power disruptions on the ground, orbital activity was seriously affected—59% of scientific missions were interrupted, astronauts had to take refuge in specially shielded areas of the International Space System, and a number of satellites were lost.[3]

The Australian Space Weather Plan, first developed in 2003 by the Australian Academy of Sciences,[4] states that the main areas where critical effects due to space weather phenomena will be felt are: satellite operations, human spaceflight, radiation exposure during air travel, navigation systems, electricity grid balancing and distribution systems, high frequency communications, short wave communications, radar surveillance systems, geophysical explorations, corrosion of pipelines and the insurance industry.

Space systems also play a vital role in researching these phenomena and warning against them. For example, the most important "warning buoy" for solar storms is NASA's Advanced Composition Explorer (ACE) launched in 1997, which can give advanced warning 15 or 45 min before a front of solar activity reaches Earth. Other dedicated solar monitors include the Solar and Heliospheric Observatory (SOHO), DSCOVR or Probe-2, but there are a number of satellites performing other missions, which are equipped with radiation monitors and other relevant instruments. NASA's experience has demonstrated that combinations of missions enable larger scale investigations. Figure 6.1 provides a depiction of some the missions.[5]

Table 6.3 was compiled by the Space Weather Prediction Centre in the US as a non-exhaustive list of specific economic costs borne by specific actors which use the services of this Centre to manage their space weather risk exposure. We can infer from the table that most of the damage is registered on the ground, not just resulting from the disruption of critical space systems with which there is a direct chain of dependency, but also from indirect effects on market expectations and on tertiary infrastructures.

The SWPC has more than 650 clients subscribing to its services, all of them corporations or governmental entities with a significant exposure to space weather phenomena. It has its own infrastructure for delivery of the files, but also has a partnership with the US National Oceanic and Atmospheric Administration (NOAA), which is the institution that drafted the severity scales for space weather phenomena and their effects on Earth and on technical systems. More than 500 million files are transferred each month, with 500 thousand new files created every month containing real time data revisions being accessed by 176 different secondary services providing data and early warning to 400 thousand smaller clients in 120 countries.

Should a Carrington event recur today (which has happened in the past, having missed Earth due to striking another region of space), the US National Research Council (NRC) estimated that the result would be damage valued at $2 trillion in the first year for the US alone, and recovery times between four and ten years without also

---

[2]https://www.nasa.gov/topics/solarsystem/features/halloween_storms.html.

[3]http://www.thespacereview.com/article/1549/1.

[4]http://www.sws.bom.gov.au/docs/Space_Weather_Plan.pdf.

[5]https://www.nasa.gov/content/goddard/heliophysics-system-observatory-hso.

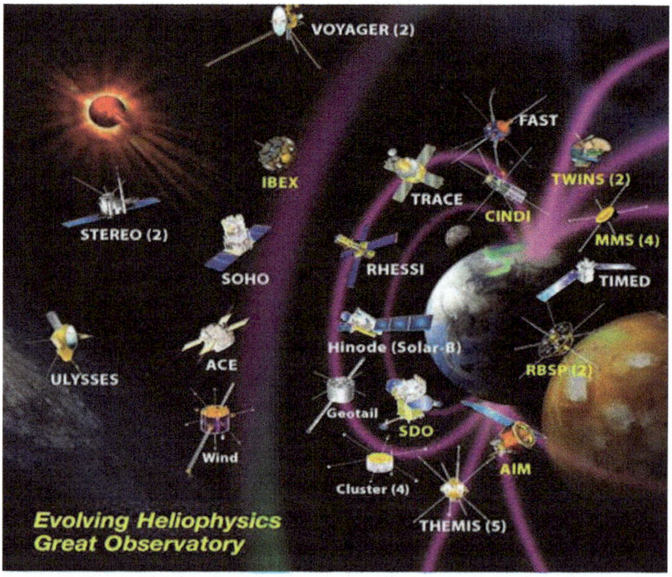

Fig. 6.1 Missions involved in studying space weather phenomena

taking account of damages to electricity grids in Europe, lost economic opportunity and so on. Other key terrestrial infrastructure, such as land-based communication systems, the underlying computer systems of financial markets, and pretty much any infrastructure dependent on a reliable supply of electricity and communications can also be disrupted. Such events could also exploit extant weak links in infrastructure systems to inflict even greater damage, with the NRC estimating that, due to vulnerable and ageing transformer stations, over 130 million consumers in the US alone would be deprived of electricity for more than a few hours.

Table 6.4 presents a list with the most well-known sources of up-to-date information on current space weather phenomena and anticipated levels (Hapgood 2010). The services with an asterisk are part of the SWENET system developed and operated by the European Space Agency. The users of these services are private companies in various fields, governments, research groups and National armed forces. Examples are provided in Table 6.4.

Examples of activities and systems affected by space weather include satellite operations, manned space flights, radiation in space and aviation, navigation systems, general electrical control systems, high-frequency communications, shortwave broadcasting, surveillance radars, geophysical exploration, electricity power grid distribution, and long pipeline corrosion.

Satellite damage and loss is one of the major effects of space weather. This can take place through radiation damage, single event effects (SEEs) and electrical charging. During the solar event of 2003 more than 30 satellite anomalies were reported. Satellite disruption will have a direct impact on the ground operations, including

Table 6.3 Economic costs to different organizations that use weather prediction centre

Impact area	Customer (examples)	Action (examples)	Cost (examples)
Spacecraft (Individual systems to complete spacecraft failure; communications and radiation effects)	• Lockheed Martin • Orbital • Boeing • Space systems Loral • NASA, DoD	• Postpone launch • In orbit—Reboot systems • Turn off/safe instruments and/or spacecraft	• Loss of spacecraft ~ $500M • Commercial loss exceeds $1B • Worst case storm—$100B
Electric power (Equipment damage to electrical grid failure and blackout conditions)	• U.S. Nuclear Regulatory Commission • N. America Electric Reliability Corp. •Allegheny power • New York power authority	• Adjust/reduce system load • Disconnect components • Postpone maintenance	• Estimated loss ~ $400M from unexpected geomagnetic storms • $3–6B loss in GDP (blackout)
Airlines (communications) (Loss of flight HF radio communications) (Radiation dose to crew and passengers)	• United airlines • Lufthansa • Continental airlines • Korean airlines • NavCanada (air traffic control)	• Divert polar flights • Change flight plans • Change altitude • Select alternate communications	• Cost ~ $100k per diverted flight • $10–50k for re-routes • Health risks
Surveying and navigation (Use of magnetic field or GPS could be impacted)	• FAA-WAAS • Dept. of Transportation • BP Alaska and Schlumberger	• Postpone activities • Redo survey • Use backup systems	• From $50k to $1M daily for single company

broadcasting of television and radio programmes, positioning, search and rescue, etc. Table 6.5 presents significant satellite losses and outages.

Space weather can degrade the accuracy of the navigation systems and can cause loss of the satellite signal (e.g., see Table 6.5). Recently, the GNSS have been strengthened by augmentation services, which generate ionospheric correction data and enable satnav receivers to measure aircraft altitudes with accuracy of 10 m. However, in the 2003 space weather event, the 50 m vertical limit set up by the FAA was exceeded. The loss of satellite navigation can occur in severe space weather conditions, such as strong ionospheric scintillation, where the signal varies very quickly so the receiver cannot maintain lock. Solar radio bursts act like a natural jamming signal, as happened in 2006, when the guided approach system used by airlines was unavailable for 15 min (Hapgood 2010).

Table 6.4 Providers of up-to-date information space weather phenomena

Universal services, offering access to data and predictions regarding space weather conditions in a scientific format which needs to be interpreted by experts	
• SWPC—Space Weather prediction centre (SUA) • SWENET—Space Weather European network (ESA) • IPS radio and space services (Australia) • ISES—international space environment service	• www.swpc.noaa.gov • www.esa-spaceweather.net/swenet • www.ips.gov.au • www.ises-spaceweather.org
Specialized services—aviation	
• SolarMetrics—professional Space Weather services for aerospace • QinetiQ atmospheric radiation model	• www.solarmetrics.com • www.qarm.space.qinetiq.com
Specialized services—energy systems	
• GIC Now! • GIC Simulator • Solar wind monitoring and induction modeling for GIC • Metatech Corporation, applied power solutions division & geomagnetic storm forecasting services • Prototype GIC forecast service	• www.aurora.fmi.fi/gic_service/english/index.html • www.spaceweather.gc.ca/se-gic-eng.php • www.geomag.bgs.ac.uk/gicpublic • www.metatechcorp.com/aps/apsmain.html • www.lund.irf.se/gicpilot/gicforecastprototype
Specialized services—oil and mineral prospecting	
• BGS geomagnetism applications and services	• www.geomag.bgs.ac.uk/services.html
Specialized services—pipelines	
• Space Weather service for pipelines	• www.spaceweather.gc.ca/se-plp-eng.php

The following are some of the actors most affected by solar weather phenomena:

- **Electric Power Grid Operators** use geomagnetic storm detection and warning products to maximize power grid stability and to mitigate power grid component damage and large-scale blackouts
- **Spacecraft Launch Operators** use radiation products to avoid electronic problems on navigation systems, preventing launch vehicles from going off course and being destroyed
- **Spacecraft Operations and Design** staff rely on space weather products to avoid electronic problems. Space weather effects on satellites vary from simple repairs to total mission failure
- **Manned Spaceflight** activities are altered to avoid or mitigate effects of radiation storms impacting crews and technological systems
- **Navigation Systems** depend on space weather information to ensure the integrity and safe use of electronic navigational systems, such as GPS

Table 6.5 Selected significant satellite losses and outages, as adopted from Cannon (2013)

Date	Event	Satellite	Orbit	Cause (probable)	Effects seen
8 March 1985		Anik D2	GEO	ESO	Outage
October 1989	CME-driven storm	TDRS-1	GEO	SEE	Outage
July 1991		ERS-1	LEO	SEE	Instrument failure
20 January 1994	Fast solar wind stream	Anik El	GEO	ESO-note: all three satellites were of same basic design	Temporary outage (hours)
20 January 1994	Fast solar wind stream	Anik E2	GEO	ESO-note: all three satellites were of same basic design	6 months outage, partial loss
20 January 1994	Fast solar wind stream	Intelsat K	GEO	ESO-note: all three satellites were of same basic design	Temporary outage (hours)
11 January 1997	Fast solar wind stream	Telstar 401	GEO	ESO	Total loss
19 May 1998	Fast solar wind stream	Galaxy 4	GEO	ESD	Total loss
15 July 2000	CME-driven storm	Astro-D (ASCA)	LEO	Atmospheric drag	Total loss
6 Nov 2001	CME-driven storm	MAP	Interplanetary L2	SEE	Temporary outage
24 October 2003	CME-driven storm	ADEOS/MIDORI 2	LEO	ESO (solar array)	Total loss
26 October 2003	CME-driven storm	SMART-1	HEO	SEE	Engine switch-offs and star tracker noise
28 October 2003	CME-driven storm	DRTS/Kodama	GEO	ESD	Outage (2 weeks)
14 January 2005		Intelsat 804	GEO	ESD	Total loss
15 October 2006	Fast solar wind stream	Sicral 1	GEO	ESD	Outage (weeks)
5 April 2010	Fast solar wind stream	Galaxy 15	GEO	ESD	Outage (8 months)
13 March 2012	CME-driven storm	Spaceway 3	GEO	SEE?	Outage (hours)
7 March 2012	CME-driven storm	SkyTerra 1	GEO	SEE/ESD?	Outage (1 day)
22 March 2012	CME-driven storm	GOES15	GEO	ESD?	Outage (days)

- **Aviation Operators** use crucial information on space weather impacts—such as communication outages, potentially harmful radiation, and navigation errors—to adjust routes and altitudes
- **Communications Operators** anticipate and react to space weather activity to mitigate impacts occurring over a wide range of communications frequencies used by emergency management officials, search and rescue systems, and many others
- **Surveying and Drilling Operations** rely on accurate and timely space weather products for safe and efficient high-resolution land surveying and sea drilling. A growing number of customers are realizing societal and economic benefits from space weather products and services. Expect this trend to continue as the world becomes increasingly dependent on space-based systems and other technologies vulnerable to hazardous space weather.

The following is a collection of scenarios of impact on various activities and infrastructures should an exceptionally strong solar weather phenomena manifest itself:

- Electricity grid—the reasonable worst-case scenario would have a significant impact on national electricity grids. The time to repair would be between weeks and months. In addition, current estimates indicate a potential for some local electricity interruptions of a few hours. Because most nodes have more than one transformer available, not all these failures would lead to a disconnection event.
- Satellites—some satellites may be exposed to environments in excess of typical specification levels, so increasing microelectronic upset rates and creating electrostatic charging hazards. Because of the multiplicity of satellite designs in use today there is considerable uncertainty in the overall behavior of the fleet but experience from more modest storms indicates that a degree of disruption to satellite services must be anticipated. Fortunately, the conservative nature of spacecraft designs and their diversity is expected to limit the scale of the problem. The best engineering judgement, based on the 2003 storm, is that up to 10% of satellites could experience temporary outages lasting hours to days as a result of the extreme event, but it is unlikely that these outages will be spread evenly across the fleet since some satellite designs and constellations would inevitably prove more vulnerable than others. In addition, the significant cumulative radiation doses would be expected to cause rapid ageing of many satellites. Very old satellites might be expected to start to fail in the immediate aftermath of the storm while new satellites would be expected to survive the event but with higher risk thereafter from incidence of further (more common) storm events. Consequently, after an extreme storm, all satellite owners and operators will need to carefully evaluate the need for replacement satellites to be launched earlier than planned in order to mitigate the risk of premature failures.
- Aircraft passenger and crew safety—passengers and crew airborne at the time of an extreme event would be exposed to an additional dose of radiation estimated to be up to 20 mSv, which is significantly in excess of the 1 mSv annual limit for members of the public from a planned exposure and about three times as high as the dose received from a CT scan of the chest. Such levels imply an increased cancer

risk of 1 in 1,000 for each person exposed, although this must be considered in the context of the lifetime risk of cancer, which is about 30%. No practical method of forecast is likely in the short term since the high energy particles of greatest concern arrive at close to the speed of light. Mitigation and post event analysis is needed through better onboard aircraft monitoring. An event of this type would generate considerable public concern.

- Ground and avionic device technology—solar energetic particles indirectly generate charge in semiconductor materials, causing electronic equipment to malfunction. Very little documentary evidence could be obtained regarding the impact of solar energetic particles on ground infrastructure and it is consequently difficult to extrapolate to a solar superstorm. More documentary evidence of normal and storm time impacts is available in respect to avionics—no doubt because the operating environment has a higher flux of high-energy particles. Our estimate is that during a solar superstorm the avionic risk will be ~1,200 times higher than the quiescent background risk level and this could increase pilot workload. We note that avionics are designed to mitigate functional failure of components, equipment and systems and consequently they are also partially robust to solar energetic particles.
- Global navigation satellite systems (GNSS)—assuming that the satellites—or enough of them—survived the impact of high energy particles, it stands to reason that a solar superstorm might render GNSS partially or completely inoperable for between one and three days. The outage period will be dependent on the service requirements. For critical timing infrastructure it is important that holdover oscillators be deployed capable of maintaining the requisite performance for these periods. There will be certain specialist applications where the loss or reduction in GNSS services would be likely to cause operational problems; these include aircraft and shipping. Today, the aircraft navigation system is mostly backed up by terrestrial navigation aids; it is important that alternative navigation options remain available in the future.
- Cellular and emergency communications—a study has concluded that the UK's commercial cellular communications networks are much more resilient to the effects of a solar superstorm than those deployed in a number of other countries (including the US) since they are not reliant on GNSS timing. In contrast, the UK implementation of the Terrestrial European Trunked Radio Access (TETRA) emergency communications network is dependent on GNSS. Consequently, mitigation strategies, which already appear to be in place, are necessary.
- High frequency (HF) communications—HF communications are likely to be rendered inoperable for several days during a solar superstorm. HF communications is used much less than it used to be; however, it does provide the primary long-distance communications bearer for long distance aircraft (not all aircraft have satellite communications and this technology may also fail during an extreme event). For those aircraft in the air at the start of the event, there are already well-defined procedures to follow in the event of a loss of communications. However, in the event of a persistent loss of communications over a wide area, it may be necessary to prevent flights from taking off. In this extreme case, there does not appear to be a defined mechanism for closing or reopening airspace once communica-

tions have recovered. Mobile satellite communications During an extreme space weather event, L-band (~1.5 GHz) satellite communications might be unavailable, or provide a poor quality of service, for between one and three days owing to scintillation. The overall vulnerability of L-band satellite communications to superstorm scintillation will be specific to the satellite system. For aviation users the operational impact on satellite communications will be similar to HF.
- Terrestrial broadcasting—terrestrial broadcasting would be vulnerable to secondary effects, such as loss of power and GNSS timing.

This section summarizes some of the agencies, and nations, involved in space weather forecasting. It does not attempt to cover all such organizations. For instance, Asia is rapidly developing its space activities, Europe contains many smaller agencies in addition to ESA. USA has a centre for space environment but has many small groups and businesses operating in the space weather area.

- Europe—European effort in space weather monitoring and analysis has until recently lacked overall coordination. Various aspects of space weather are monitored by different organizations: solar observations are made through the French Meudon Observatory, radio propagation warnings are prepared by BAE systems p/l and a wide range of forecasts is made by the University of Oulu. Currently, the European Space Agency (ESA) is seeking ways to coordinate the present groups into a cohesive European Space Weather Centre but this is at an embryonic stage
- USA—the three principal agencies undertaking significant space weather activity are the civilian Space Environment Center of the National Oceanic and Atmospheric Administration (NOAA), National Aeronautical and Space Agency (NASA) and the US Air Force (Air Force Weather Agency). NOAA is supported by the US Department of Commerce and works closely with the USAF under a National Space Weather Plan
- Russia—the Russian network of space weather monitors is still one of the largest in the world, but its capability is limited by poor support and by lack of coordinated effort. The majority of the space weather monitoring is operated by the Institute of Applied Geophysics in Moscow in cooperation with the military
- Asia—in Japan the Communications Research Laboratory operates a space weather centre outside Tokyo. Data and analysis are carried out and distributed to organizations that use them. Kyushu University hosts the Space Environment Research Centre and Nagoya University has a solar terrestrial environment laboratory both using some real-time data from Australia. In China, there is a space weather centre at the Beijing solar observatory and at the Chinese Research Institute for Radio Propagation. Although the Chinese network of monitors is extensive, the communication of data in near-real time from many sites is not available and therefore warnings of space disturbances are of limited value. Korea, Indonesia and possibly Thailand and Malaysia are developing space weather programs
- Africa—South Africa is establishing a space weather forecast centre with the help of Australian expertise. South Africa has the longer-term objective of providing assistance to rural communications across the whole of Africa

- International bodies—under the auspices of ICSU (International Council for Science) COSPAR (Committee for Space Research) is the prime organization for coordinating international space science programs, including space weather, while SCOSTEP (Scientific Committee for Solar-Terrestrial Physics) specializes in Sun-Earth relations. The International Space Environment Service (ISES) provides a rapid exchange of space weather information. The International Union for Radio Science (URSI) sponsors symposia on subjects that affect radio and consequently space weather is now a discussion topic within URSI. The Cosmic Ray Commission of the International Union of Pure and Applied Physics (IUPAP) sponsors biennial international cosmic ray conferences that include the radiation aspects of space weather
- Future directions—major efforts are being made, especially in the U.S.A and Japan, to understand the hazards of space environment in preparation for the use of space for research, industrial and military activities. Space programs are in place for monitoring and modelling the Sun-Earth environment and many more are planned as international cooperative efforts.

At the present time there are the following issues in relation to space weather:

1. Future technical systems will require greater prediction accuracy than currently possible
2. Technology is moving ahead rapidly into areas influenced by space weather
3. Developed countries are very dependent on satellites and spacecraft for information, communications, and remote sensing
4. Space weather activities, even when they are at international standards, are widely dispersed among government, university and the private sectors
5. A strategy is needed to maximize the benefits of space weather activities for industry and social infrastructure.

What can be done to mitigate the risks of space weather? Firstly, the proliferation of early warning systems hints at the benefits of being forewarned—systems can be shut down to avoid the worst effects of the geomagnetic induced current, flights can be grounded or rerouted, the use of grids can be adjusted to bypass the most vulnerable sections, space activities and experiments can be postponed, and back-up systems can be initiated in time, as well as a host of other protection measures.

Another means of increasing the resilience of the system-of-systems to the effects of space weather on any of its component infrastructures is to maintain redundancies, extra system capacity or extra assets. The robustness of the asset in question is also very important, as the proper shielding and other precautions can reduce the detrimental effects of space weather. On the one hand, there is an associated cost with such preparations, both financial for the entity ordering the satellite system as well as a cost of opportunity in terms of weight, number of instruments and quantity of fuel on-board. The current constraints of accessing space lead to limited payloads and having to choose, to a certain extent, between security and efficiency. The experience of the past few decades with space weather has led to significant advances in shielding and asset hardening, but there is still a non-trivial chance of approximately 5% for a space asset to spontaneously malfunction during its lifetime due to the ambient stresses like radiation. This is especially important since replacements are difficult and time consuming, as well as a not very valid investment on the part of a profit driven company to whom preparations for a remote and unlikely event are considered sunk costs which should be minimized. This is a pattern that repeats itself throughout Western economies, where private companies own and operate between 60 and 80% of their nations' critical infrastructures and relationships with authorities can become strained, due to different objectives and incentives.

Other means of protecting critical infrastructures, including terrestrial ones, from space weather are being developed. For instance, key transformer stations can be hardened against space weather effects. Electronics can use triple redundant circuitry wherever possible. To reduce the risk of cascading disruptions throughout the CI chain because of the dependency on CSI of secondary and tertiary critical infrastructures, decision makers can explore the replacement or supplementation of space services with ground-based variants—signal amplifying stations, ground-based atomic clocks, higher quality signal receivers, either a switch to or maintaining alternative systems which are more easily and less expensively shielded.

In the long-term, a viable international framework and legislation must be developed in order to enable the involvement of multiple stakeholders, meaning not just nations, but also businesses, Universities, representatives of civil society and international organizations, in the risk governance process. Technical recommendations can be advanced to ensure a minimum standard of security for space systems, the wide dissemination of best practices and information and informing the public regarding the new resources. There are significant obstacles facing such a vision, especially if we extrapolate from the situation today, where technical recommendations by the United Nations Committee on the Peaceful Uses of Outer Space can be safely ignored, especially by the countries that matter, because there is no mechanism for encouraging compliance.

Ultimately, space weather is a powerful threat to the functioning of the critical infrastructures, both terrestrial and situated in space, on which the prosperity, security and safety of our nations are built. By its very nature, however, in accordance with the concept of 'anti-fragility' developed by Taleb (2014), the constant stresses posed by the ambient and low-level space weather phenomena to space systems have served as a constant reminder of the risk of a powerful event, spurring the development and implementation of protective measures. In this respect, whatever resilience the 'critical space infrastructures' have today is the result of prior efforts encouraged by the certainty of future crises afforded by the occasional damage, destruction and disruption suffered by space systems, as well as terrestrial infrastructures.

6.2.2 Space Debris

Space debris represents space pollution and encompasses expended launch vehicles and damaged satellites. It reduces the expected value of space activities and increases the probability of damaging existing satellites or other space vehicles. Approximately

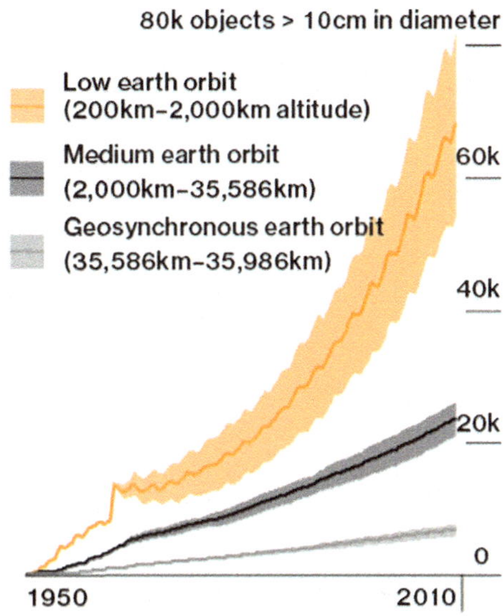

Fig. 6.2 The evolution of space debris in the three main orbits (https://www.bloomberg.com/news/articles/2014-01-16/space-junk-in-earth-orbit-the-cleanup-problem)

49% of extant satellites are in LEO, 6% are in MEO, and 41% are in GEO, the remainder being in elliptical or other orbits.[6]

The US Air Force currently tracks about 23,000 pieces of space debris and is able to see objects the size of a basketball at a 15-mile distance. But there are about half a million more debris bits it cannot see. That has raised fears about the Kessler syndrome, the scenario put forth by the NASA scientist Donald Kessler in the 1970s in which space becomes so overpopulated that collisions set off bigger and bigger chain reactions, making it nearly impossible to navigate.[7]

According to ESA (2017), of 42,000 total and whole objects placed in space by 5,250 launches, 23,000 are still present and only 1,400 are still active (UCS estimates are higher, as we have seen, taking into account the proliferation of cubesats).

The following is a breakdown of space debris by source.[8] Figure 6.2 is provided to illustrate increasing debris in space.

- 42%—disintegration of space systems, especially satellites
- 22%—whole but not functional space systems
- 19%—from specific activities of space missions
- 17%—debris from the launchers that placed space systems outside Earth's atmosphere.

The following is a breakdown of active satellites by orbital occupancy, underlining the most probable areas for debris accumulation[9]:

[6]https://www.ucsusa.org/nuclear-weapons/space-weapons/satellite-database#.WyCAsSAnbIU.

[7]See the Donald Kessler Interview, http://webpages.charter.net/dkessler/files/KesSym.html.

[8]http://www.orbitaldebris.jsc.nasa.gov/library/EducationPackage.pdf.

[9]https://www.ucsusa.org/nuclear-weapons/space-weapons/satellite-database#.WyCB0yAnbIV.

- 49% in low orbit;
- 6% in medium orbits;
- 41% in geosynchronous orbit;
- 4% in other orbits, including elliptical.

The estimated time for reentry of orbital debris in Earth's atmosphere, where it is safely disposed of, varies according to altitude, in the absence of an exterior force to provide it with another vector (Whitlock et al. 2008):

- A few days, for less than 125 miles altitude;
- A few years, if the space debris is in the 125–370 miles altitude band;
- A few centuries, for more than 500 miles;
- At GEO, the timeframe expands so much that it can be safely stated that the debris is in permanent orbit.

Some estimates suggest that China (40%), USA (27.5%) and Russia (25.5%) have generated the most orbital debris[10]:

- China—40%;
- United States—27.5%;
- The Russian Federation—25.5%.

Debris can damage or destroy communication, weather, navigational, governmental, and military satellites. There have been several instances when debris approaching the ISS presented a threat to the station. Astronauts put on their space suits and moved to the Soyuz capsules so that they have an opportunity to start moving towards the Earth if needed.[11] The coating of the American shuttle spacecraft has been damaged twice. In 2006, a tiny space fragment collided with a satellite in its orbit, as a result of which residents of the Far East were left without a TV signal for a while. Taking into account that the planet's technologies are increasingly linked to space, orbital debris can at any moment interrupt the usual course of life.

Besides expended launch vehicles, debris is created by satellites, as they reach the end of their productive lives and break up, or as a result of impact with debris or with other satellites (Taylor 2006). The intentional destruction by the Chinese of their weather satellite (FengYun 1C) in January 2007, and the accidental collision between the Cosmos 2251 and Iridium 33 satellites in February 2009 are such examples. The destruction of the FengYun weather satellite generated 2,316 trackable objects (larger than a golf ball) and over 150,000 additional pieces of debris over 1 cm in size, at an altitude approximately 540 miles above the surface of the earth (Pardini and Anselmo 2007). The Cosmos-Iridium collision generated at least 2100 additional pieces of trackable debris, at an altitude approximately 490 miles above the earth's surface.

---

[10] http://www.spacemart.com/reports/Space_junk_endangers_mankinds_usual_course_of_life_999.html and http://www.cbc.ca/news/technology/space-debris-1.4545467.

[11] See the US Orbital Debris Mitigation, http://www.orbitaldebris.jsc.nasa.gov/library/USG_OD_Standard_Practices.pdf.

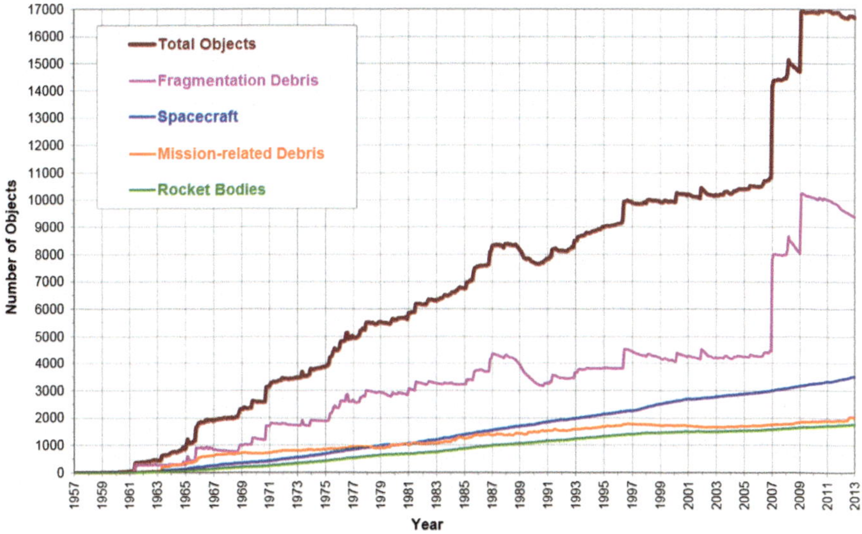

Fig. 6.3 The evolution of number of debris number in space, breakdown by origin

In its 2007 FengYun-1C test, in which an inactive weather satellite was destroyed by China, the equivalent of 12% of the entire stock of orbital debris was created in one strike.[12] This was only the second kinetic ASAT test in history, after the inaugural one in 1983 performed by the Americans. The year after the Chinese test, in operation 'Burnt Frost,' the USA tested a new generation of ASAT weaponry.[13] This whole episode shows the destructive power of ASAT weaponry and the potential for escalation in space conflicts, with devastating global results, which is why the non-militarization of space, the non-proliferation of space weaponry and the control of their use is on the agenda for major global policy initiatives, like the Conference for Disarmament or various UN legislative projects in the field of space.

The evolution of debris number in space is illustrated by Fig. 6.3.

A single fleck of paint, travelling at orbital speeds, can destroy a satellite with a force that puts a high-powered bullet to shame. The International Space Stations regularly performs evasive maneuvers to avoid collisions with large pieces of debris though the photo below shows that even the smallest ones, which are harder to track and avoid, still pose incredible danger. NASA's Space Surveillance System only tracks debris as large as 5 cm, whereas the 7-mm chip in an ISS viewport was caused by a piece a few thousandths of a millimeter large[14] (see Fig. 6.4).

[12]See 'Evolution of the Debris Cloud Generated by the FengYun-1C Fragmentation Event,' http://issfd.org/ISSFD_2007/10-4.pdf.

[13]https://www.ucsusa.org/sites/default/files/legacy/assets/documents/nwgs/a-history-of-ASAT-programs_lo-res.pdf.

[14]http://thescienceexplorer.com/technology/space-debris-hit-international-space-station-causing-small-crack-window.

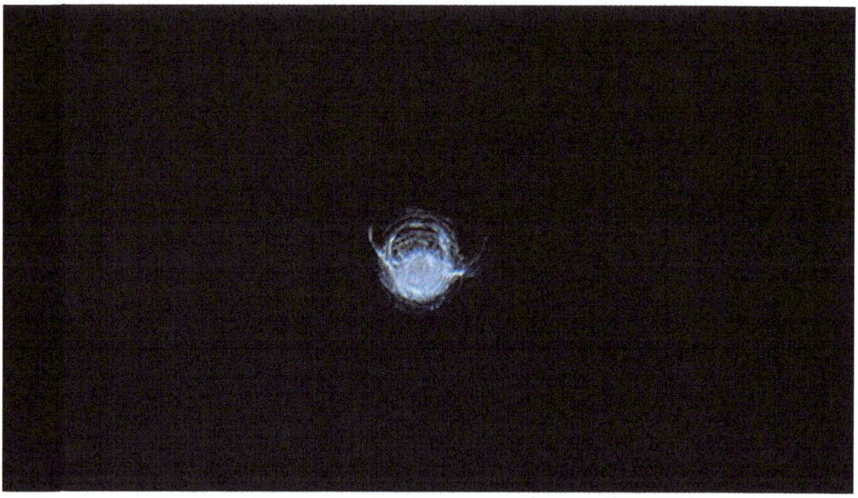

Fig. 6.4 Mark left in ISS viewport by impact with very small debris

Where possible, satellites also perform evasive maneuvers, but at the cost of their usable lifespan, since their limited supply of fuel for thrusters is required for station keeping, which is the maintenance of their position throughout their operational lifetime. Collisions are not just with tiny debris, but also with pieces of satellites or entire non-functioning satellites, as happened during the collision of telecommunications satellite Iridium 66 with an inactive Russian satellite, Cosmos 2251, in 2009. Even though space is, proverbially, very big, the most useful orbits are so crowded, that satellites routinely pass within proximity of each other at hypervelocity, while Iridium 66 and Cosmos 2251 were estimated to pass at around 500 m distance.[15] There is no current means of refueling and repairing satellites in space, though a few start-ups are working on solutions and NASA has, in the past, undertaken a risky manned mission to repair the irreplaceable Hubble telescope. The sort of treatment reserved for Hubble and the ISS is not available to other space assets, not if the industry is going to be financially sustainable.

The biggest threat when it comes to space debris is not just the potential for crippling or destroying satellites, or even endangering astronauts. The biggest threat, in the wider picture, is that every collision creates new debris, compounding the problem and increasing the statistical likelihood of future collisions, which would create even more debris. This has the potential to turn into a cascading phenomenon dubbed 'The Kessler Syndrome,' after NASA scientist Donald Kessler.[16] He theorized that, at a certain concentration of space debris, consecutive collisions will make the likelihood of immediate impact almost certain, rendering Earth orbit an uninhabitable minefield. No unshielded rocket will be able to pass, no new satellites placed, and

[15]https://www.space.com/19450-space-junk-worst-events-anniversaries.html.

[16]See the Donald Kessler Interview, http://webpages.charter.net/dkessler/files/KesSym.html.

no manned mission can take place. This is not a theoretical scenario, so much as a prediction for the end state of this rapidly degrading natural environment.

Currently, there are approximately 21,000 human-generated (radar tracked) pieces of debris measuring over 10 cm, 600,000 (untracked) pieces of debris between 1 and 10 cm, and over 100,000,000 (untracked) pieces of debris between 0.1 and 1 cm in earth orbit.[17]

Debris greater in size than 10 cm can be tracked by earth stations, and satellites can, given sufficient warning, engage in evasive maneuvers to potentially avoid a collision. A collision with a piece of debris greater than 10 cm, however, will likely destroy a satellite and generate significant amounts of additional debris. Debris greater in size than 1 cm but less than 10 cm cannot be earth-tracked, but can destroy a satellite (Adilov et al. 2015). Thus, these pieces of debris are particularly dangerous (since they are not currently tracked). Debris less than 1 cm in size can damage or potentially destroy a satellite, although additional shielding, which increases production costs, can protect the satellite from damage (Adilov et al. 2015). Salter (2015) calculated that the mass of the debris inventory in orbit is 6,100, 2,300 of which are situated in Low Earth Orbit. Orbital debris has degrees of persistence, as mentioned before. Thus, at very low altitudes (less than 125 miles) space is quickly self-cleansing. However, peak debris density in LEO occurs at 550 miles, which suggests centuries would pass before the region is self-cleaned (assuming no additional debris is added during that time) (Whitlock et al. 2008). Several debris collisions occurred in the past 20 years, according to Union of Concerned Scientists[18]:

- 1991 Inactive Cosmos 1934 satellite hit by catalogued debris from Cosmos 296 satellite
- 1996 Active French Cerise satellite hit by catalogued debris from Ariane rocket stage
- 1997 Inactive NOAA 7 satellite hit by uncatalogued debris large enough to change its orbit and create additional debris
- 2002 Inactive Cosmos 539 satellite hit by uncatalogued debris large enough to change its orbit and create additional debris
- 2005 U.S. rocket body hit by catalogued debris from Chinese rocket stage
- 2007 Active Meteosat 8 satellite hit by uncatalogued debris large enough to change its orbit
- 2007 Inactive NASA UARS satellite believed hit by uncatalogued debris large enough to create additional debris
- 2009 Active Iridium satellite hit by inactive Cosmos 2251.

---

[17]http://www.unoosa.org/pdf/limited/AC105_C1_2011_CRP14E.pdf.

[18]https://www.ucsusa.org/sites/default/files/legacy/assets/documents/nwgs/securing-the-skies-exec-summary.pdf.

6.2.2.1 Risk Governance for Space Debris

As outlined in the previous section, space debris has become a recognized, immediate and permanent threat to the functioning of critical space systems, which is why it has also received the most attention at all levels, compared to other threats, such as "space weather". The following section outlines the complexities of the ad hoc risk governance system that is crystalizing out of disparate efforts by different actors to ameliorate the problem. It also highlights the deficiencies of current approaches, which are fragmented and stifled by issues of jurisdiction, sovereignty and uncertain liability. To a certain extent, this is a reflection of the problems facing the entire framework for governance of risks to space systems and their day to day operations.

To simplify this vast subject matter, we have developed the following schematic (Fig. 6.5) which highlights the most important criteria for classification of space debris risk governance efforts.

Translating the schematic into a narrative is equally daunting with the limitation of space. We should start from the issue of solutions to the space debris problem. The solutions are:

- To minimize the creation of space debris during launch—proper disposal of launch system components through reentry, robust design to prevent issues such as the formation of small debris from the physical damage of the launch procedure
- To minimize vulnerability to space debris during satellite operation (to prevent accidents and to avoid the creation of space debris)—hardening the satellite with

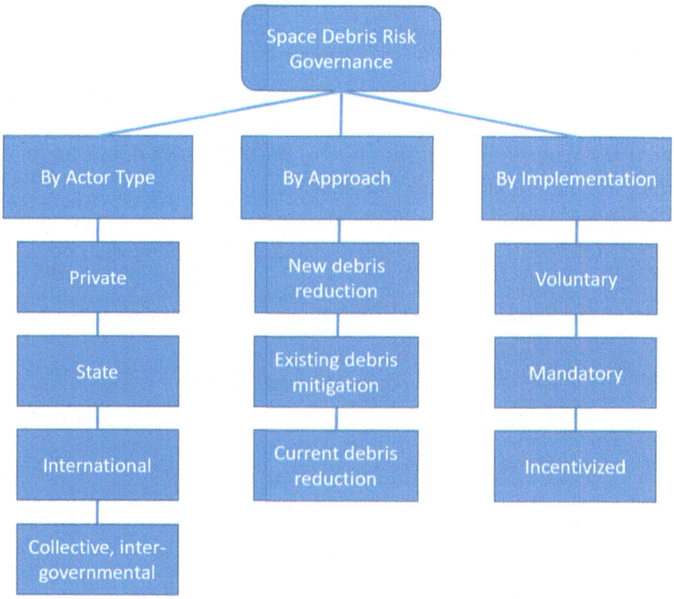

Fig. 6.5 Schema for Classification of Space Debris Risk Governance Efforts

shielding, which may still create new debris, even if the satellite may be saved, through the damage to the shield; mobbing the satellite out of the way using the limited reserves of fuel normally used for station-keeping, reducing the useful lifespan of the satellite
- To minimize the creation of space debris after the useful lifespan of an asset has ended—maintaining enough fuel in reserve (as a sacrifice of useful lifespan) to nudge the deactivating satellite into the atmosphere and decrease the likelihood of collisions, or the satellite acting as the progenitor of new space debris.

Governmental bodies are helping, both in Academia and in the Armed Forces, by improving the capacity for detecting debris in orbit and maintaining permanent overwatch on behalf of the owners and operators of the 'critical space infrastructures.' The solution of cleaning up orbits through the use of technological means for collecting space debris or inducing reentry is still in the realm of fiction. However, the solutions detailed above are quite possible and are, to a certain extent, implemented where safety counts for the most and cost counts the least—in the powerful militaries of spacefaring nations, where even satellites constellations such as GNSS systems (invariably military controlled, with the exception of the European Galileo system) have spare units in orbit ready to replace a damaged unit in order to maintain coverage or the quality of the signal. In another section, we mentioned that most military communications and procedures are still routed through civilian satellites. For them, technical standards relating to robustness, resilience, proper launch and disposal are pursued at varying governmental levels. Should the operator in question be a private actor, as most terrestrial critical infrastructures and an increasing share of space systems are owned and operated by private entities, then security will be viewed as a sunk cost inflating the budget allocated to the space system and diminishing its returns. Therefore, regulation takes the place of incentive for sustainable and security conscious behavior. It is possible for national regulatory agencies to recommend technical standards and also have the authority to impose them, but the most important drive comes at supranational levels.

First, there is an Inter-Agency Debris Commission which brings together the most powerful space agencies for the explicit goal of mutual cooperation and coordination on these issues. Secondly, but at a more important level because of its steady organization (as opposed to the IADC's infrequent reunions), there is the United Nations Committee on the Peaceful Uses of Outer Space which has Subcommittees developing various aspects of space governance, including such technical standards. The problem with this body is that its recommendations are applied on a voluntary basis, with no means of enforcement or penalty, unless an authoritative National body were to adopt the same standards and make them mandatory for companies and organizations registered in the specific country (Joyner 2005). Different treaties are supported by a mosaic of states, which are at various stages of adopting them, while other treaties lack the support of the most powerful space players, who are holding out for a framework that is to their specific advantage.[19] Of course, there is

[19]The 'Legal Subcommittee of the UN Committee on the Peaceful Uses of Outer Space,' which was approved during 54th session in Vienna, 13–24 April 2015.

also a positive means of enforcing compliance, should the body in question be in a position to dictate terms in the market by virtue of occupying an important position as authorized builder, significant purchaser or significant operator of space systems. Then, it could enforce preferences which, for the sake of simplifying the complexities of manufacture, space system producers would find it more advantageous to internalize and apply across their whole product line.

There are other ways to incentivize sustainable behavior with regards to space debris—through the use of generic market mechanisms. The gains in security and resilience from the adoption of robust design and sustainable practices could be converted to financial value and have some accrue to the respective operator. For instance, banks considering whether to finance the space venture would consider the already risky business less so should the space system adhere to recognized standards of protection. This might translate into more accessible financing, to reflect the lower risk of failure. Insurance companies, as well, may lower their premiums if the insured asset is hardened against impact. The collective effect of the reduction in new debris creation and the gradual removal by natural means of existing debris may lower the debris density of certain orbital bands, leading to lower risk of an impact occurring. Again, this would reduce insurance premiums. Another way to incentivize sustainable debris creation levels is to figure out and assign liability for the destruction of an asset through the impact with a trackable piece of debris originating from an improperly disposed asset or a launch vehicle. Rather than risk paying damages and lost profits to another company for the results of its debris creation, a company would be incentivized to apply the latest standards in robustness, disposal of assets and debris reduction. Since it is not possible to track every piece of debris, but it is possible to estimate how much debris was created during the course of an impact event or a routine process, one could also imagine an international mechanism for levying a fine from debris producers as a behavior incentive, as part of a mechanism for financing research, development and deployment of space debris solutions or even to partly cover the losses of affected parties. There is a large knowledge barrier in the way of the proper functioning of such mechanisms, especially to understanding and pricing space debris risk. An interesting and recent development in this regard is the proposal to compile a space infrastructure vulnerability cadaster (Gheorghe and Yuchnovicz 2015), which would work very much like a land cadaster, in that it would provide a visual representation of the severity of the agglomeration of space debris in a certain orbital band. Relying the environmental science term of 'critical load,' the space cadaster would define intervals of impact risk based in relation to this critical load (which would represent the certainty for being hit and is the theoretical threshold for the commencement of the Kessler Syndrome), from little to no risk, to acceptable risk, then moderate, then high and then unacceptable. The resulting map would highlight the degree of degradation in the orbital environment and would also guide decision makers and business planners. It would be a basis for defining insurance premiums, for levying taxes for clean-up or for charging users for the sustainable utilization of an increasingly scarce resource in the respective orbital band (just like congestion pricing in city centers reduces traffic).

This sounds very good in theory, but the road to such solutions is arduous, even when attempted by private actors. For instance, there was an initiative on the part of The International Institute for Private Law (UNIDROIT) which serves as an adequate example to highlight the incipient struggle for establishing a legislative and administrative framework for space activity, especially with regards to commercial space development. For a number of years, it was engaged in formulating a Space Assets Protocol to aid in formalizing and removing the uncertainty from space entrepreneurship.[20] This protocol began development within the context of the Cape Town Convention on International Interests in Mobile Equipment, which opened on the 16th of November 2001 to provide a generic legal framework for registration of ownership and security interests, as well as legal remedies for default. There are three specific sector protocols: aircraft equipment, railway rolling stock, space assets. The first two passed, but the third ultimately failed due to concerted opposition from both spacefaring nations and private industry, which found it inadequate. In addition to defining the term 'space asset,' the Protocol would have had, among others, the following components[21]:

- Debtor's rights and assignments of debtor's rights and rights of use;
- Public service exemption from default remedies;
- Insurer's salvage interest;
- An international registry to record and determine priority (according to when registered) among creditor rights in space assets. It would include transponder leases, conditional sales agreements and security interests.

The parallel with maritime law and custom is obvious. Its failure notwithstanding, the efforts of UNIDROIT underline the need for creating an appropriate framework to support commercial space activities, the kind of framework that can then be employed not just for encouraging the growth of the space industry and the diversification and reduction in cost of services, but also as a foundation for regulatory measures aimed at sustainable use of the most valuable space in proximity to Earth.

One final aspect to the space debris risk governance outline is to mention the deliberate threats involving space debris. The destructive nature of the phenomenon makes its imitation attractive for actors interested in damaging space assets. In the first section of the article, it was mentioned how states with ASAT capabilities are less likely to employ them because of mutual dependency issues. Kinetic ASAT capabilities, such as those demonstrated by the US and China, have the power to turn space into a veritable minefield since the obliteration of a single satellites by the Chinese increased the number of known debris by 12%. The potential for a cascading destruction phenomenon in orbit is very high in case of such an attack. Other forms of space warfare are also potential contributors to the space debris issue—cyber-attacks, jamming, laser blinding or other damage and intentional 'frequency fratricide' may lead to a loss of control over the trajectory of the satellite, which could veer off course and become a hazard for other systems, in addition to dangers to itself. There was

[20] http://www.unidroit.org/overview-2012-space-assets.

[21] See http://www.hfw.com/UNIDROIT-Draft-Space-Assets-Protocol.

the case of the Galaxy 15 satellite that stopped answering commands of the ground station, for about 4 months, in 2010 (Ministry of Defence 2012). In 1998, another satellite from Galaxy constellation, this time number 4, situated in geostationary orbit, was similarly affected, leading to the loss of functionality of 90% of pagers from U.S. and problems with TV broadcasts, for a couple of days (Rinaldi et al. 2001). The Conference on Disarmament is one venue where the issues of preventing an arms race in space and the militarization of space have been discussed with a view towards transforming it into legislative reality. At the same time, rogue states, terrorist and organized crime groups are excluded from these kinds of incentives to abstain from space attacks and, increasingly, have the means to attempt them. The proliferation of ASAT technology is an issue, but not the biggest one, since even a hacker with a laptop can find countless ways of disrupting or subverting the functioning of a satellite (Gheorghe and Vamanu 2007). By definition, this is a low risk, low cost, high reward attack, which makes it very attractive. With regards to ASAT capabilities of intermediate complexity, there is also the possibility of launching a rocket full of pre-made debris into the desired orbit or using maneuvering satellites to displace the intended space system without creating debris.

6.2.3 Deliberate Threats Against Space Systems

Deliberate human threats to critical space infrastructures are many, varied and highly efficient. Their development dates back to the beginning of the American and Russian space programs, which had an important military component and a 'dual-use' philosophy regarding the development of new technologies. Nowadays, many actors in the field of space develop anti-satellite systems of legitimate technologies for protection that can be modified to become efficient anti-satellite weapons. A general taxonomy of anti-satellite capabilities consists of the following categories:

- Cybernetic attacks
- Laser attacks
- Signal jamming
- Kinetic attack with missiles
- Electromagnetic attacks
- De-orbiting by attacking with maneuver satellites;
- Passive attacks, by "mining" the space where that system orbits.

The vulnerability of space systems is predicated on:

- The predictability of CSI trajectory
- The orbital dynamics of CSI relative to various regions of Earth, that allows regular visual contact from many regions, including some that can be considered sources of hostile elements
- The difficulty of CSI replacement, from the perspective of the technical efforts, the costs and the amount of time needed

- The efficiency of these attacks calculated as a ratio between the damages caused and the costs of an attacker for fulfilling such a mission (including the costs of failed attempts). A weather satellite can provide vital information for millions of beneficiaries and thousands of users, especially in emergency situations, when permanent monitoring is imperative
- Reduced costs for implementing certain attack methods—a cybernetic attack upon a communication satellite requires only a computer, an Internet connection and a trained user, therefore capable of being implemented as easily from a command center as well as a suburb from an emerging country. The required components for jamming signals between a command center and a satellite can be bought, legitimately and cheaply, from various sources. Even the technological and cost barriers for highly conventional military attacks, such as rockets or lasers, have been substantial reduced in the past decades. China is noted for the resources dedicated to its space program, with an acute military component
- The ascension of non-state actors which, as in the case of nuclear weapons, are not receptive to the logic of classic deterrence in the same way that a country would (Gheorghe and Vamanu 2007). Therefore, a country will develop anti-satellite capacities, but is unlikely that it will use them offensively, because of the substantial risk of retaliation against its own critical space systems. Also, in the nuclear domain, the number of countries that tend to obtain nuclear weapons is reduced, yet the intelligence services are preoccupied with the attempts of non-state actors, such as terrorist groups, of obtaining weapons of mass-destruction, and their decentralized and hyper-ideological nature makes them immune to a sovereign state's motivation to protect its population or to maintain the integrity of its territory[22]
- Anti-satellite attacks are o good way of causing substantial economic loss but evading direct loss of human lives. Therefore, it is considered a more humane offensive method and less reprehensible from the perspective of population perception. In this way, space systems become targets of opportunity for actors that do not wish to rise to a new level of infamy that would close doors in case of eventual negotiations or political projects.

Attacks against critical space systems have become more facile and more numerous, as the required technology advances, as well as the motivations of various actors in risking a possible retaliation for such an attack. The system of peaceful exploitation of outer space, after the Cold War, has been based on international cooperation in the context of space hegemony of U.S. The rise of China as a space actor, in the circumstances of U.S.A denying a potential partnership with China, because of a justified fear of technology transfer, produces what some call aa new space race' that is not restricted only to civilian applications. China aggressively developed anti-satellite capabilities, both through the potential of cybernetic attacks, as well as developing weaponry capable of destroying satellites in orbit. The US is the country with the most

[22]See https://www.stimson.org/content/anti-satellite-weapons-deterrence-and-sino-american-space-relations.

advanced thinking in the domain of dependence on 'critical space infrastructures,' their military applicability and defense doctrine against such attacks.

Its vulnerability is emphasized by the fact that it owns and operates half of the 270 existing military satellites, plus several hundred civilian, private or 'dual-use,' satellites that also have a military applicability. Among them is the first global positioning and navigation system, GPS, which still is a project of American Department of Defense. That branch of the US government still reserves the right to prioritize its needs to the detriment of its allies or civil users. The dependence on space systems is inexorably bound to the ascent of cyberspace as an environment for business, politics and social interactions, as well as military operations. A single Global Hawk drone that flies over the Middle East, consumes more transmission bandwidth than was consumed during the entire Gulf War in 1991, and 90% of the military traffic passes through civilian satellites, many with a private owner, and not through systems constructed to be resilient to various means of interrupting their functioning. Furthermore, 68% of American ammunition used in Iraq was guided through satellites, while only 10% was guided in the same manner during the Gulf War. Already, American strategists have stopped talking about the 'fog of war' and have started talking about the 'cloud of electrons' and about the fact that space systems are an 'Achilles' heel' for the USA (Easton 2009).

Military exercises even from 1990s or early 2000s, like Army After Next, Navy Global, and Air Force Global Engagement, Space Game 2, Schriever 1 and 2, or other simulations from private domain, DEADSATS, confirmed the fact that 'politicians, economists and company chiefs ignored the fact that space losses can affect national, economic and social security, not just in the United States, but also in the entire world.' American experts concluded that even major military powers, like U.S.A, could be 'taken hostage by the unknown elements of a new type of war.' Nowadays, these fears are directed towards China, and another military exercise, 'Pacific Vision,' led by American Air Forces from the Pacific theater of operations, demonstrated the vulnerability of commercial communications satellites on which they depend, as well as the vulnerability at cybernetic level and radar systems. Referring to the ASAT test of China from 2007, the first one after the US experiments 20 years earlier, General Harnel from Space and Missile Systems Center, declared that 'losing asymmetric advantage in space will regress the American war machine from the informational age to the industrial age' in favor of the adversary (Easton 2009).

6.2.3.1 Electronic and Cybernetic Attacks

Since useful orbits are not many and the majority of satellites will have a mobility limited by the architectural needs, their trajectories around Earth can be easily identified, in order to start attacks at the moment of their passing above certain areas, taking into account the fact that is impossible to monitor all the activities at such a large scale. At least, these approaches, as a concept, do not lead to the creation of new space debris.

6.2.3.2 Signal Jamming

Attackers could use signal jamming technologies in order to interrupt communications between satellites and ground stations. Interference in radio communication between a satellite and ground users can be implemented on uplink (data transfer from ground station to satellite) or downlink (data transfer from satellite to ground station), which is more vulnerable than command and control links between control centers and satellites. In the case of downlink jamming, the attacker interferes not only with the satellite, but also with the receiving of its transmissions at ground stations. For example, since GPS signal format is public, it would be relatively easy for a ground or air-based transmitter to generate a jamming signal. These types of systems, although illegal, are available in a wide variety on the free market.

Both United State and Russia possess jamming capabilities efficient up to geostationary orbits, especially against non-military targets, which are less shielded against such types of attacks. In 2002, the US initiated the Counter-Communications System, about which there are not many publicly available details. Non-military satellites are deigned to be used by an increased number of varied users, therefore their design philosophy is based on the ease of access, not on the obstruction of jamming. Most of the interferences are unintentional, resulting from human errors, but there were cases where different countries jammed foreign satellite transmissions in their own territory or cases where even non-state actors blocked commercial communication. The equipment necessary for jamming operations is similar to the equipment of legitimate satellite communications, and this method is not very challenging from a technical standpoint. For jamming a GPS signal, an attacker only needs to produce a signal with enough intensity to overwhelm the legitimate one. Ground jamming is more favorable in these situations, because it is simpler for the attacker to be close to ground receiving stations than the sources of transmissions, situated in the geostationary orbit, at an altitude of 20,000 km. Locating the attacker is simple, therefore it facilitates a counter-attack.

It is more complicated to jam an uplink, because the attacker would not have the advantage of reduced distance, as in the case of downlink, and he must know the exact direction and reception frequency of the satellite transponder which receives the signal. Since communication satellites have a large number of users, each with their own frequency, the attacker needs to know the exact frequency of the users he wants to target, or to dedicate important resources for jamming all the users, including ones that he would not want to jam. Taking into account the diversity of users of a space system, it is very likely that an attacker without discernment would be able to induce direct and indirect damages to his own country. Of course, a hacker can take control over the satellite and stop bothering with interference procedures, if he does succeed in passing through the defense of control systems.

Successful jamming is hard to be confirmed from the attacker's perspective, but it has certain advantages like:

- It is more difficult for the responsibility for the attack to be assigned to a specific actor, compared with other attacks

- It does not generate new space debris
- It is temporary and reversible.

Counter-jamming techniques, among which are information encrypting and rapid, unexpected and frequent changes of operating frequencies, are efficient, but reduce the maximum data transmission capacity through the communication link.

The Center for Developments, Concepts and Doctrine within the Ministry of Defense of Great Britain published a report, warning about 'frequency fratricide,' both unintentional and deliberate. Adjacent satellites cannot operate on the same frequency, without the generating some interference. The International Telecommunication Union coordinates the common use of radio spectrum and promotes international cooperation in ensuring uninterrupted satellite orbits. Occasionally, the loss of control over a satellite can lead to the modifying its orbit with the potential to intersect with other satellite orbits, leading to possible radio interferences. There was the case of the Galaxy 15 satellite that stopped answering commands of the ground station, for about 4 months, in 2010 (Ministry of Defence 2012). In 1998, another satellite from Galaxy constellation, this time number 4, situated in geostationary orbit, was similarly affected, leading to the loss of functionality of 90% of pagers from U.S. and problems with TV broadcasts, for a couple of days (Rinaldi et al. 2001).

6.2.3.3 Laser Attacks

It is possible to use lasers for 'blinding' the satellites, since the imaging components of a reconnaissance platform are naturally susceptible to damage, temporary or permanent, in case of a light wave which is strong enough being focused in the optical system. The creation, maintenance and usage of such a platform for space assaults, or at distances close to space, involve a high level of resource commitment.

Stationary ground lasers can interfere with the satellite sensors or can cause equipment damages. Low intensity lasers can impair the functioning of the high-resolution reconnaissance satellites, inhibiting the surveillance of large surface areas. Just as the ability of a satellite's data reception capacity can be overwhelmed by a jamming signal, a light source that is stronger than the target observed by the satellite will overload the sensors. The commercially available lasers have the ability to temporarily blind a satellite, but the presence of several filters and detectors require from the attacker detailed knowledge about the satellite in order to stop the operations, despite these contingencies and redundancies. On higher intensities, lasers can partially blind the satellite, through the destruction of part of the main sensor. The advantage in this case is that the laser does not have to be trained on the target; if the target is not damaged, the effects last only so long as the laser is trained on the device, presumably above the area whose reconnaissance must be prevented. A powerful enough laser, trained on a satellite, may overheat it or damage the structure and main components.[23]

[23]https://www.ucsusa.org/nuclear-weapons/space-security/a-history-of-anti-satellite-programs#.WyCYt6dKjIU.

Attacker advantages through laser assaults:

- Reaching the target with light speed
- From the tactical viewpoint, the system flexibility makes it possible to calibrate in order to produce reversible effects, as well as significant and permanent damage
- The concentration of high amounts of energy in a very thin beam and in a limited range of frequencies, allowing theoretically for targeting a specific part of the satellite. In practice, there are constraints on the part of the equipment used and on the part of atmospheric effects.

Ground lasers have significant operational disadvantages, including:

- Large distance from the target to be blinded, while the laser power is inversely proportional to the square of the distance
- The presence of the Earth's atmosphere causes distortions of the beam, making targeting difficult, and certain weather phenomena, such as fog and rain, limit the power that can be transmitted by the laser
- Laser ground stations can be the target of a counter-attack
- Laser ground station can use a laser only against a target in direct line of sight
- Protecting a satellite is relatively simple, involving the use of reflective, conductive or absorptive materials.

The research on these technologies continues in the United States and other countries. These include Russia, which resumed, in 2003, the Soviet program Sokol Eshelon (Сокол-Эшелон) to achieve a laser mounted on an aircraft of Beriev A-60 type, in order to temporarily blind enemy reconnaissance satellites. One of the key points of the project was the attempt to paint the AJISAI Japanese satellite as a target at an altitude of 1,500 km in 2009.[24]

An example of a laser ASAT military program is the American project Airborne Laser (ABL), whose aim was to create a megawatt laser which would fit in the cone of a plane in order to destroy missiles which were in space flight phase. The technology is useful for anti-satellite purposes, being even better adapted to them, because the missiles could easily be endowed with countermeasures such as reflective paint and the aircraft equipped with ABL is vulnerable to attack while trying to intercept missiles. Not being stationary, the aircraft could travel anywhere on Earth in order to target a specific space system at any given time. However, the program was stopped by the Missile Defense Agency in 2011.

In 2006, China has used, at least once, a laser to illuminate an American satellite from the ground, as an evidence of the performance of Chinese technology. Many other countries have the ability to track the motion of a satellite with a low intensity laser, because these capabilities are used in basic scientific research (using the satellite reflections of a laser to measure the gravity field of Earth, movement of plate tectonics, etc.). The International System of satellite measurements with the help of lasers consists in 40 ground stations in 23 countries and 30 cooperative satellites—the

[24] See for example: http://russianforces.org/blog/2012/11/russia_to_resume_work_on_airbo.shtml.

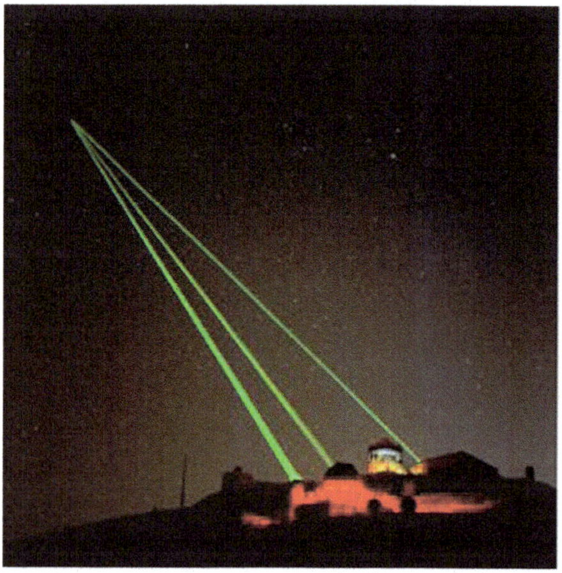

Fig. 6.6 Directed Energy directorate starfire from the Air Force Research Laboratory at Kirtland Air Force Base, USA (*Source* http://www.thelivingmoon.com/45jack_files/03files/Kirtland_AFB_Starfire_DEW.html)

equipment itself can be used to illuminate a different satellite, which is not part of the network.

A research program of the American Air Force derived from the Starfire project foundations, managed to combine a system sensitive enough to keep up with the satellite movement, within an optical system that compensates for atmospheric effects. There are also projects for equipping high altitude aircraft with such lasers, reducing the distance to the satellite and improving the performance and power consumption of the laser.[25] For objects in low orbit, laser attacks will have become a credible threat in the near future. This appears to be the case when one considers the research of Air Force Research Laboratory Directed Energy Directorate (see Fig. 6.6).

Permanent damages can be prevented by the satellite operators through:

- Closing the optical device iris, temporarily interrupting the provision of services, and possibly achieving the attackers' objective
- Relocating the system, which, given the current inability of refueling in space, will shorten the lifetime of any component, except the International Space Station.

Depending on the tasks assigned to the laser, the energy demand increases, which means that these lasers will be more complicated, larger and less mobile. Such lasers need cooling equipment; powerful energy sources and, in some cases, exhaust systems. The MIRACL system developed by the United States in the 1990s is powered by chemical reactions of missile type and is hosted in a large facility. A very powerful

[25]See for example (National Research Council 2013) and http://www.airforcemedicine.af.mil/News/Display/Article/1365970/directed-energy-weapons-research-a-new-frontier-for-air-force-medicine/.

anti-satellite laser weapon requires advanced optics. Thus, depending on the severity of damages which the mission profile requires, lasers have a key disadvantage in terms of needed investment and expertise, compared to other offensive systems.[26]

6.2.3.4 Electromagnetic Pulse Attacks

Low orbit satellites are at risk of becoming collateral damage to nuclear explosions at high altitudes, which generates an electromagnetic pulse (EMP), as a result of gamma radiation produced during nuclear reactions. High altitude electromagnetic pulses are a cause of significant concern for worldwide military systems due to their ability to permanently stop the operation not only of the space systems, but also of the ground systems, which form the basis of the critical infrastructures (Foster et al. 2008). A nuclear device will, upon detonation, radiate a portion of its total yield as X-rays, with the fraction realized a function of weapon design and attached delivery system. For a detonation above a (spectrally dependent) threshold altitude, X-rays emitted horizontally or upward will propagate to large distances virtually unattenuated by the atmosphere. X-rays emitted downward will be absorbed over ranges of tens of kilometers upon reaching sufficiently dense air. Neutrons and gamma rays emitted by a detonation similarly propagate upward great distances into space for detonations above threshold altitudes. Major atmospheric attenuation of these energetic emissions occurs at altitudes below approximately 40 km. For detonations up to a few hundred kilometers altitude, a majority of the kinetic yield of the weapon may be converted to ultraviolet (UV) photons. The combined flux of energetic photons (X-ray, gamma, and UV) and neutrons irradiates a vast region of space, diminished by spherical divergence, as shown in the Fig. 6.7 (Foster et al. 2008).

The actual size of the hazard zone depends on weapon yield, detonation altitude, and the degree of satellite hardening against disruption or harm. Damage to satellite structures and to coatings on solar panels and sensor optics occurs when X-ray and UV fluxes exceed critical thresholds. Electronics damage similarly ensues when X-ray and gamma pulses induce destructive electric currents in circuit elements and when energetic neutrons penetrate solid-state circuitry.

Compared to ground detonations, several effects are relatively unique to high altitude bursts[27]:

- Electromagnetic pulse (EMP) is important only for high altitude bursts. For such detonations, ionization of the upper atmosphere can produce a brief intense pulse of radio frequency radiation which can damage or disrupt electronic devices. For explosions above most of the atmosphere, EMP can affect large areas
- Ionization of the atmosphere from explosions in the atmosphere can interfere with radar and radio communications for short periods

[26]https://www.ucsusa.org/nuclear-weapons/space-security/a-history-of-anti-satellite-programs#.WyCdAKdKjIU.

[27]http://www.johnstonsarchive.net/nuclear/hane.html.

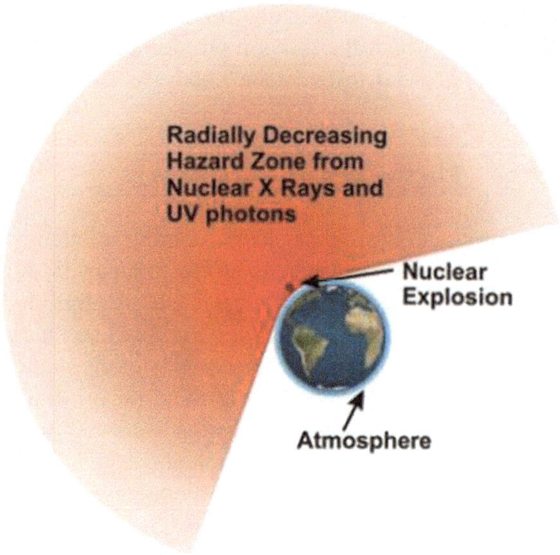

Fig. 6.7 Areas of space irradiated by photons and neutrons, adapted from Foster et al. (2008)

- Charged particles produced by explosions above the Earth's atmosphere can be captured by the Earth's magnetic field, temporarily creating artificial radiation belts that can damage spacecraft and injure astronauts/cosmonauts in orbit.

From 1958 to 1962, the U.S. and U.S.S.R. conducted over a dozen nuclear tests in the Earth's upper atmosphere or in space–the highest at an altitude of 540 km. Table 6.6 lists information on all tests above 20 km altitude.

6.2.3.5 Kinetic ASAT: Anti-satellite Weapons Attacks

These types of weapons were essential in helping the army for the last four decades and their feasibility has been successfully demonstrated by the USA in 1985 (Lieggi and Quam 2007), when they were used to destroy a defective research satellite. China has repeated the experiment in 2007. The situation is worrying for the army systems commanders, which must acknowledge the possibility that an opponent can take the strategic initiative through various technical means. One of these possibilities is the use of an anti-satellite weapon against a critical space system in a mission that lasts only a few minutes, which makes it impossible to be intercepted. An example of such a weapon can be considered the intercontinental ballistic missiles (Gheorghe and Vamanu 2007).

The first country to test a satellite destruction system was the USA, which, in 1985, destroyed a defective solar flares research satellite. On 11 January 2007, China destroyed a defective weather patterns monitoring satellite (Fengyun 1C), at 865 km altitude. The incident raised questions about the Chinese military develop-

Table 6.6 High-altitude nuclear explosions as suggested by Robert Johnston (http://www.johnstonsarchive.net/nuclear/hane.html)

Nation	Test name	Date	Alt. (km)	Location	Yield (kt)	Warhead
USA	HARDTACK I–Yucca	28 Apr 1958	26.2	Pacific Ocean	1.7	W-25, plutonium implosion?
USA	HARDTACK I–Teak	01 Aug 1958	76.8	Johnston Island	3800	W39, thermonuclear, 50% fission
USA	HARDTACK I–Orange	12 Aug 1958	43	Johnston Island	3800	W39, thermonuclear, 50% fission
USA	Argus I	27 Aug 1958	200	South Atlantic Ocean	~1.5	W-25, plutonium implosion?
USA	Argus II	30 Aug 1958	240	South Atlantic Ocean	~1.5	W-25, plutonium implosion?
USA	Argus III	06 Sep 1958	540	South Atlantic Ocean	~1.5	W-25, plutonium implosion?
USSR	#88 Groza ("Joe 79")	06 Sep 1961	22.7	Kapustin Yar	10.5	R-12?
USSR	#115 Grom ("Joe 98")	06 Oct 1961	41.3	Kapustin Yar	40	R-5
USSR	K PROJECT–#127 K-2 ("Joe 109?")	27 Oct 1961	150	Kapustin Yar	1.2	R-12?
USSR	K PROJECT–#128 K-1 ("Joe 105?")	27 Oct 1961	300	Kapustin Yar	1.2	R-12?
USA	DOMINIC I/FISHBOWL–Starfish Prime	09 Jul 1962	399	Johnston Island	1400	W-49, thermonuclear
USA	DOMINIC I/FISHBOWL–Checkmate	20 Oct 1962	147	Johnston Island	<20	XW-50X1
USSR	K PROJECT–#184 K-3 ("Joe 157")	22 Oct 1962	290	Kapustin Yar	300	R-12?
USA	DOMINIC I/FISHBOWL–Bluegill Triple Prime	26 Oct 1962	48.2	Johnston Island	400	W50, boosted fission
USSR	K PROJECT–#187 K-4 ("Joe 160")	28 Oct 1962	150	Kapustin Yar	300	R-5 M, thermonuclear
USSR	K PROJECT–#195 K-5 ("Joe 168")	01 Nov 1962	59	Kapustin Yar	300	R-12?
USA	DOMINIC I/FISHBOWL–Kingfish	01 Nov 1962	96.3	Johnston Island	400	W50, boosted fission
USA	DOMINIC I/FISHBOWL–Tightrope	04 Nov 1962	21	Johnston Island	<20	W31

ment, although the republic has denied the intent to engage in a space arms race. What is important about these two incidents, especially the last one, is that the destruction method used was a kinetic strike (a frontal collision at high speed between a target and the intercepting object). The kinetic strike produces many distinct items—for the 750-kg Chinese satellite, NORAD has detected 2,316 new objects that they could monitor (they had the size of a golf ball or they were larger), with the likelihood of 100,000 smaller objects, up to a centimeter long, but equally dangerous. This single incident has increased the number of monitored orbit debris by 12%. Ironically, China was one of the promoters of space weapons ban, which President Bush was presenting at the time as a legitimate option for American defense.

An important component of a state's ASAT capabilities especially that of the United States, is the dual-use technology for the missile systems. The intercontinental ballistic missiles travel at speeds and altitudes comparable to those of some satellites, so anti-ballistic systems could also have ASAT functions—with minor modifications. Furthermore, weight constraints of space systems limit their ability to take countermeasures, unlike ballistic missiles, increasing the ASAT efficiency of the dual-use system. In addition, satellites follow predictable paths and programs, giving the attacker a chance to plan the attack in advance. Thus, the target has fewer defensive options than a rocket in flight.

USA developed a system of interceptors called 'Ground-based Midcourse Defense' (GMD) that was placed in Fort Greely in Alaska and in Vandenberg Air Force Base in California. GMD consists of three stage rockets that can carry into space a maneuvering-capable interceptor module with infrared sensors. Target destruction is made by direct impact. If it were launched against low orbits satellites, the presence of fuel on board would allow him to travel thousands of miles in this orbit, overcoming the limitation of ground ASAT systems (which are useful only against targets in the space above them). A GMD could target a significant proportion of the overall number of low orbit satellites.[28]

In February 2008, USA demonstrated the ASAT capability of the AEGIS naval defense system, by destroying a defective US reconnaissance satellite, located at 240 km altitude. The operation was called 'Burnt Frost' and the USA-193 satellite would have returned in the atmosphere anyway. USA denied any connection to China's ASAT experiment from 2007, claiming the presence of toxic materials on board as a reason to prevent the reentry in the atmosphere. Satellite destruction also prevented its recovery for technology theft by another country. Another ASAT-capable US system is the 'Terminal High Altitude Area Defense' interceptor.[29] Figure 6.8 depicts a launch of an Aegis SM-3 interceptor, similar to the one that destroyed the USA-193 satellite in 2009.

---

[28] See for example: https://www.ucsusa.org/nuclear-weapons/space-security/a-history-of-anti-satellite-programs#.WyGA1yAnbIU.

[29] See https://www.ucsusa.org/sites/default/files/legacy/assets/documents/nwgs/a-history-of-ASAT-programs_lo-res.pdf.

Fig. 6.8 Image of Aegis SM-3 interceptor (See 'Improving Tactical Training in the Surface Community,' https://idrivewarships.wordpress.com/2013/07/03/improving-tactical-training-in-the-surface-community/)

AEGIS systems are used not only by USA, but also by its allies. For example, Japan has a co-research program of the system, together with the United States, and among potential customers we find South Korea and various European countries.

Amongst the countries with significant kinetic ASAT technology potential, we find:

- China—has developed anti-satellite technologies since 1980s, including for intercontinental ballistic missile functions. The historical test of 2007 was the first in 20 years and was the largest producer of space debris ever recorded. Shortly after, the United States resumed the ASAT weapons ground testing. The first test against ballistic missiles was successfully held in January 2010, but the target was sub-orbital and has no longer produced any space debris. China is also developing an ASAT missile to be launched from a submarine and that can target geostationary orbit satellites
- India—in 2010, the general director of the Indian Defence Research and Development Organization said that a project to create an anti-satellite weapon located in an outer atmospheric vehicle and a laser sensor is undergoing, and that the project was initially developed to serve against ballistic missiles
- Israel—is developing the Hetz 3 ('arrow') missile together with the United States. The project is a cooperation between the Israeli Aerospace Industries and Boeing, and the missile is intended for outer space ballistic missiles interception, with a success rate of 99%. For this reason, Hetz 3 could also be used as an ASAT weapon

- Russia—in 2009, Russia resumed a project started in the 1980s—the use of a MiG-31 fighter plane as a platform for launching satellite interceptor missiles. In parallel, a project involving an aerial platform for an ASAT laser was resumed.

These collateral damages can have demoralizing purposes, like the attacks against states exhibiting proven confidence in critical space systems, or they can be the secondary objective for some troublesome states that wish to cause the greatest possible damage. Such an attack is available to players disposing of a moderate budget, basic flight control systems, intermediate locations and the will to endure retaliation. Finally, it is not necessary to attack a particular space system in space. One insidious way to generate a cascading destruction of the space systems is the mere launch of a rocket whose load is to be comprised of numerous bearing balls, which afterwards would be left on predefined target trajectory (Primack 2002).

Considering Kessler Syndrome, it is certainly possible that a situation in which the concentration of objects in orbit will be so great that a collision would generate enough new objects for producing, immediately other collisions, and causing new space debris formation. The result is a chain reaction that would eventually transform the space debris into the maximum number of items, each with minimum weight. The final effect is a minefield effect that could make low-Earth orbit, the center of human activity in space, uninhabitable for generations, while enhancing, at the same time, the danger of transit for any vehicle that would like to leave the vicinity of the Earth.

The devastating effects on infrastructure makes threatening with Kessler Syndrome fall under the MAD label '*mutually assured destruction*,' and could be a particularly effective deterrent factor, the second after the nuclear threat, complying with the definition offered by the Rand Corporation (Long 2008). A space attack would be a pyrrhic victory, but Rand Corporation warns that this deterrent is among the strongest advantages in geostrategic games. In such games, the big players are always the more responsible, because they also have the most to lose. The "contests" regarding the number of nuclear weapons owned are a matter of prestige, because the least expensive deterrent is a single weapon of mass destruction and the lack of doubts regarding the intention of using it in case of an attack, or even for preemptive purposes. The same principle can be applied to the space games.

We can even imagine a scenario where a country such as North Korea (or even a terrorist organization with access to private technologies) instead of trying to develop intercontinental ballistic missiles with nuclear capabilities, would decide to take people hostage in a much less expensive way—a simple and low-tech missile whose load to be made up of thousands of tiny metal balls that could be scattered in orbit (Primack 2002). If the missile were to be aimed at one of the densely populated areas mentioned above, the result could be catastrophic, but a 'blind' launch is also likely to cause major damages.

Fig. 6.9 Space debris created after the testing and usage of ASAT weapons

6.2.3.6 Satellite Maneuverability

Research is ongoing for anti-satellite technologies with reduced complexity, which do not use explosives or create new space debris (see Fig. 6.9). Such a product is represented by the maneuvering satellites, which are able to approach and even touch a satellite without the owner's permission. Both civilian actors, interested in commercial applications like in situ repair, and military players were concerned about the development of technologies for this purpose, such as capabilities for intersection and proximity operations or autonomous mission planning. NASA developed a program called DART: *Demonstration for Autonomous Rendezvous Technology* in 2005, that involved launching a satellite that could approach a civilian satellite without assistance from the ground. DART collided with that satellite, but an American Air Force program, XSS 11: *Experimental Satellite System 11* had more success. Russia, Japan, China and the European Space Agency are in various stages of developing such capabilities.[30]

Such a satellite would have several advantages as an anti-satellite weapon. Its gun would not need much power or sophistication; therefore, it would not be too heavy, and it could produce zero space debris. Such a satellite would also be difficult to detect, especially if it were carried in space by a legitimate launch mission as a

[30]https://www.ucsusa.org/nuclear-weapons/space-security/a-history-of-anti-satellite-programs#.WyHtoadKjIU.

secondary payload. Not needing high relative speeds to destroy targets, the satellite could maneuver in very close proximity to the target, enhancing its chances for a successful mission. Also, placing it into orbit could be done anytime, giving the tacticians the possibility to act before a rival could suppress the capacity to launch or to avoid the risk of launching in a crisis situation and under adverse conditions. The disadvantages are that satellites, once identified as such, would be vulnerable to attacks by various means, and the degradation of the satellite would diminish the confidence in a successful mission, when it would be decided that it is time to activate the satellite's offensive capabilities.

The BX-1 microsatellite lot, that was tested and released by China, can be used for inspecting rival satellites, or even to launch various kinds of co-orbital attacks. An interesting incident occurred during the first test, when the BX-1 satellite passed nearby the International Space Station, simulating a real operation (Easton 2009).

6.2.4 Anti-ASAT Administration Efforts

The way to prevent an arms race in outer space is a long-time topic on the agenda of the Conference on Disarmament, the main international forum for negotiating arms control treaties. A group for space weapons was established in 1985 and the group continued to meet until 1994, but with very little success. The UN General Assembly has also tried to promote this goal since 1983, when the first attempt to adopt a resolution in favor of preventing an arms race in outer space was made. The United States has voted at least eight times against this project, especially after they amended the National Space Policy, in 2006, in order to prohibit the formation of any statutory approval or any other mechanisms that would restrict the use of outer space by the United States—also for military purposes. In theory, the United States policy is counterproductive, because it prevents the building of a constraining framework for the space technology militarization rate in those countries that have emerged as US rivals, especially China.

Critics of the USA highlight the proliferation of the anti-satellite weapons (ASAT) and therefore, predict that the USA will have to accept higher costs just to maintain the current security level of its space assets, beyond other initiatives in the respective field. ASAT systems include space weapons designed to incapacitate or destroy satellites for strategic military purposes. Arguably, one must observe the potential dual-use of space technologies. This issue might as well be highlighted as indicated (Gheorghe et al. 2018):

> *At 5:28 p.m. EST January 11, 2007, the People's Republic of China successfully destroyed a defunct Chinese weather satellite, FY-1C. The destruction was carried out by a modified medium-range ballistic missile with kinetic ASAT warhead. FY-1C was a weather satellite orbiting Earth in polar orbit at an*

altitude of about 537 miles (865 km), with a mass of about 750 kg. Launched in 1999, it was the fourth satellite in the Feng Yun series.

This event triggered interest in domain (see e.g., Gheorghe et al. 2018; Gheorghe and Vamanu 2007) model for hitting targets using readily available tools. In creating a simulation, Gheorghe and Vamanu (2007) use Google Earth is used to supply maps and geographical information. Prior to I/O interface for ASAT simulator, one needs create an ad hoc interface to Stoff's 'Orbitron' which enables one to select a target to be hit—or again, get close to the target as possible. Figure 6.10 shown a customized purposeful interface and an input-output interface for an ASAT simulator that used in for hitting a specific.

Since all systems are governed by certain laws, familiarity with laws and equations for unmanned flight conics is handy. For example:

$$r = p/(1 + e\cos(\theta - \theta_0)) \tag{6.1}$$

may conveniently be rephrased as

$$r = p/(1 + (p/r_0 - 1)\cos\theta - p\sin\theta/(r_0 \mathrm{tg}\,\alpha)) \tag{6.2}$$

where

$$p = r_0^2 v_0^2 \mathrm{tg}^2\alpha/(KM_{\mathrm{E}}(1 + \mathrm{tg}^2\alpha)) \tag{6.3}$$

A careful examination of the above suggests that one can have:

r_0 (m) is the initial distance of the object from the reference focus of the conic

r (m) is the current distance of the object from the reference focus of the conic

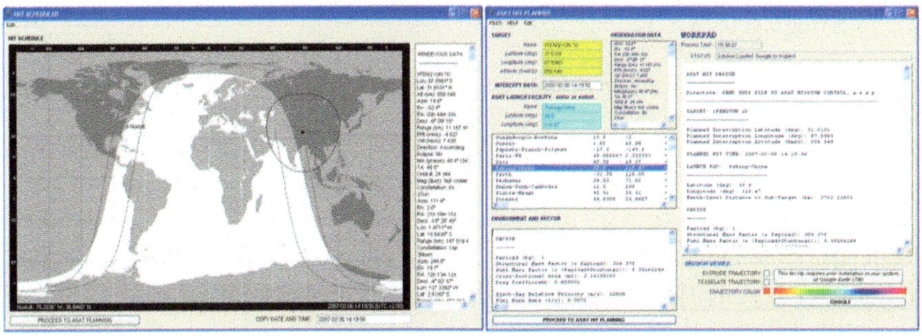

Fig. 6.10 A simple interface developed for a simulation in which a satellite is a target

θ_0 (m)	is the initial polar angle of the object from the reference focus of the conic
θ (m)	is the current polar angle of the object from the reference focus of the conic
v_0 (m/s)	is the initial velocity imposed upon the object, at the origin of the ballistic flight
α (rad. from the local vertical)	is the initial tilt angle, at the origin of the ballistic flight
$M_E(\text{kg}) = 5.9798e^{24}$	is the earth's mass
$K(\text{N m}^2/\text{kg}^2) = 6.673e^{-11}$	is the universal gravitational (Newton) constant.

And when applied in Eqs. (6.1) and (6.2), the trajectory of an ASAT weapon can be described using r_0 and θ_0 (polar coordinates of the launch site) and (r, θ) can be considered as the current coordinates of the rocket. If one assumes that the rendezvous point, which is shared by target and rocket, the polar coordinates r_1, θ_1 can be obtained by processing *Orbitron*'s output. And since these coordinates obey trajectory Eq. (6.3), the following algebraic condition for the tilt angle and launch velocity must be fulfilled:

$$\begin{aligned} &\text{tg}^2\alpha(KM_E r_1(1 - \cos\theta_1) + r_0 r_1 v_0^2 \cos(\theta_1 - r_0^2 v_0^2) - \text{tg}\alpha\, r_0 r_1 v_0^2 \sin\theta_1 \\ &\quad + KM_E r_1(1 - \cos\theta_1)) = 0 \end{aligned} \tag{6.4}$$

In theory, Eq. (6.4) provides a 'solution' for altitude, velocity, and tilt angle during the controlled flight phase. It this becomes possible to compute for compute the difference between the actual speed and the ideal speed required to hit target, boost the rocket to get to the ideal speed as fast as feasible, and cut off the engines. The total vehicle mass m (kg) is (textbook-routinely) assumed to vary with time T (s) as:

$$\text{m} = \text{m}_0(1-\delta T) \tag{6.5}$$

with δ (1/s)—a constant burnout rate. The gravity constant g (m/s^2) depends on altitude via the polar radius r, as:

$$g = KM_E/r^2 \tag{6.6}$$

where the polar radius can obviously be written as:

$$r = R_E + z \tag{6.7}$$

with the Earth radius taken as $R_E = 6.37e6$ m, and z (m)—the geometric altitude. Essentially depending on altitude is also the air density, ρ_a (kg/m^3), described in this exercise as suggested by Richard Shelquist in his scholarly-informative web pages (Shelquist 2006):

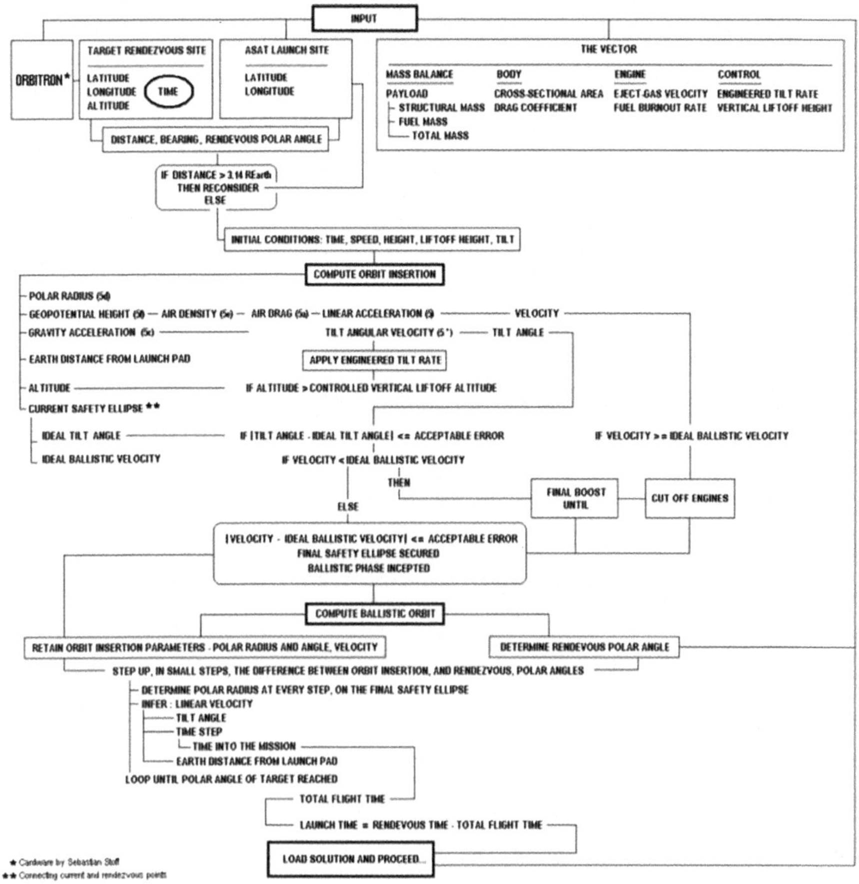

Fig. 6.11 A 'simple' programming approach for ASAT (Gheorghe and Vamanu 2007)

$$\rho_a = \left(\frac{a_1 - H}{a_2}\right)^{\frac{1}{a_3}} \tag{6.8}$$

here H (km) is the geopotential altitude in kilometers, relating to the common, geometric altitude, z, as:

$$H = zR_E/(z + R_E) \tag{6.9}$$

With this time, Earth's radius (R_E) expressed in kilometers, the coefficients in Eq. (6.7) are $a_1 = 44.3308$, $a_2 = 42.2665$, and $a_3 = 0.234969$. Figure 6.11 is a representation describing an inferred algorithm from the above description.

Simple applets can be developed for live support of an ASAT simulator and numerical experiment have since shown that one can reach the sought rendezvous points with relative accuracy. Table 6.7 is a sequel excerpts of a few relevant lines from an ASAT mission documentation, as provided by the simulator. While such a 'game' involves trading off between accuracy and computational time, it is possible to optimize onboard fuel mass so as to minimize several factors including dead weight, safety, and loss of control. As we gain knowledge, perhaps 'serious gaming' to the issue might just the new normal for space warfare involve space systems.

Clearly, any party researching new technological systems for civilian purposes (e.g., maneuvering satellites and space debris de-orbiting systems) will also acquire (inevitably) military capabilities, even if it does not declare as such (de Zwart et al. 2018). Their conversion would take very little time, in the event that such a capability would be required—in a crisis situation. There is also the possibility of avoiding the stipulations of these legal frameworks, by using unrestricted ground-level military technologies. It is noted that the AEGIS missile systems could damage satellites, same as the intercontinental ballistic missiles. Furthermore, the legal provisions regulating the placement of the offensive systems in space are easily bypassed by launching rockets from the ground or from aircrafts, by using ground lasers, or even by detonating nuclear explosions in the atmosphere in order to produce an electromagnetic pulse.

In 2006, the European Union created a draft for a Code of Conduct, in order to propose behavioral rules for the responsible space-interested parties. The purpose of this document is to improve the safety of space operations, to protect the space environment—especially regarding the formation of new space debris—and to promote the idea that space systems should not be considered legitimate targets for aggression and conflict. Signatories of such document might need to discuss and even agree on key issues, such as, (1) not placing weapons in orbit, (2) not threaten or use force against space systems, and (3) not encourage or assist others in such activities.

Major omissions of this program are: ground development of the ASAT weapons is not prohibited, launching "dual-use" satellites—which could be used as weapons—is also not prohibited and the lack of a proposal for some means of verifying the compliance with these provisions.

In 2008, Russia and China submitted a proposal that was discussed in the Conference on Disarmament, for limiting the legitimate uses of ASAT weapons but which, in line with their stated interests, is not blocking the development of ASAT technologies.

The Americans were critical about the European initiative and avoided to make counter proposals, signaling the end of this project through non-involvement. However, the National Space Policy of 2010 emphasized a greater disposition towards cooperation, transparency and arms controls, as long as it is demonstrable that the security of the United States would be enhanced by such a project. The United States has not taken a leadership position for these discussions and the US maintains that it is just an observer of the negotiations that would be initiated in the Conference on Disarmament.

However, several groups have attempted to codify and disseminate good practices in the space domain through voluntary agreements. These include the technical subcommittee recommendations of the United Nations Committee for Peaceful Uses of Outer Space or the Code of Conduct for responsible space actors of the Stimson Cen-

Table 6.7 An excerpt from an ASAT mission documentation (Gheorghe and Vamanu 2007)

ASAT hit design
Directive: send this file to ASAT mission control, a.s.a.p.
Mission summary
Target: **'FENGYUN 1D'**
Planned interception latitude (deg): 31.6101 Planned interception longitude (deg): 87.5983 Planned interception altitude (kmAG): 858.548 *** Planned hit time: 2007-02-06 14:19:56** *** Required launch time: 2007-02-06 14:04:48** *** Flight time to impact: 0 h. 15 m. 7 s.**
Launch pad: Peking-China
Latitude (deg): 39.9 Longitude (deg): 116.47 Earth-level distance to sub-target (km): 2702.44429
Insertion data
Distance from launch pad (km): 58.7903947 Insertion altitude (km): 86.7781608 Insertion tilt angle (deg. from vertical): 43.91255 Insertion velocity (m/s): 5268
Orbital data
Maximum altitude (mAG): 1191498.7 Minimum velocity (m/s): 3120.4757
Impact data
Impact velocity (m/s): 3831.81685 Impact tilt angle (deg. from vertical): 58.4910387 Flight time to impact: 0 h. 15 m. 7 s. Visibility circle radius (km): 3128.21451 Distance error (%): 0.65460801 Altitude error (%): −0.7262643
Vector
Payload (kg): 1 Structural mass factor (x Payload): 304.375 Fuel mass factor [x (Payload + Structural)]: 5.58166189 Cross-sectional area (m^2): 3.14159265 Drag coefficient: 0.000001 Eject-gas relative velocity (m/s): 10,000 Fuel burn rate (1/s): 0.0075 Controlled tilt (deg/s): 2 Vertical liftoff height (m): 100
Environment: ISA—the 1976 International Standard Atmosphere
Sea level standard pressure (Pa): 101325 Sea level standard temperature (K): 288.15 Temperature lapse rate (K/km): 6.5 Gas constant (J/(mol K)): 8.31432 Molecular weight of dry air (g/mol): 28.9644 Air density law coefficient a1: 44.3308 Air density law coefficient a2: 42.2665 Air density law coefficient a3: 0.234969

(continued)

Table 6.7 (continued)

Constants
Number pi: 3.14159265 Earth radius (m): 6,378,388 Earth mass (kg): 5.9798e24 Gravity constant (N m^2/kg^2): 0.6673e−10 Insertion phase time step (s): 0.5
Flight time distance altitude tilt target speed target vector mass *(s) (m from pad) (mAG) (deg) (deg) (m/s) (m/s) (kg)*
Controlled flight to insertion
0.5 0.99792983e−15 16.2979692 0 43.197666 32.5959384 5408 2002.33797 1 0.29981117e−14 48.964497 0 43.3171403 65.3330556 5408 1994.80094 1.5 0.60049012e−14 98.0707182 0 43.4064731 98.2124425 5408 1987.2639 2 0.10022688e−13 163.688319 0 43.4908328 131.235202 5408 1979.72688 . . . . 50 46770.776 83666.8951 43.3802 44.188849 4245.05207 5274 1256.17188 50.5 48260.4179 85218.302 43.8365 44.069641 4301.57943 5271 1248.63484
Ballistic flight to target
52.7332 58790.3947 86778.1608 43.91255 NA 5268 NA 1241.09781 54.4747 67798.5423 96252.7052 44.01205 NA 5250.8275 NA 1241.09781 56.2244 76806.69 105708.041 44.11247 NA 5233.68393 NA 1241.09781 . . . . 904.965 2761234.69 858548.0 58.7151134 NA 3819.38797 NA 1241.09781 907.631 2770242.84 852312.672 58.4910387 NA 3831.81685 NA 1241.09781 NA—not applicable

ter. Even if such voluntary agreements are easier to negotiate, the formally ratified ones last longer, and they include mechanisms for ensuring the compliance with the provisions, having greater success in the long term. They include the establishment of an institutional group which will have the authority to mediate and resolve disputes.

In 2010, the European Union returned to the idea of creating a Code of Conduct for Outer Space Activities, advancing the proposal that any deliberate attack should be considered legitimate only in case of self-defense, safety of its property and preventing the formation of new space debris. In 2012, the United States refused to sign the document, but the US has stated its willingness to work with the European Union and other nations in order to develop an International Code of Conduct for Outer Space Activities.[31]

[31] More information can be found here: https://swfound.org/media/166384/swf_draft_international_code_of_conduct_for_outer_space_activities_fact_sheet_february_2014.pdf.

Final Remarks

In this chapter, the concept of Critical space infrastructure Protection (CSIP) is proposed. CSIP is a concept that relates to the preparedness, readiness and response to serious incidents that involve the space infrastructure, key resources, and asserts.

On March 9, 1999, Deputy Defense Secretary John Hamre warned the United States Congress of a cyber terrorist 'electronic Pearl Harbor' saying, "It is not going to be against Navy ships sitting in a Navy shipyard. It is going to be against commercial infrastructure." Later this fear was qualified by President Clinton after reports of actual cyber terrorist attacks in 2000: "I think it was an alarm. I don't think it was Pearl Harbor. We lost our Pacific fleet at Pearl Harbor. I don't think the analogous loss was that great."[32] These comments highlighted a need 'national' framework for protection of infrastructure systems. And these there, many approaches have been developed at location, regional, national, and international levels. However, a new playing ground has emerged: space. And the question we pose might sound obvious: what is the CSIP for any nation? Such frameworks might as well address the obviously risks and threats including, space weather, space debris, ASAT, and electromagnetic pulse attacks. Comprehensive solutions might revolve around elements of space infrastructure assurance: analysis and assessment, remediation, indicators and warning systems, mitigation, and incident response, and reconstruction. Along with such approaches, policy and diplomacy cannot be understated along with axiological argumentation.

References

Adilov, N., Alexander, P. J., & Cunningham, B. M. (2015). An economic analysis of earth orbit pollution. *Environmental & Resource Economics, 60*(1), 81–98. https://doi.org/10.1007/s10640-013-9758-4.

Cannon, P. (2013). *Extreme space weather: Impacts on engineered systems and infrastructure*. London: Royal Academy of Engineering. Retrieved from https://www.raeng.org.uk/publications/reports/space-weather-full-report.

de Zwart, M., Hays, P., Jakhu, R., Jaramillo, C., Meyer, P., Stephens, D., et al. (2018) *Space security index 2018: Executive summary*. ISBN: 978-1-927802-21-2. Retrieved from http://spacesecurityindex.org/wp-content/uploads/2018/06/SSIExecutiveSummary2018.pdf.

Easton, I. (2009). *The great game in space: China's evolving ASAT weapons programs and their implications for future U.S. strategy*. Arlington, Virginia: Project 2049 Institute.

ESA (European Space Agency). 2017. *Space debris by the numbers*. Retrieved from http://www.esa.int/Our_Activities/Operations/Space_Debris/Space_debris_-_by_the_numbers.

Foster, J. S., Gjelde, E., Graham, W. R., Hermann, R. J., Kluepfel, H. M., Lawson, R. L., et al. (2008). *Report of the commission to assess the threat to the United States from electromagnetic pulse (EMP) attack* (No. A2473—EMP). Washington D.C.: EMP Commission. Retrieved from http://www.empcommission.org/docs/A2473-EMP_Commission-7MB.pdf.

[32]Remarks by the president in photo opportunity with leaders of high-tech industry and experts on computer security: https://fas.org/irp/news/2000/02/000215-secure-wh1.htm.

Gheorghe, A. V., & Vamanu, D. V. (2007). Risk and vulnerability games. The anti-satellite weaponry (ASAT). *International Journal of Critical Infrastructures, 3*(3/4), 457–470. https://doi.org/10.1504/ijcis.2007.014120.

Gheorghe, A. V., & Yuchnovicz, D. (2015). The space infrastructure vulnerability cadastre: Orbital debris critical loads. *International Journal of Disaster Risk Science, 6*(4), 359–371. https://doi.org/10.1007/s13753-015-0073-2.

Gheorghe, A. V., Vamanu, D. V., Katina, P. F., & Pulfer, R. (2018). Critical infrastructures, key resources, and key assets. Cham, Switzerland: Springer International Publishing. https://doi.org/10.1007/978-3-319-69224-1_1.

Hapgood, M. (2010). *Space weather: Its impact on Earth and implications for business*. London: Lloyd's 360° Risk Insight.

Joyner, C. C. (2005). *International law in the 21st century: Rules for global governance*. Lanham, Md: Rowman & Littlefield Publishers.

Lieggi, S., & Quam, E. (2007). China's ASAT test and the strategic implications of Beijing's military space policy. *Korean Journal of Defense Analysis, 19*(1), 5–27. https://doi.org/10.1080/10163270709464125.

Long, A. (2008). *Deterrence—From Cold War to Long War: Lessons from six decades of RAND deterrence research*. Santa Monica, CA: RAND Corporation.

Ministry of Defence. (2012). *Space: Dependencies, vulnerabilities and threats*. Wiltshire: The Development, Concepts and Doctrine Centre. Retrieved from https://assets.publishing.service.gov.uk/government/uploads/system/uploads/attachment_data/file/33689/20120313mne7_space_vulnerabilites.pdf.

Pardini, C., Anselmo, L., (2007) *Evolution of the debris cloud generated by the FengYun-1C fragmentation event*. Pisa, Italy: Space Flight Dynamics Laboratory, Istituto di Scienza e Tecnologie dell'Informazione Alessandro Faedo. Retrieved from http://issfd.org/ISSFD_2007/10-4.pdf.

Primack, J. (2002). *Debris and future space activities*. Santa Cruz: University of California. Retrieved from http://physics.ucsc.edu/cosmo/Mountbat.PDF.

Rinaldi, S. M., Peerenboom, J., & Kelly, T. K. (2001). Identifying, understanding, and analyzing critical infrastructure interdependencies. *IEEE Control Systems, 21*(6), 11–25. https://doi.org/10.1109/37.969131.

Salter, A. W. (2015). *Space debris—A law and economics analysis of the orbital commons*. Mercatus Center, George Mason University, SUA. Retrieved from https://www.mercatus.org/system/files/Salter-Space-Debris.pdf.

Shelquist, R. (2006). Air density and density altitude calculations. Retrieved January 1, 2017, from https://wahiduddin.net/calc/density_altitude.htm

Taleb, N. N. (2014). *Antifragile: Things that gain from disorder* (Reprint ed.). New York: Random House Trade Paperbacks.

Taylor, M. W. (2006). *Orbital debris: Technical and legal issues and solutions* (Masters). Montreal: McGill University. Retrieved from http://digitool.library.mcgill.ca/R/?func=dbin-jump-full&object_id=99558&local_base=GEN01-MCG02.

Whitlock, D. O., Stansbery, E., & Johnson, N. L. (2008). *History of on-orbit satellite fragmentations* (No. NASA/TM-2008–214779). Washington D.C., United States: National Aeronautics and Space Administration. Retrieved from https://www.hsdl.org/?abstract&did=722495.

Chapter 7
Crisis and Emergency Situation Management

Abstract The purpose of this chapter is to provide a forward-looking summary to critical space, undersea, and underground systems. Proposed areas of research at the methodological, epistemology, ontological, and nature of man are then presented.

7.1 Space Crisis Management

Modern societies are highly dependent on the continuous operation of critical infrastructures (CI) that ensure the supply of critical goods and services. According to Moteff et al. (2003), they are 'infrastructures so vital that their incapacitation or destruction would have a debilitating impact on defense or economic security.' These include among others power generation and transmission systems, drinking water supply, information and communication grids, and transport and industrial systems (Katina and Keating 2015). Disruptions resulting from natural disasters, system attrition or premeditated acts may have rapid repercussions for the population's security and livelihood, on business continuity, as well as on investor confidence, and can affect other critical infrastructures through domino effects as described by, among other, Hokstad et al. (2012). Examples abound in recent history of widespread infrastructure disruption in geographic areas—Zimmerman and Restrepo (2009) analyzed a large-scale power blackout which also disrupted systems dependent on electricity, like water supply, telecommunications, and rail transport. Cities, as engines for growth and concentrated accumulations of critical infrastructures, are a natural experiment in critical disruption propagation, with effects during the interruption of power supply ranging from impacts on food supply and healthcare systems, to the functioning of public services and the maintenance of public order.[1]

Space systems, such as remote sensing satellites, global navigation satellites systems and communication satellites, have become key enablers for a wide variety of commercial, scientific and military applications. The rapid growth of their capabili-

[1]Public Safety and Emergency Preparedness Canada—Incident Analysis, 'Ontario–U.S. Power Outage—Impacts on Critical Infrastructure: http://cip.management.dal.ca/publications/Ontario%20%20US%20Power%20Outage%20-%20Impacts%20on%20Critical%20Infrastructure.pdf.

© Springer Nature Switzerland AG 2019

A. Georgescu et al., *Critical Space Infrastructures*,
Topics in Safety, Risk, Reliability and Quality 36,
https://doi.org/10.1007/978-3-030-12604-9_7

ties has offset the high cost barrier of the required investment and new developments promise an even greater reduction in the cost of space infrastructure.[2] As such, some of the extant space systems have become deeply embedded in the functioning of advanced societies, supporting economies, lifestyles and governance processes. Mureșan and Georgescu (2015) have argued that the increasing dependence on certain space systems places them firmly in the area of critical infrastructure or as components in the wider critical infrastructure system-of-systems. One of the most important applications of space system capabilities is in the field of crisis and emergency situation management. A review of extant literature highlights the potential involvement of a wide array of space capabilities in such processes, from command, coordination and control to Earth Observation, positioning and navigation. The sheer depth of the potential toolkit at the disposal of the security decision maker is belied by the fragmentary nature of the actual use of these capabilities, which rely on actor sophistication, access to services and a host of other factors. In this chapter, a generic model of crisis and emergency situation management that more neatly divides the issue into various processes for which the authors argue the potential or realized benefits of the use of space systems is provided. The authors conclude by analyzing some key aspects of the dynamic of space sector development as it pertains to crisis and emergency situation management, as well as the complexities inherent in the growth of reliance on them, including the creation of new risks, vulnerabilities and threats.

7.2 A Space Based Paradigm

The proposed space-based model divided, chronologically, into before, during and after phase (Fig. 7.1), for which a number of processes have been identified with the potential for space system contribution. The chronological order should not suggest a linearity of crisis and emergency situation phases. Rather, the model is more adequately thought of as a circular collection of processes. The final process in the post-crisis phase will transition, in some ways seamlessly, into the first processes of the pre-crisis phase.

The paradigm must also be completed conceptually through the re-acknowledgement of a significant aspect of Critical Infrastructure Protection theory. The concept, as stated by Rinaldi et al. (2001), is that interdependencies between critical infrastructure systems lead to the risk of cascading disruption of their functioning. This interdependency is not linear, because of the complexity of the system-of-systems. Every infrastructure is influenced, to varying degrees by many other infrastructures (with critical dependencies being either permanent or circumstantial). This suggests a network of infrastructure nodes, in which the cascading disruption, failure, or decline feeds on itself through this web of relations, amplifying

[2] https://www.esoa.net/Resources/Why-Satellites-Matter-Full-Report.pdf.

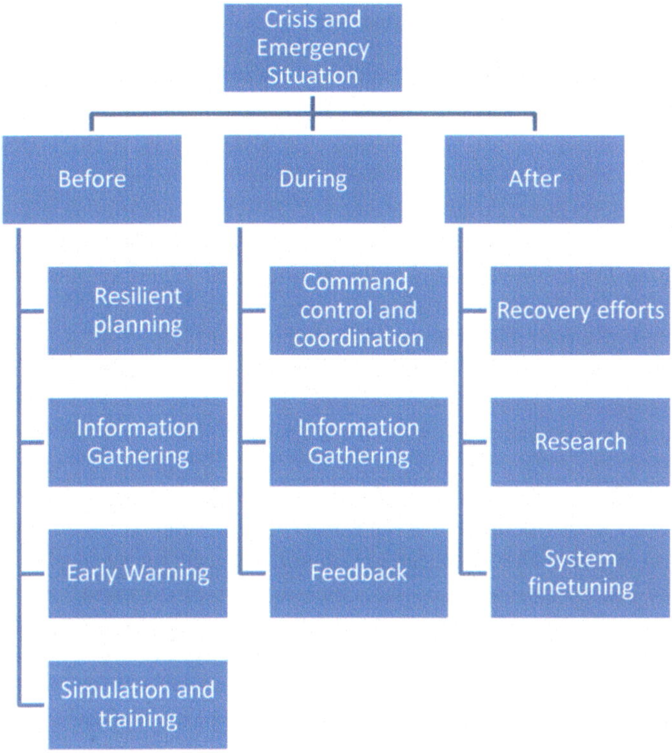

Fig. 7.1 A proposed space-based model for emergence management

dysfunction, leading to greater human and material losses (Calida and Katina 2012; Gheorghe et al. 2018; Katina and Hester 2013).

7.2.1 *The Before Phase*

For a generic threat or threat spectrum, the responsible stakeholders will not just address its materialization, but seek to create the conditions for the minimization of their chance to appear and the prior establishment of damage mitigation mechanisms. This is a key facet of resilience governance, and space systems can play an important role in the processes described below, not just in a supporting role, but also by opening up new avenues for development.

7.2.1.1 Resilient Planning Component

Decision makers can plan not just for the materialization of a negative event and place countermeasures, but also for the intrinsic resilience and robustness of the critical infrastructure whose security must be assured, which begins in the planning stages of the building of new infrastructure. Space systems play an important role in both of these instances. As pointed out by Bouchon (2006), the generational accumulation of new critical infrastructures leads to bewildering complexities in interactions between multiple long-lived systems. This build-up of infrastructure 'sedimentation' at the bedrock of societal development and functioning means that every new infrastructure must be carefully planned with regard to its impact on the system-of-systems, and to possess certain qualities, such as resilience, robustness, adaptability, capability of graceful decline and so on (Katina and Hester 2013). This is true not just for technical assets, like power plants, pipelines and electricity grids, but also for organizational infrastructures and other socio-technical systems, like public administrations and military organizations, which are defined as critical infrastructures in most formal taxonomies of infrastructures, like the European and American ones.

Resilient planning should be thought of umbrella term for a complex set of tasks, such as the gathering of assorted data, the selection of sites and routes for placement of new infrastructure, risk assessments under an all-hazards approach and so on. An example which illustrates the potential usefulness of space systems and the inability for stakeholders to neglect their use because of costs, since shirking such responsibilities could have dire repercussions, is the placement of new nuclear power plants or, indeed, of nuclear waste repository sites. These infrastructures function on some of the longest imaginable timeframes, from 50 years for reactors (80 or more with new developments), to hundreds and thousands of years for nuclear waste sites. Space capabilities are a valuable addition to ground-based research into the geology of an area, its weather patterns, its hydrology, the vegetation, as well as the scope of human activity, habitation and impact on the environment. Moreover, regular data gathering by space systems provides a less onerous means of creating datasets for analysis of long-term patterns, crucial to any long-lived critical infrastructures. This provides a much needed fourth dimension to resilient planning.

7.2.1.2 Information Gathering Component

Space systems have added new dimensions to the continuous monitoring of critical infrastructure functioning and the adjacent environments, bringing a wealth of new data into the analysis and decision process. With the help of Earth Observation capabilities, experts can analyze the surrounding environment, whether natural or manmade, as well as the interaction between the critical infrastructure and its environment, such as the flow of potential pollutants. Space systems do not just offer new datasets or new business models for conducting security analysis, but can also fundamentally shift the paradigm, through the possibility of constant monitoring leading to historical datasets from which new information can be gleaned on risks and vulnerabilities. Ideally, such inquiries should be paired with regular monitoring on the ground, both cursory and in-depth, but the use of space systems adds much needed flexibility to the security manager's disposition of assets and resources. An example of constant monitoring, especially as a replacement for more onerous and less useful ground surveillance, is the provision of security for a nuclear waste repository, which may be under threat not just from natural elements, but also from terrorist and criminal elements seeking to leverage the understandable isolation of these sites for easing the theft of materials (Haeme 2003).

7.2.1.3 Early Warning

Space systems are becoming valuable tools in early warning against the materialization of a risk or a threat. The potential regularity of space surveillance enables rapid observation of developing crises and can provide decision makers with valuable additional time to marshal resources and implement mitigation measures. For the general public, this is most often seen in the build-up to extreme natural phenomena, like hurricanes, whose close monitoring enables timely evacuation. With regards to anthropic threats, space capabilities are also finding increasing utilization in the nuclear non-proliferation efforts. They are becoming rapidly established components of the Comprehensive Test Ban Treaty's International Monitoring System for a number of different reasons—space systems enable discrete and non-invasive monitoring of problem areas, provide ample early warning for the preparation of nuclear tests and can quickly provide political decision makers and the public with undeniable evidence of malfeasance, spurring a response.[3]

An area of development made possible only through the development of space systems is the early warning capability for 'solar storms,' a particular subset of 'Space Weather' phenomena towards which the technological progress of humanity has rendered it extremely vulnerable. Every critical infrastructure, especially 'critical space infrastructure', is vulnerable to disruption because of coronal mass ejections of plasma or geomagnetically induced currents, if only because of their potential effects on electricity generation and transmission systems. On 13–14 September 1989, one of the largest solar storms since the Carrington Event in 1859, when humanity's state of development made it least susceptible to such effects, resulted in the loss of contact with numerous space assets for over a week and left 6 million inhabitants of the Canadian province of Quebec without electricity for several hours, with many planes being grounded or rerouted (Cannon 2013). The 2003 'Halloween storm' was another peak of solar activity[4]—alongside power disruptions on the ground, orbital activity was seriously affected—59% of scientific missions were interrupted, astronauts had to take refuge in specially shielded areas of the International Space System, and a number of satellites were lost (Butt 2010). Another such event, occurring by chance

---

[3]For example, see: https://nanopdf.com/download/b-jasani-role-of-satellites-in-strengthening-nuclear-security_pdf.

[4]https://www.nasa.gov/topics/solarsystem/features/halloween_storms.html

when the Earth is at the optimal point for advanced societies to suffer full exposure to its effects, has been estimated by the US National Research Council (NRC) to result in damages valued at 2 trillion dollars in the first year for the US alone, and recovery times between four and ten years (Baker et al. 2011). The cited report, however, does not take into account the damage to electricity grids in important trading partners such as those in Europe, lost economic opportunity, the evaporation of value in mark-to-market systems on the basis of investor confidence and so on. In an interconnected world, the propagation of infrastructure disruption risks makes the potential damages, as well as the time to recovery, inestimable. Land based communication systems, the underlying computer systems of financial markets, and any other infrastructure dependent on a reliable supply of electricity and communications are also under threat. Weak links in infrastructure systems, because of bad planning or mismanagement, can be triggered by space weather events to inflict even greater damage—the NRC estimates that the electricity grid of the United States is critically reliant on just a few dozen central transformer stations which are unshielded and obsolete, because of which over 130 million consumers in the US alone would be deprived of electricity for more than a few hours (Baker et al. 2011).

Space systems in deep space are used not just to study the various 'space weather' phenomena, but also to provide early warning between 15 and 30 min in advance of an event occurring. It might not seem like sufficient warning, but it is enough for certain critical systems to be shut down, for operators to initiate mitigation measures and other back-ups, and for economic agents relying on space systems to land airplanes, to not attempt sensitive operations or to switch to other forms of navigation. The most important 'warning buoy' for solar storms is NASA's Advanced Composition Explorer (ACE) launched in 1997; other dedicated systems include the Solar and Heliospheric Observatory (SOHO), DSCOVR or Probe-2, as well as a host of satellites and probes pursuing other missions but with useful on-board instrumentation.

7.2.1.4 Simulation and Training

The training of personnel, the development of procedures and their testing, as well as the simulation of scenarios to ensure preparedness and discover any gaps in security are vital functions of resilience governance, with an important impact on crisis and emergency situation management. If space systems are adopted as a component of the resilience governance mechanism and of crisis management in particular, then it stands to reason that they, or a simulated presence, will be involved in the simulation and training activities to familiarize operators and decision makers with their use and the interpretation of their data, as well as with their impact on the wider management process, since, as we shall see in future sections, reliance on space systems, in itself, poses a risk of disruption. A complementary approach to training is serious gaming. Serious gaming is discussed in Chap. 8.

7.2.2 *The During Phase*

This is the phase during the materialization of the negative event. The management of a crisis event is a main area of contribution for space systems, where the earliest areas of positive impact were identified, and solutions developed to harness the potential of space capabilities. The proposed model sees three main components to space system use during a crisis—coordination, information gathering and feedback.

7.2.2.1 Command, Control and Coordination

Natural disasters, terrorist attacks and other events may interrupt terrestrial communication systems, especially land-based, or even subvert them. This may prevent proper coordination of field assets and their response to the crisis situation and the new developments, as well as hamper the implementation of procedures. In case of a terrorist attack, this would be by design, in order to maximize the chaos and the damage, but natural disasters and other negative occurrences may have the same effect. Space systems offer a useful back-up for communications, suitable not just for response activities, but also for public morale, by enabling the collection of valuable secondary information, like the names of survivors. The post-disaster analysis of the Fukushima Daiichi nuclear plant meltdown and the tsunami that caused it in 2012 highlighted the role of the Kiku-8 and Kizuna satellites in coordinating responders, both governmental and municipal, and in maintaining outside lines of communications to notify families when their loved ones had been saved.

7.2.2.2 Information Gathering

Real-time reconnaissance by satellites is crucial for the modern approach to crisis and emergency situation management process, though it should still be supplemented, to the extent possible, by "boots on the ground". The use of space capabilities can be tailored to the requirements of individual crisis situations, resulting in relevant and useful applications for the decision maker's toolkit. In the case of the Fukushima dual-disaster, space systems played several valuable roles in information gathering—they provided real-time monitoring of flooded areas, surveillance of damage to the Fukushima reactors (radiation, temperature etc.), surveillance of other critical infrastructure damage, environmental monitoring and so on. The Fukushima disaster also highlighted two key issues with regards to how the competent authorities dealt with the consequences of space systems becoming critical infrastructures for the crisis and emergency situation management activities.[5] The Fukushima disaster provides a vivid example of the reliance on space systems and how this growing dependency creates new risks, vulnerabilities and threats, which will be touched on in a later section—the main Japanese Earth Observation satellite whose configura-

[5]For example, see: http://www.unoosa.org/pdf/pres/stsc2012/tech-11E.pdf.

tion made it suitable for the needs highlighted by this event, ALOS, spontaneously malfunctioned just after the disaster, depriving the decision makers of its critical services. Secondly, the loss of the provision of space services was made up from several different sources—from other Japanese satellites, but primarily by international partners sharing information gathered by their own satellites, either on an ad-hoc or a pre-organized basis which will be described in detail in the next section.

7.2.2.3 Feedback

A composite process which includes elements of the previous entries in this particular section but was highlighted separately because of its conceptual relevance for a crisis or emergency situation. Space systems are becoming an important backbone of the feedback loop that keeps decision makers up-to-date with minute changes in the crisis environment and the consequences of their prior actions and policies. This kind of information is vital in a fast-developing crisis scenario, where decision makers shift resources, change tactics, implement procedures at the exact required moment, correct errors and react to unforeseen challenges in a timely manner.

7.2.3 Post-crisis Phase

Should not be considered limited to reestablishing normal business operations and regaining quality of life. This must also include processes of analyzing the critical infrastructure failures and the performance of crisis management policies and activities in order to minimize the chances of a future crisis occurring and better prepare the system-of-systems for rapid recovery from disruption and for better damage mitigation. The resilience of the critical infrastructure system-of-systems is dependent on these improvements. For all of these considerations, space systems offer potential benefits.

7.2.3.1 Recovery Efforts

The urgency of the crisis situation gives way to a potentially arduous recovery process, in which decision makers must contend with multiple demands on their resources, often with severe gaps in certain critical infrastructure capabilities. Some of these can be compensated for by space systems, such as damaged communication landlines or damaged data collection systems. Crisis situations are often a drain on military resources, which are used to conduct search and rescue, handle logistics, debris clearing, maintain public order and so on. Space systems can assist them in these activities but can also assist decision makers in monitoring borders and maritime areas, providing early warning for consequent, maybe related threats, resulting from the activity of foreign militaries or organized crime groups like smugglers. Therefore, space systems offer important short-term capabilities for the maintenance of national security to compensate for temporary shortcomings.

7.2.3.2 Research

The cost-effective monitoring and data collection enabled by space systems, over long periods of time, make them ideal facilitators of research into the aftermath of natural disasters and other crises, as well as similar events, to add to the body of knowledge regarding these threats and the response of the system-of-systems to their materialization. Following the Fukushima disaster, Japanese researchers began collaboration with Ukrainian counterparts for the study of the Chernobyl Nuclear Power Plant area to discern the consequences of the presence and spread of radioactive particles in air, soil and water[6] (Yoshimoto et al. 2013). Such research may prove important for the recovery prospects of the Fukushima area and the improvement of civil protection capabilities.

7.2.3.3 System Finetuning

Decision makers utilize the results of previous processes to finetune both the resilience governance and crisis management systems, as well as the characteristics of the system-of-systems and its components, thereby improving security outcomes. Coming to terms with the failure of the ALOS satellite at its moments of pressing need, and the undeniable utility of space applications during the Fukushima Daiichi Crisis, the Japanese authorities launched a constellation of eight nanosatellites. Each had specific functions and instruments, but also some redundancies, acting as a resilient collective surveillance system with a more advantageous security profile than a single asset solution[7] (Yoshimoto et al. 2013). As will be shown in the next section, Japan updated its space capability build-up paradigm to take into account new developments in system architecture, pricing and so on. Given the challenging security environment of the Japanese archipelago, Japan will probably lead the way in promoting a more intricate interplay between terrestrial and space systems, as well as other components of the Japanese national disaster management paradigm, for crisis and emergency situation management. The development of space capabilities in service of resilience governance will continue, with other actors being able to learn from this experience and Japanese best practices.

[6]For example, microsatellites where used for environment monitoring during Fukushima and Chernobyl, http://www.nanosat.jp/images/report/pdf/NSS-05-0104.pdf.

[7]http://www.nanosat.jp/images/report/pdf/NSS-05-0104.pdf.

7.3 Critical Considerations on Space Services Provisioning

Japan's experience with supplementing its provisioning with critical space services during and after the Fukushima disaster and the loss of the ALOS satellite exemplifies the potential for international accords, collective charters and sharing agreements to bypass the issue of momentary or permanent service scarcity, especially during a crisis period. The political tenability of a nation free riding on the 'critical space infrastructure' of another without constructing its own is a relatively unexplored subject, though World Bank projects involving free access to space data for economically or environmentally disadvantaged areas have proven both successful and capable of maintaining political good will. Certainly, the extent to which a small number of spacefaring nations, led by the United States, account for a vast majority of space systems that supply space services, some of them critical, to billions of consumers, shows that there is ample scope for a country with no space assets of its own to maintain a high degree of sophistication which would make it reliant on space systems. The US alone represents, according to the open sources satellite database of the Union of Concerned Scientists, 859 of the 1,886 known satellites, or 45.5%.[8] Nevertheless, Japan followed up on the loss of the ALOS satellite by adopting a nimble formula of more numerous, but smaller, satellites with greater redundancy. Until then, it relied on a number of sources for space services related explicitly to crisis and emergency situation management needs:

- The International Disasters Charter—a response mechanism for pooling satellite capabilities and other resources and making them available free of charge to the affected parties in a timely manner. It was set up in the year 2000 by the European Space Agency, the French Space Agency CNES and the Canadian Space Agency, and has 14 members today. During the Fukushima Crisis, numerous satellites provided data or were on call to do so, belonging to US actors (Landsat 5 and 7, EO-1, IKONOS, GeoEye, Quickbird-2, Worldview 1 and 2), India (Cartosat-2), ESA (Envisat), Canada (Radarsat-2), South Korea (Kompsat 2), Germany (Terra SAR X, RapidEye), France (Spot 4 and 5) and China (HJ). Other such arrangements predate this Charter, like SERVIR. A newer model for cooperation comes from the European Union, which initiated the 'Copernicus Emergency Management Services' in 2012 which is tied to the Charter specifically for improving the quality of gathered data
- Sentinel-Asia—a highly effective model for regional cooperation involving 78 groups dedicated to disaster reduction, which served to make up for the loss of the ALOS satellite. Thailand's Geo-Informatics and Space Technology Development Agency contributed the use of its THEOS satellites, while Taiwan contributed Formosat-2
- Individual offers—countries that have developed capabilities which are not included in pre-existing sharing arrangements for various reasons but offered the

[8]This number is based on statistics of the Union of Concerned Scientists open-source satellite database statistics: http://www.ucsusa.org/nuclear-weapons/space-weapons/satellite-database.html#.Vg0BUCvkVTB.

Japanese governments the use of their capabilities. Italy, Russia, Spain and Dubai are numbered among them.

These types of cooperation are increasingly necessary and common throughout the entirety of the space services industry, because of the basic characteristics of space systems as an infrastructure: the significant economic barriers to their creation and launch, their very low numbers when compared to the number of users and beneficiaries, their stressful environment beset also by the rising risk posed by space debris, their weight limitations, which require compromises on redundancy, shielding, lifetime duration, extra instruments, interoperability etc., the grey areas and uncertainties in the still forming legal and organizational framework for the exploitation of space and the rising anti-satellite capabilities (Gheorghe and Vamanu 2007) not just of nation-states, but also of non-state actors like terrorist groups and organized crime (cyber-attacks, in particular, have a low threshold cost for initiation and potentially the highest cost to benefit ratios). Though it is not the purview of this particular paper, the issue of space systems as critical infrastructures implies that there is an inherent risk in critical dependencies on such vulnerable and scarce assets. Dependencies for crisis and emergency situation management are even more critical, because the disruption of services may occur (intentionally or accidentally) when they are most needed, compounding the damage and loss of life. Rather than being a 'silver bullet' for whatever ground-based crisis situation they address, using space systems involves accepting and planning for a host of new risks, vulnerabilities and threats, from technical to political. This continuous churn of risk profiles because of security decisions and policies is a defining element of resilience governance, according to Mureşan and Georgescu (2015), who compared it to a tightrope balancing act in which decision makers strive to keep their security ledger in the black.

There are developments which may assist in easing the issue of reliable provisioning of space services for many activities, including crisis and emergency situation management—technological and economic advances working towards reducing system and launch costs, institutional developments and clearer risk profiles enabling the better financialization of the industry (involving also insurance, financing development, assigning liability etc.), better instruments reducing the weight demands for satellites and so on. The latest State of the Satellite Industry Report, launched in September 2015 to account for the year 2014, noted not just the impressive growth in space services revenue (230% over a decade), but also a powerful dynamic in terms of new launches—208 in 2014, compared to 107 in 2013, driven mostly by the growth in CubeSats (63% of 2014 total).[9] These are smaller systems aiming for efficient use of launch space to share costs, which have become much more viable both through standardized designs contributing to economies of scale, but also through improvements in instrument sensitivity. Also, of note is the rise of new business models for accessing space services, such as the aforementioned World Bank projects, or

[9]Based on the report: '2015 State of the Satellite Industry Report' commissioned by Satellite Industry Association: http://www.sia.org/wp-content/uploads/2015/06/Mktg15-SSIR-2015-FINAL-Compressed.pdf.

the European Space Agency's business model for the Copernicus program of Earth Observation satellite launches, involving free and unrestricted access to data up to a certain quality.[10] The ESA also decided on not including in the terms of use for the Galileo global navigation satellite system a provision enabling arbitrary suspension or degradation of service for users, similar to that of the American GPS and Russian GLONASS systems. Galileo can also improve its signal quality through interoperability with both GPS and GLONASS, which are not interoperable with each other, increasing systems resilience.[11] In a nod to crisis and emergency situation management concerns, Galileo also features a "Search and Rescue" function that enables the accurate location of the source of distress signals.[12]

With regards to the challenging security environment, there have been advances in the required collective approach to improving the security outcomes in the very international orbital environment. International bodies such as the UN's Committee on the Peaceful Uses of Outer Space propose technical standards for shielding space systems from the effects of "space weather" and reducing the production of new space debris by deorbiting satellites at the end of their useful lifespan and minimizing debris production during launch. Steps are also being taken towards preventing the militarization of space and the targeting of space systems in a conflict, which may lead to a logic of MAD: mutually assured destruction (Long 2008).

Final Remarks

Space systems are emerging as important tools for crisis and emergency situation management with regards to critical infrastructure disruption. In a wider sense, they are becoming a foundational element of critical applications for resilience governance, such as information gathering and command, control and coordination. This is due to a unique combination of capabilities and advantages, as well as their complementarity to the classical toolkit for crisis management. Using an illustrative, generic model for viewing the requirements of crisis management in a wider sense, the authors contend that space systems provide significant services in all phases and processes of response to a crisis, thereby supporting security experts and decision makers in long-term efforts to also increase the resilience and robustness of societies. The benefits to quality of life and business continuity are self-evident.

The resulting paradigm suggests an optimal state for fully utilizing space systems at their current and predicted levels of capability but is not representative of factual reality. In reality, even the most advanced nations utilize a patchwork toolset that may lean heavily towards the use of some systems in some phases and processes, while ignoring others. Further developments of a technological, economic or orga-

[10]http://copernicus.eu/data-access-satellite.

[11]From European Space Agency: http://www.esa.int/Our_Activities/Navigation/Galileo_and_EGNOS.

[12]http://ec.europa.eu/growth/sectors/space/galileo/sar_en.

nizational nature in this field may increase access to space applications. With the advent of regional and global infrastructure systems, as identified not just in the scientific literature, but in the reality of actual governance efforts, such as the European Programme for Critical Infrastructure Protection, the promotion of new instruments for managing risks and minimizing disruptions gains new urgency. The interdependencies that cross-national borders and the jurisdiction of the competent national authorities ensure that security for the whole is skewed by the lowest common dominator, regardless of the security outcomes for individually managed components in the system-of-systems. It is in the interest of the most advanced and space-realized actors to promote adoption by other countries of these new capabilities.

At the same time, there is a growing realization that space systems, themselves, are becoming critical infrastructures. Since criticality is also heavily context dependent, this means that a reliance on space systems for crisis and emergency situation management will automatically designate these as ‘critical space infrastructures’, imparting a new risk of disruption that must be taken into account. As shown by individual examples, the materialization of these new risks is not without precedent, meaning that security decision makers must plan for maximizing the reliability of access to space services and of substitutive capacity. International cooperation plays an increasing role in ensuring the provisioning of critical space services for national authorities in an emergency situation. Ultimately, this new critical dependence has materialized from the obvious benefits of utilizing space systems. This means that the security trade-off of using space systems in resilience governance, in a wider sense, and emergency management, in a narrower sense, to deal with other risks, vulnerabilities and threats, while engendering new ones, is a positive sum game or can be developed as such.

References

Baker, D., Balstad, R., Bodeau, J.M., Cameron, E., Fennell, J.F., Fisher, G.M., et al. (2011). *Severe space weather events: Understanding societal and economic impacts*. Washington, DC: National Academies Press. Retrieved from http://lasp.colorado.edu/home/wp-content/uploads/2011/07/lowres-Severe-Space-Weather-FINAL.pdf.

Bouchon, S. (2006). *The vulnerability of interdependent critical infrastructures systems: Epistemological and conceptual state-of-the-art* (No. EUR 22205 EN). Ispra, Italy: Institute for the Protection and Security of the Citizen, European Commission Joint Research Center. Retrieved from https://publications.europa.eu/en/publication-detail/-/publication/e3c8da0d-9d5a-408d-a098-38f6b624508d/language-en.

Butt, Y. M. (2010, Feb 1). The EMP threat: fact, fiction and response (part 1). The Space Review. Retrieved from http://www.thespacereview.com/article/1549/1.

Calida, B. Y., & Katina, P. F. (2012). Regional industries as critical infrastructures: A tale of two modern cities. *International Journal of Critical Infrastructures, 8*(1), 74–90. https://doi.org/10.1504/IJCIS.2012.046555.

Cannon, P. (2013). *Extreme space weather: Impacts on engineered systems and infrastructure*. Londaon: Royal Academy of Engineering. Retrieved from https://www.raeng.org.uk/publications/reports/space-weather-full-report.

Gheorghe, A. V., & Vamanu, D. V. (2007). Risk and vulnerability games. The anti-satellite weaponry (ASAT). *International Journal of Critical Infrastructures, 3*(3/4), 457–470. https://doi.org/10.1504/IJCIS.2007.014120.

Gheorghe, A. V., Vamanu, D. V., Katina, P. F., & Pulfer, R. (2018). *Critical infrastructures, key resources, and key assets*. Cham, Switzerland: Springer International Publishing. https://doi.org/10.1007/978-3-319-69224-1_1.

Haeme, T. (2003). *Use of satellite and airborne remote sensing in the safeguards of a nuclear waste repository site* (No. STUK-YTO-TR–199). Espoo: Finland: VTT Information Technology. Retrieved from http://inis.iaea.org/Search/search.aspx?orig_q=RN:34068266.

Hokstad, P., Utne, I. B., & Vatn, J. (Eds.). (2012). *Risk and interdependencies in critical infrastructures: A guideline for analysis*. London: Springer. Retrieved from www.springer.com/us/book/9781447146605.

Katina, P. F., & Hester, P. T. (2013). Systemic determination of infrastructure criticality. *International Journal of Critical Infrastructures, 9*(3), 211–225. https://doi.org/10.1504/IJCIS.2013.054980.

Katina, P. F., & Keating, C. B. (2015). Critical infrastructures: A perspective from systems of systems. *International Journal of Critical Infrastructures, 11*(4), 316–344. https://doi.org/10.1504/IJCIS.2015.073840.

Long, A. (2008). *Deterrence—From Cold War to Long War: Lessons from six decades of RAND deterrence research*. Santa Monica, CA: RAND Corporation.

Moteff, J. D., Copeland, C., & Fischer, J. (2003). *Critical infrastructures: What makes an infrastructure critical?* (pp. 1–17). Washington, DC: The Library of Congress.

Mureşan, L., & Georgescu, A. (2015). The road to resilience in 2050. *The RUSI Journal, 160*(6), 58–66. https://doi.org/10.1080/03071847.2015.1123948.

Rinaldi, S. M., Peerenboom, J., & Kelly, T. K. (2001). Identifying, understanding, and analyzing critical infrastructure interdependencies. *IEEE Control Systems, 21*(6), 11–25. https://doi.org/10.1109/37.969131.

Yoshimoto, S., et al. (2013). *Environment Monitoring of Fukushima and Chernobyl Areas using a Constellation of Earth Observation Microsatellites*, Universitatea din Tokio, parte a Programului "Japan-Ukraine Cooperation Technical Demonstration Program for Supporting Aftermath Responses to Accidents at Nuclear Power Stations". Retrieved from http://www.nanosat.jp/images/report/pdf/NSS-05-0104.pdf.

Zimmerman, R., & Restrepo, C. E. (2009). Analyzing cascading effects within infrastructure sectors for consequence reduction. In *2009 IEEE Conference on Technologies for Homeland Security* (pp. 165–170). https://doi.org/10.1109/THS.2009.5168029.

Chapter 8
Serious Gaming and Policy Gaming

Abstract Over the last few decades, practitioners and management scholars increasingly criticized the conventional strategy making methods, arguing that rapidly changing environments require emerging and creative approaches. Serious gaming (simulation game or gaming, used interchangeably within the text) discipline is found to be increasingly useful within mainstream strategy literature involved with former strategy making approaches. Serious gaming as an approach for understanding infrastructure is disused in this chapter. This discussion includes concepts game cycle, data negation, scenario development as well as model for estimating consequences and probabilities.

8.1 Serious Gaming

Serious gaming (simulation game or gaming, used interchangeably within the text) discipline is found to be increasingly useful within mainstream strategy literature involved with former strategy making approaches (Geurts et al. 2007). A definition of gaming simulation is given as a representation of a set of key relationships and structure elements of a particular issue or a problem environment, where the behavior of actors and the effects of their decision are a direct result of the rules guiding the interaction between these actors (Ancel 2011; Wenzler 2005).

Serious gaming is an activity where two or more independent decision-makers seek to achieve their objectives within a limited context. Serious games allow researchers to model problems with societal aspects which can often be found in next generation infrastructure transition efforts. The advantage of simulation gaming over traditional computer simulation models is that the stakeholders do not have to be represented by mathematical formulations; instead, they are played by the participants themselves (Bekebrede 2010). Representing complex systems with serious gaming models save the model builders the need to build in the psychological assumptions since they are represented by the stakeholders. Greenblat and Duke (1975) suggest that "the participants (or the players) of the game perform a set of activities in an attempt to achieve goals in a limiting context consisting of constraints

© Springer Nature Switzerland AG 2019

A. Georgescu et al., *Critical Space Infrastructures*,
Topics in Safety, Risk, Reliability and Quality 36,
https://doi.org/10.1007/978-3-030-12604-9_8

and of definitions of contingencies" (p. 106). These games are labeled 'serious' because their primary objective is educational and/or informative as opposed to pure entertainment.

Simulation games have many different forms. The aim is to provide insights for various goals. However, commonality is in that reality can be simulated through the interaction of role players using non-formal symbols as well as formal, computerized sub-models when necessary. This approach allows the group of participants to create and analyze future worlds they are willing to explore. Lately, large organizations reported serious gaming simulation uses for their organizational change management efforts (Ancel 2011). Duke (1974) argues that formal complexity communication methods are inadequate when it comes to problems of the future due to their exponentially increasing complexity. Gaming simulation techniques can handle "many variables" and are distinguished from other techniques by being relatively uncalibrated and intuitive (Duke 1974, p. xv). Each serious game is situation specific; consequently, they should only be performed within the intended and designed context. Failure to do so will result in poor results.

Serious games are developed to serve several different purposes. However, the most important contribution of gaming methods is their ability to enhance communication among various actors. This tends to lead to utilization of gaming methods in complex system exposition where complex systems with social aspects can be examined (Bekebrede et al. 2005; Duke 1974). Policy-gaming exercises carry various objectives like understanding system complexity, improving communication, promoting individual and collective learning, creating consensus among players, and motivating participants to enhance their creativity or collaboration (Geurts et al. 2007). Policy games are often used in understanding complex infrastructure systems which will be covered within the next section.

8.2 Infrastructure Serious Gaming

The discrepancies associated with such infrastructure transitions are related to the lack of understanding of societal aspects of these systems. For that reason, several serious gaming exercises are developed to assist decision makers, experience system complexity and train stakeholders. Serious games can represent the multi-level system architecture by proprietary rules at the player level, interaction of the players, and the system levels. The complexity associated with infrastructures (both the technical/physical and social- political levels) is integrated within the gaming platform for stakeholders to experience an abstract representation of the system and make informed decisions (Mayer 2009). Several infrastructure systems are represented using serious games.

Unlike hard-system methods, the gaming and simulation approach is quite flexible and easily adaptable to other quantitative methods, scenarios, and computer models (Mayer 2009). Policy gaming methods can help both participants and modelers understand the big picture and identify critical elements of the complex problem at hand. Because of the iterative and experimental nature of these gaming and simulation environments, participants are able to test different approaches within both a safe environment and a condensed timeframe (Wenzler 2008). INFRASTRETEGO is an example of a serious gaming-based decision-making tool, aimed at encapsulating the Dutch electricity market. Game developers used the game to examine strategic behavior in a liberalizing electricity market while examining the effectiveness of two main types of regulatory regimes. Strategic behavior is the use of administrative and/or regulatory processes such as stalling, delaying, or appealing interconnection negotiations, engaging in anti-competitive pricing, or other methods that can be encountered within liberalization of utility industries. Empirical research indicates that strategic behavior may affect the level playing field and public values in a negative way. Overall, the game was able to identify the undesirable, unintended and unforeseen effects of strategic behavior phenomena. Serious gaming enabled monitoring and measurement of strategic behavior as it occurred since participants did not have any fear of litigation and were able to report the development of the strategic behavior which cannot be observed in real-world situations (Kuit et al. 2005; Wilson et al. 2009).

8.3 Serious Gaming and Data Generation

There are a limited number of studies regarding use of serious games as a data generation method. For example, Rosendale (2016) employed role-playing as a data generation method about the use of language in speech act situations. The gaming method, used to gather data in platonic and romantic situations, a limited application, as authentic interactions cannot be observed without violating privacy of participants. However, Rosendale states that the role-play method has been shown to be a valid and reliable method, the limitations of using this method brought up questions about its validity and ability to represent real world interactions between humans (Rosendale 2016). Demeter (2007) also suggested using role-play as a data collection method related to apology speech acts by analyzing how apologies take place in different situations. Participants, chosen from English majors at a university from Romania, were engaged in a role-playing environment and asked to apologize within the scenarios presented to them. The naturally occurring discussions were collected and compared against another method called discourse completion tests. The author concluded that in some instances, role-playing produced more realistic data since it allowed participants to actually speak instead of writing their responses and they were more authentic since a natural setting was created by the scenarios (Demeter 2007). Another qualitative study using role-playing to generate data was conducted by Halleck (2007). The gaming method was used to evaluate a nonnative speaker's oral efficiency using simulated dialogues. The biggest advantage of using role-playing is given as its ability to simulate a real conversation environment without violating participants' privacy. Besides generating data for speech act studies, the only study related to data elicitation was the REEFGAME, simulating the marine ecosystems in order to learn from different management strategies, livelihood options and ecological degradation (Cleland et al. 2012). The data generation ability of the game was limited to the decision-making processes of the stakeholders (fishers) which can be categorized under collective learning regarding complexity, and it was not elaborated on any further. Considering the studies above, the literature survey did not provide any intensive data generation study conducted with serious gaming approaches, demonstrating the uniqueness of the study at hand.

8.4 Gaming Cycles

During the pre-gaming phase, it is necessary to collect all the gaming variables depending on the modeled system. Such variables include scenarios, stakeholders and their interactions, historical data regarding the system and information on the parameter(s) upon which the success of the transition process will be measured. The computer-based simulation mechanism keeps track of the process throughout the gaming exercise. Depending on the application, the computer-based simulation can evaluate risk or reliability of an infrastructure system or keep track of generation capacity or throughput of a certain utility. Once the adequate numerical simulation mechanism and all the supporting data are collected, the game is developed.

Developing the game is an iterative process where versions are often tested by playing with several groups and then fine-tuning. The gaming cycle includes the execution of the gaming exercise with the participation of experts. The game usually starts with the presentation of the scenario to the participants. Participants are asked to perform according to their predetermined roles. Considering the new information, they have been presented, participants are asked to make collective decisions about the investigated parameters. The decisions are taken as the input variables for the computer assisted simulation mechanism where initial conditions for the next step are calculated. The iterative process enables participants to experience and shape the future phases of the transition process. The presence of participants (preferably experts or real stakeholders) social values, norms and beliefs provides the realistic input for the social interaction and the decision-making process.

The post-gaming phase of the methodology involves data collection and analysis which surfaced during the gaming cycle. At this level the elicited data are arranged and presented back to the participants for further analysis and feedback. And although not performed, it is possible to use several other types of COTS software to organize and analyze the data.

8.5 Scenario Development

Scenarios constitute one of the main elements of complex models, simulations or serious games. The scenario allows the modelers to create the environment in which the particular system operates. Similar to other gaming parameters (e.g. the technological advancement contributions or their timelines), the scenario to be investigated is often determined by subject matter experts. In order to support strategic planning efforts, workshops, serious games, think- tanks or other behavioral expert elicitation techniques are often organized (Jacobs and Statler 2006). Elicitation techniques include, among others, Delphi Method and Nominal Group Technique. An example for such a study was conducted in 1997 by the National Research Council (National Research Council 2001). Experts from academia, aviation related public and private sectors, scientists, consultants, the armed forces, and government agencies were able to determine five scenarios with great depth for the next 15–25 years. Out of five chosen scenarios, the aviation industry experienced three of the predicted futures, including the 9/11 attacks, a steep increase in fuel prices, and the effects of the global market in aviation.

When constructing a scenario, the overall goal must be formed during the initial stage of the process. Goal formation includes the determination of the expected results, system boundaries, knowledge base, stakeholder functions, etc. Once the goal is clarified, the scenario is constructed in an iterative manner. On the other hand, developing scenarios for gaming purposes is highly dependent on the type of the simulation's goals. If the purpose of the game is to offer policy recommendations or implications, the scenario has to play a dominant role. If the purpose of the game is educational, the scenario must be able to promote creative thinking and imagination in its participants (deLeon 1975).

8.6 Serious Gaming as a Way of Understanding Infrastructures

Luna-Reyes et al. (2005) state that the presence of social and organizational factors can cause up to 90% of the information system project failures, resulting in not delivering the expected benefits. For that reason, it is crucial to integrate such societal factors into the design of large-scale infrastructure design processes. Thus far, it is evident that serious gaming methods proving insights in planning efforts for large-scale sociotechnical systems. In fact, Ancel (2011) suggests that serous gaming's ability to integrate the social and technical aspects of infrastructure development delineates these methods as the most appropriate candidate for creating a venue combining computer assisted stakeholder interaction. In this way, serious games provide insights into how to address issues arising from the interaction of players, roles, rules and scenarios. Mayer (2009, p. 845), describes serious gaming derived applications as "a hard core of whatever the computer model incorporated in a soft shell of gaming (usually through some form of role- play)."

8.6.1 Rapid Risk Assessment Model

One of such methods is the rapid risk assessment model. Rapid Risk Assessment Model (RRAM) serves as to estimate and quantify of risk values, comprised of separately calculated accident probabilities and their respective consequences. The probabilities within the model are estimated via the Probability Number Method (PNM), and the consequences are approximated via numerical manipulations. The RRAM is supported by historical and expert elicited data as well as the gaming to numerically generate the risk values throughout the methodology.

The RRAM was used as the risk simulation mechanism selected for the case study. However, depending on the problem at hand, this model can be replaced with any adequate software, method, or an existing study measuring aspects like network capacity, throughput, financial status, etc. The adaptability of the gaming method allows developers to switch and/or combine different approaches which will provide a systemic view of the problem.

8.6.2 Basics of RRAM

The RRAM was created through the joint effort of the International Atomic Energy Agency (IAEA), the United Nations Environment Programme (UNED), the United Nations Industrial Development Organization (UNIDO), and the World Health Organization (WHO) under the umbrella of the United Nations. The model and the associated method were developed as an affordable solution for a quick turn-around needed to determine risks associated with handling, storage, processing and transportation of hazardous materials. The risk assessment methodology (including the PNM approach) was supported by an extensive database containing various types of substances (e.g., flammable, toxic, etc.), safety precaution measures, population densities and environmental factors (International Atomic Energy Agency 1996). However, as opposed to answering questions such as the maximum number of fatalities or effect of distance, the PNM induced risk assessment methodology was more focused on prioritization of actions in the field of emergency preparedness.

8.6.3 Consequences and Probabilities

Risk is defined as the product of the probability of an accident and its respective consequences (Bedford and Cooke 2001). The IAEA study estimates probabilities and consequences separately. The consequences of an accident (e.g. an event caused by storage or transportation of certain hazardous materials) are calculated via simple numerical manipulations, taking into consideration the characteristics of the substance and correcting factors regarding the area, population density, accident geometry, etc. The required data to form the components of the equation is obtained through previous modeling efforts and expert opinions. On the other hand, the probabilities are estimated via PNM where the probability of a certain accident happening is calculated via a dimensionless 'probability number' N, which is in turn transformed to actual probabilities. The probability number is adjusted/updated according to the various correcting factors. The relationship between the probability and N is given via $N = |\mathbf{long}_{10}\boldsymbol{P}|$. The risk is defined as the product of the consequences and the probabilities of unwanted outcomes (hazardous events). For the NextGen case study, the adequate interpretation and calculation steps were adopted and integrated which generates the risk simulation mechanism.

8.6.4 Estimation of the Societal Risks

The previous section provided calculations of human casualties (fatalities) associated with an accident, along with the probabilities of such accidents occurring. The risk to the public from these activities is estimated by combining these two values. The consequences are categorized with respect to the fatalities and the probability classes are categorized by the order of magnitude of the number of accidents per year (e.g. societal risk operational instrument). The consequence-frequency (x-y) diagram is created. The main goal is to obtain a list of activities whose risks have to be further analyzed before others. The risk matrix representation is one of the primary outcomes of the method.

8.6.5 Post Gaming (Data Analysis)

Throughout the gaming effort, the discussions and possible negotiations within the opposing parties are important findings that can lead to different problem-solving approaches. The results of a game run are analyzed to examine if the gaming exercise influenced the beliefs, intentions, attitudes, and behavior of participants, yielding to a better understanding of complexity (Joldersma and Geurts 1998). The serious gaming exercise serves both as an individual and collective learning platform for the stakeholders, leading to an elevated level of knowledge over the system (Wilson et al. 2009). The individual learning takes place during the decision-making process where each stakeholder group represents its respective point of view. The reflective conversations between the participants enable feedback and help participants build informed judgments. Therefore, the presence of realistic interactions among players helps the testing and evaluation of NextGen related technologies in the future (Joldersma and Geurts 1998). Also, like individual learning, collective or organizational learning provides insight into the system at hand (i.e. NextGen aviation safety values).

Expert feedback is a main contributor in all phases of the methodology. Experts from all stakeholder groups help shape possible scenarios, provide numerical data regarding the future technological enablers and also assist in evaluating the developed methodology in different categories. Expert participation in all three phases of the gaming-based elicitation methodology is prominent since it allows game developers to constantly modify the gaming components by taking participant comments and recommendations into consideration. Due to the large level of the system, no one expert is sufficient for gathering all the data needed to develop gaming based on the given methodology. This methodology provides a common elicitation aggregation opportunity for next generation infrastructure development including space systems.

To support the case study applications, concise RRAM approach is provided as follows:

First, risk is defined as the product of the probability of an accident and its respective consequences. The methodology calculates probabilities and consequences separately (See International Atomic Energy Agency 1996).

8.7 Consequences

The consequences of a certain accident are constructed with respect to the characteristics of the substance and its correcting factors, among others, area, population density, and accident geometry and indicated in Eq. (8.1):

$$\boldsymbol{C_{a,s} = A \cdot \delta \cdot f_A \cdot f_d \cdot f_m} \tag{8.1}$$

where,

$C_{a,s}$	is consequences per accident, and per substance (fatalities/accident)
A	is the calculated area based on the various inputs based on each class and amount of substance
δ	is population density within the impact area
$f_A \cdot f_d$	is a correcting factor for the impact area geometry
f_m	is a correcting factor for the cases if people within the distance have shelter, opportunity to flee, or a way to find out about the event before it happens (e.g. warning from odor, tanks that explode one by one)

Each component of the formula is entered from the existing tables where data is collected from extensive calculations, modeling, and expert opinions. The formula outputs a casualty number for each type of accident which will be matched with its respective probability of occurrence.

8.8 Probabilities

The probability of each substance installation (e.g. flammable liquid stored in a processing plant) is calculated via the following formula. The average probability number (used as a starting point) for each installation is determined by $N^*_{1,s}$ and varies by each type of substance and installation. The rest of the components in the equation are used to adjust this initial estimation in the presence of various correcting factors and indicated in Eq. (8.2):

$$\boldsymbol{N}_{1,s} = \boldsymbol{N}^*_{1,s} + \boldsymbol{n}_l + \boldsymbol{n}_f + \boldsymbol{n}_o + \boldsymbol{n}_p \tag{8.2}$$

where,

$N_{1,s}$ is the probability number for an accident to happen for a substance within an installation
$N^*_{1,s}$ is the average probability number for the installation and the substance
n_l is the correction parameter for the frequency of loading/unloading operations
n_f is the correction parameter for the safety systems associated with flammable substances
n_o is the correction parameter for the organizational and management safety
n_p is the correction parameter for wind direction towards the populated area

Based on the correction factors, the average $N_{1,s}$ value is recalculated and converted back into probability values based on the logarithmic relationship given above. In the same stream of thought, a similar approach can be used to estimate the risks associated with transportation of hazardous substances (e.g., see Vamanu et al. 2016) or other systems including space systems. Specifically, for risks of transportation of hazardous substances, $N_{t,s}$, the formula is provided as follows:

$$\boldsymbol{N}_{t,s} = \boldsymbol{N}^*_{t,s} + \boldsymbol{n}_c + \boldsymbol{n}_{t\delta} + \boldsymbol{n}_p \tag{8.3}$$

where,

$N^*_{t,s}$ is the average probability number for the transportation substance
n_c is the correction parameter for the safety conditions of the transportation system
$n_{t\delta}$ is the correction parameter for the traffic density
n_p is the correction parameter for wind direction towards the populated area

Final Remarks

Calculations of human casualties are associated with 'accident,', along with the probabilities of occurrence of accidents. The risk to the public from these activities is estimated by combining these two values. Consequences are categorized with respect to fatalities, and the probability classes are categorized by one order of magnitude of the number of accidents per year. The results obtained from the consequence and probability calculations are represented on the risk matrix, providing an overall picture of the risk. The interpretation of the risk matrix provides the acceptability criteria for the societal risk. The thresholds for acceptable risks can be based on accident frequency/probabilities as indicated in Fig. 8.1, consequences in Fig. 8.2, or a combination of both, which is based on the ALARA (As Low As Reasonably Achievable) principle and as suggested in Fig. 8.3.

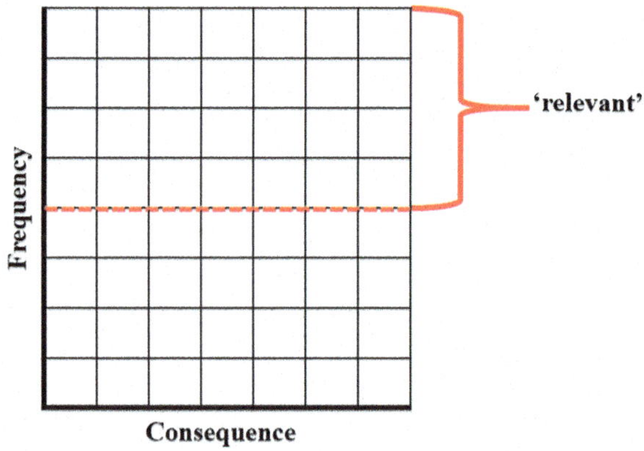

Fig. 8.1 Acceptability determined by frequency

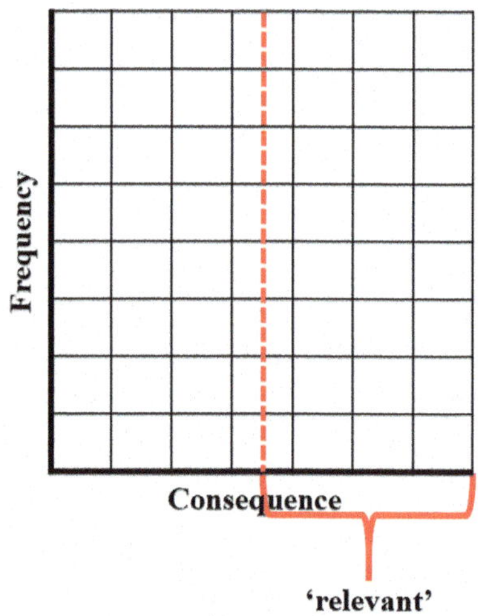

Fig. 8.2 Acceptability based on consequences

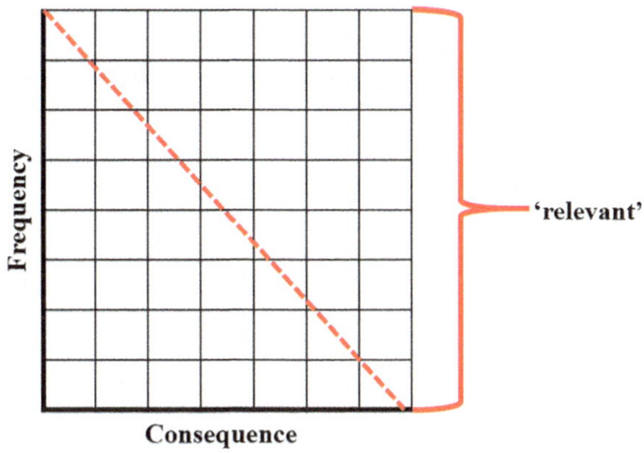

Fig. 8.3 Acceptability based on frequency and consequences

References

Ancel, E. (2011). *A systemic approach to next generation infrastructure data elicitation and planning using serious gaming methods* (Ph.D.). Old Dominion University, United States—Virginia. Retrieved from http://search.proquest.com.proxy.lib.odu.edu/docview/896960555/abstract/C4484A36FA44018PQ/5.

Bedford, T., & Cooke, R. (2001). *Probabilistic risk analysis: Foundations and methods*. New York, NY: Cambridge University Press.

Bekebrede, G. (2010). *Experiencing complexity: A gaming approach for understanding infrastructure systems*. Delft University of Technology, Delft, The Netherlands. Retrieved from http://resolver.tudelft.nl/uuid:dae75f36-4fb6-4a53-8711-8aab42378878.

Cleland, D., Dray, A., Perez, P., Cruz-Trinidad, A., & Geronimo, R. (2012). Simulating the dynamics of subsistence fishing communities: REEFGAME as a learning and data-gathering computer-assisted role-play game. *Simulation & Gaming, 43*(1), 102–117. https://doi.org/10.1177/1046878110380890.

deLeon, P. (1975). Scenario designs: An overview. *Simulation & Games, 6*(1), 39–60. https://doi.org/10.1177/003755007561002.

Demeter, G. (2007). Symposium article: Role-plays as a data collection method for research on apology speech acts. *Simulation & Gaming, 38*(1), 83–90. https://doi.org/10.1177/1046878106297880.

Duke, R. D. (1974). *Gaming: The future's language*. Beverly Hills, CA: Sage Publications. Retrieved from https://trove.nla.gov.au/version/45243363.

Bekebrede, G., Mayer, I., van Houten, S.P.A., Chin, R.T., & Verbraeck, A., (2005). How serious are serious games? Some lessons from infra-games. In *Presented at the Proceedings of Digital Games Research Association (DiGRA) Conference: Changing Views—Worlds in Play.* Vancouver, BC: DiGRA. Retrieved from http://www.digra.org/wp-content/uploads/digital-library/06278.53186.pdf.

Geurts, J. L. A., Duke, R., & Vermeulen, P. A. M. (2007). Policy gaming for strategy and change. *558*. Retrieved from https://repository.ubn.ru.nl/handle/2066/157314.

Greenblat, C. S., & Duke, R. D. (1975). *Gaming simulation: Rationale, design and applications*. New York: Wiley Inc.

Halleck, G. (2007). Symposium article: Data generation through role-play—Assessing oral proficiency. *Simulation & Gaming, 38*(1), 91–106. https://doi.org/10.1177/1046878106298268.

IAEA. (1996). *Manual for the classification and prioritization of risks due to major accidents in process and related industries*. Vienna: International Atomic Energy Agency. Retrieved

from http://www-pub.iaea.org/books/IAEABooks/5391/Manual-for-the-Classification-and-Prioritization-of-Risks-due-to-Major-Accidents-in-Process-and-Related-Industries.
Jacobs, C. D., & Statler, M. (2006). Toward a technology of foolishness: Developing scenarios through serious play. *International Studies of Management & Organization, 36*(3), 77–92. https://doi.org/10.2753/IMO0020-8825360304.
Joldersma, C., & Geurts, J. L. A. (1998). Simulation/gaming for policy development and organizational change. *Simulation & Gaming, 29*(4), 391–399. https://doi.org/10.1177/104687819802900402.
Kuit, M., Mayer, I. S., & de Jong, M. (2005). The INFRASTRATEGO game: An evaluation of strategic behavior and regulatory regimes in a liberalizing electricity market. *Simulation & Gaming, 36*(1), 58–74. https://doi.org/10.1177/1046878104272666.
Luna-Reyes, L. F., Zhang, J., Ramón Gil-García, J., & Cresswell, A. M. (2005). Information systems development as emergent socio-technical change: A practice approach. *European Journal of Information Systems, 14*(1), 93–105. https://doi.org/10.1057/palgrave.ejis.3000524.
Mayer, I. S. (2009). The gaming of policy and the politics of gaming: A review. *Simulation & Gaming, 40*(6), 825–862. https://doi.org/10.1177/1046878109346456.
National Research Council. (2001). *Maintaining U.S. leadership in aeronautics: Scenario-based strategic planning for NASA's aeronautics enterprise*. https://doi.org/10.17226/5546.
Rosendale, D. (2016). Role-play as a data-generation method. *Simulation & Games*. https://doi.org/10.1177/104687818902000410.
Vamanu, B. I., Gheorghe, A. V., & Katina, P. F. (2016). *Critical infrastructures: Risk and vulnerability assessment in transportation of dangerous goods—Transportation by road and rail* (Vol. 31). Cham, Switzerland: Springer International Publishing. Retrieved from http://link.springer.com/10.1007/978-3-319-30931-6.
Wenzler, I. (2005). Simulations and social responsibility: Why should we bother? In R. Shiratori, K. Arai, & F. Kato (Eds.), *Gaming, simulations, and society: Research scope and perspective* (pp. 139–148). Tokyo: Springer Tokyo. https://doi.org/10.1007/4-431-26797-2_15.
Wenzler, I. (2008). *The role of simulation games in transformational change*. WVB, Berlin: Planspiele Fü r Die Organisationsentwicklung.
Wilson, K. A., Bedwell, W. L., Lazzara, E. H., Salas, E., Burke, C. S., Estock, J. L., et al. (2009). Relationships between game attributes and learning outcomes: Review and research proposals. *Simulation & Gaming*, *40*(2), 217–266. https://doi.org/10.1177/1046878108321866.

Chapter 9
Space Capabilities—A Critical Tool for Nuclear Issues

Abstract In this chapter, the increasing role of space systems in the functioning of the critical infrastructures which are part of the Nuclear Industry is discussed. This discussion includes the role of feasibility and environmental studies required for the approval of new reactors, as well as the security of the nuclear industry's various processes, permanent security for the waste disposal sites, non-proliferation efforts and issues of support of decision makers and investigators.

9.1 The Nuclear Issue

Since civilian and military nuclear applications predate the important technological developments which have permitted an increasingly critical role for space systems, the instances in which space applications are truly critical for the functioning of the Nuclear Industry are limited. They are used in a complementary fashion, alongside more traditional instruments. Nuclear energy may be produced and transmitted without the intermediation of space systems. However, these have emerged as potent instruments in the increasing security apparatus surrounding all facets of the Nuclear Industry, not just the production of energy. Emerging (and perhaps no so emerging issues) include:

- The mining, refining and transport of nuclear material
- The disposal of nuclear waste, which is a permanent responsibility
- The proliferation of nuclear technology and its development along military lines
- The handling of public perception of security governance in the field and the actual performance of security governance efforts, especially from human and environmental perspectives.

The critical dependence of the Nuclear Industry on space systems is mostly limited to its important role in effective crisis and emergency situation management, as well as the secondary and tertiary influences registered through the Nuclear Industry's dependence on other CI systems, which may be more dependent on CSI. For example, modern electricity grids are very complex, with a decentralized network of variable producers which must be balanced against each other to avoid overloads

© Springer Nature Switzerland AG 2019

A. Georgescu et al., *Critical Space Infrastructures*, Topics in Safety, Risk, Reliability and Quality 36, https://doi.org/10.1007/978-3-030-12604-9_9

or the interruption of supply to consumers. This is done through the synchronization services provided by atomic clocks aboard GNSS satellites. In general, the most important space services for the Nuclear Industry will be synchronization, communications and Earth Observation, with differences in criticality based on subsector, circumstances and the presence of substitutes for these services.

Ultimately, discussions of the criticality of one space system or another to a critical infrastructure are about the CIs exposure to risks affecting the CSI in question. CSIs function in the harshest environment known to man, with significant natural and man-made risks, as well as vulnerability to low cost—high impact forms of warfare (e.g., jamming and cyber warfare) which are available to poor rogue states or to non-state actors and individuals. They are also very expensive to deploy and to replace, and difficult to substitute in their functioning until a replacement is ready. Often, the criticality of a space system to a CI, including to the Nuclear Industry, will only become apparent of the CSI in question were to malfunction at precisely the worst moment for it to do so, like the Japanese ALOS Earth Observation satellite which malfunctioned just as the Fukushima disaster demanded all resources in intelligence gathering. We suggest four main fields of involvement of CSI in the functioning of Nuclear CI: feasibility and environmental studies, security for the waste disposal, non-proliferation efforts for compliance, and support of decision makers and investigators.

9.2 Feasibility and Environmental Studies

The feasibility and environmental studies are required for the approval of new reactors, as well as the security of the nuclear industry's various processes. As the demand for security governance in the field increases, so does the requirement that maximum diligence be applied in the selection of study of new sites to host facilities related to Nuclear Industry, meaning not just reactors, but also research centers, material processing and enrichment facilities, as well as transport operations. The decision regarding were to place nuclear reactors is especially important, since their long horizon for exploitation demands that the environment itself minimize the potential for catastrophic risks and disruptions, with the least potential impact on the environment and on human settlements. (Pandey et al. 2010; Komljenovic 2018). An area afflicted by a nuclear meltdown might become unfit for human habitation, and weather patterns may carry radioactive fallout all over the world, including very densely populated areas. Generally, Earth Observation satellites are used to map the following:

- The geological characteristics of the surveyed area—the threat of earthquakes, rock slides, volcanism and how the area may reach to the weight of the facility itself
- The geography of the area and the characteristics of the air flows in the lower and upper atmosphere—how the relief of the surrounding area and air currents will enable or hinder the dispersion of radioactive particles

- The characteristics of the local water supply—seasonal variations in debit, the water composition, the other consumers and users of water. The goal is not only to map risks to human communities and economic actors which are dependent on the same water supply, but also to gauge the threats to the new nuclear facility, which needs a certain quantity and quality of water to be cooled. For instance, there are 13 nuclear facilities located along the Danube River, as well as many other nearby industrial facilities. The industrial accident which released millions of liters of dense and viscous liquid byproducts from an artificial reservoir in the proximity of Lake Balaton in Hungary could have led to the contamination of the Danube. If this had happened, then the cooling of the Romanian and Bulgarian nuclear reactors downstream of Hungary might have been compromised by the physical characteristics of the water, potentially leading to accidents
- The mapping of interconnections with critical infrastructures and human settlements—proximity to military bases that provide security, adequate transportation for sensitive materials, nearby cities to house the employees of the facility, but also potential for destruction in case of accidents
- The establishment of detailed and relevant data series, as long as possible, pertaining to security characteristics for the area. The tsunami that went over the tsunami wall of the Fukushima facility was a once in a lifetime occurrence, whose severity might have been predicted and planned for had adequate data been available.

9.3 Permanent Security for the Waste Disposal Sites

While this type of activity and facility within the Nuclear Industry could have been folded in the category above, its importance is such that it warrants separation. Until useful applications are developed for nuclear waste, such as reactors that can run on them, these materials will have to be permanently safeguarded against natural and human factors. The tens of thousands of years required for their dangerous levels of radiation to subside demand, for the foreseeable future, a permanent security apparatus to guard against accidental dispersions of materials, as well as theft by criminals and terrorists. Space systems, especially Earth Observation satellites, but also communications, allow for a better management of disposal sites at national levels, with coordination of rapid response capabilities to any threats.

Of course, the selection of the waste disposal sites is subject to even more stringent criteria than that of nuclear reactors, with a preference for geologically stable and easily observable areas, far away from human habitation and difficult to reach. Transports of nuclear waste, which often involve multiple connections and international transits, also benefit from surveillance and coordination capabilities. Meanwhile, operational waste storage sites must be subject to permanent surveillance for the foreseeable future, even after they are closed, especially as nuclear proliferation advances. Factors to be taken into consideration when deciding the best instruments and exploring the local challenges are seasonal variations in temperatures, weather patterns, vegetation, the presence of snow etc. A useful example in this regard is the Olkiluoto Waste Disposal Facility in Finland, which is paired with the Olkiluoto Nuclear Power Plant. The facility's security is not just a subject for the International Atomic Energy Agency (IAEA), but also for the site operators and the Finnish authorities. They utilize space capabilities that complement the climate and the geographical characteristics of the area, to provide permanent situational awareness, regardless of threat. For instance, not just facility is under surveillance, but also its approaches, since criminal elements may decide to tunnel into the facility and extract the nuclear materials present there.

The Olkiluoto area is frequently cloudy and suffers from a long winter season, with significant temperature, weather and humidity variations between seasons. The existence of snow for large stretches of time must be taken into consideration, as well as its spectral property shift during the melting season. The area also has significant natural foliage covering it and obscuring ground-based activities. For these reasons, imaging based on SAR (Synthetic Aperture Radar) is preferred for the area.

9.4 Non-proliferation Efforts and Nuclear Compliance

Efforts are also necessary to address non-proliferation and verifying nuclear compliance of problem states or other entities. For example, the Comprehensive Nuclear-Test-Ban Treaty (CTBT) was established to forbid the detonation of nuclear weapons in Earth's atmosphere, underground or underwater. It is of the utmost importance for the proliferation of nuclear technology, which is weaponizable, to be kept under control. Countries which have civilian nuclear sectors but are suspect of trying to become nuclear powers for geopolitical gain in their respective neighborhoods must be permanently assessed for covert efforts at nuclear militarization. The threat is not only for a state to obtain nuclear weapons and possibly a delivery system, although ad hoc methods, such as covert transport, are liable to be better than ICBMs. There is also the possibility that non-state actors will obtain nuclear weapons, even in small quantities, which they will be more willing to use. According to numerous studies, including ones by the RAND Corporation, the features of non-state actors make them mostly immune to traditional deterrence logic, based on the threat of mutually assured destruction. Non-state actors have no territory or population to worry about and may even be inimical to the country on whose territory they are based. Also, their political goals are probably limited and well defined, which makes the persuasiveness of a single nuclear weapon very attractive in their probable logic.

Figure 9.1 illustrates the weapons manufacturing processes, some of which can be assessed through Earth Observation systems in orbit (Jasani 2002). From the perspective of the CTBT enforcers, the main usefulness of satellite systems is in detecting preparations for a nuclear test or the after effects of a nuclear test. The sooner this is done, the more likely it is that the illegal activity will be curtailed. As seen in Jasani's (2002) research, space systems can play a major role in the detection of certain stages of the process and therefore alerting the international community.

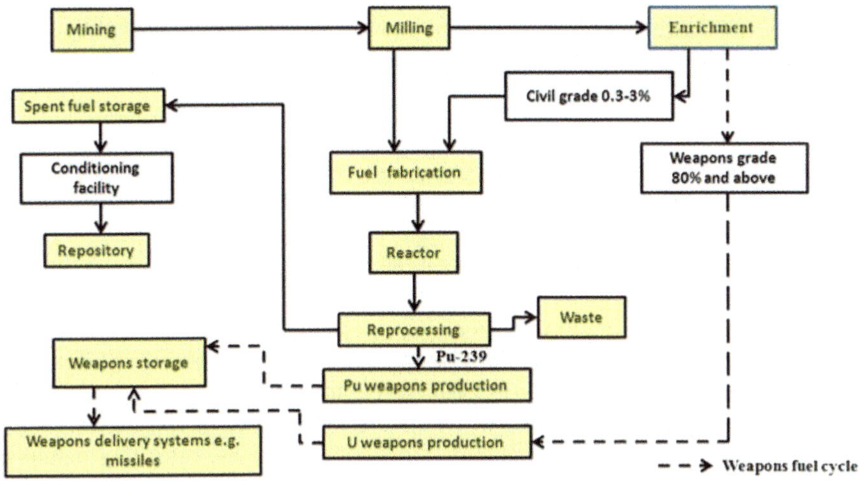

Fig. 9.1 Jasani's (2002) processes required for obtaining nuclear weapons

For these reasons, the most important space systems for non-proliferation efforts are specialized Earth Observation satellites. However, the criticality threshold has not been reached, because the CTBT has an entire system at its disposal, to whom space systems are a complementary addition, not yet critical but growing in importance. The main elements of the system are:

- The International Monitoring System—which relies on seismic sensors, hydroacoustic sensors, infrasound sensors for atmospheric and radionuclei detection stations, with satellite systems being a more recent addition. In all, the system has 321 monitoring stations and 16 research laboratories. The use of space systems does have a long history, despite their lack of centrality in the system. In the 1970, the American Corona satellites were used to assess the Russian and Chinese nuclear programs. Both the Russians and the Americans detected nuclear preparations by South Africa in the Kalahari Desert in the 1980s, and eventually persuaded the country to dismantle its military nuclear program. The same systems detected Indian tests in Rajasthan in 1995, and the examples can continue
- The International Data Centre
- The Global Communications Infrastructure—which mediates communications between the monitoring stations, the Data Centre, the national governments and so on, relying on six dedicated communication satellites and three ground centers
- Consultation and clarification processes
- On-site inspections—which are justified beforehand by evidence, including those gathered by satellite, which also define the planning for the focus of the inspectors' activity
- Trust building measures—which also rely on satellites for non-intrusive general verification of adherence to agreements.

The use of satellite capabilities has grown for a number of reasons, not including the general factors such as better costs and instrument sensitivity. From the perspective of compliance verifiers, satellites have the following advantages:

- They are passive detectors and non-intrusive, thereby not infringing on the rights of those surveilled
- They allow for permanent monitoring, especially visually, and can justify further and more intrusive methods of verifying compliance
- They provide early warning capabilities for a nuclear test
- Unlike other means of detection, satellites can pinpoint a test area precisely, giving a valuable starting point to inspectors
- Satellites also provide publicly useful evidence of illicit nuclear activity. The international public is more likely to be receptive to the message if presented with incontrovertible visual evidence of a build-up for a test or the post-testing landscape. This international community can then pressure the offending nation to reduce or renounce its nuclear ambition
- Satellites can also offer surveillance under sensitive conditions. Theoretically, it is not illegal to prepare for a nuclear test, just to perform one, and countries may try to mislead the international community with regards to the state of their nuclear program. Satellites allow for the detection of a wealth of details, some of which may be outside the acceptable limits for the country's nuclear program.

9.5 Support of Decision Makers and Investigators

There needs to be support for those involved in this type of research, ranging from decision makers to investigators. The use of space systems for the management of crisis and emergency situations is already well established, since they provide command, control and coordination capabilities to responders on the ground, as well as information to specialists assessing the severity of the event and its evolution. It was only natural that space capabilities would be compatible with the needs of nuclear accident responders and investigators, which was highlighted during the Fukushima disaster, when the Sentinel Asia initiative to which Japan belongs provided immediate surveillance of the affected area, as the crisis unfolded. During this time, the main Japanese Earth Observation satellite involved, ALOS was lost. Even though ALOS 2 and 3 have been launched, the Japanese have decided that redundancy and robustness in their Earth Observation capabilities is required, which is why a constellation of microsatellites developed by Japanese universities has been launched, with redundant instrumentation and better overall resilience. Table 9.1 presents each of the satellite along with their main functions (Haeme 2003). Haeme (2003) also provides capability associated with some of these satellites are presented in images presented in Fig. 9.2.

Other involved satellites were SPOT-5 and TerraSAR-X, the latter belonging to Airbus Defense and Space GEO-Information Services. It is interesting to note that

Table 9.1 Microsatellites for Earth Observation in nuclear issues launched by the Universities of Japan, adapted from Haeme (2003)

	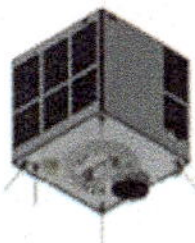		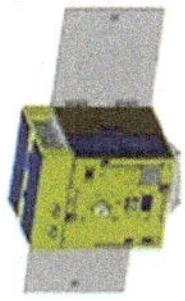	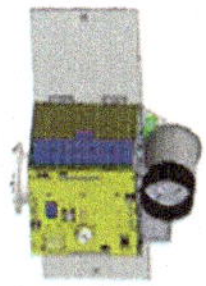
Satellite name	Hodoyoshi-1	Hodoyoshi-2	Hodoyoshi-3	Hodoyoshi-4
Operator	Univ. of Tokyo	Univ. Tohoku Univ. of Tokyo	Univ. of Tokyo	
Orbit	SSO 552 km	SSO	SSO 630 km	SSO 630 km
		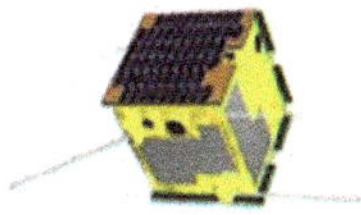		
Satellite name	Uniform-1	Chubusat-1	TSUBAME	QSAT-EOS
Operator	Univ. Wakayama Univ. of Tokyo	Univ. Nagoya Univ. Daido	Technologies Institute of Tokyo	Univ. Kyusyu
Orbit	SSO 630 km	SSO 529 km	SSO 541 km	SSO 536 km

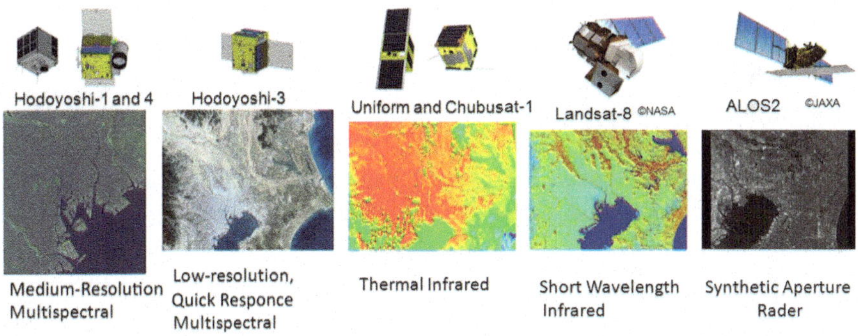

Fig. 9.2 Japanese nanosatellite capabilities

these satellites were not only used during the Fukushima disaster and in the surveillance afterwards, but also for researching the conditions at the Chernobyl NPP in Ukraine, due to increased Japanese interest in the site of the most famous meltdown and consequent total evacuation.

In addition to Earth Observation satellites, communications satellites were also heavily used to connect the authorities with command posts in cities, reestablish contact with the population and coordinate anti-meltdown and evacuation efforts. The Kiku-8 and Kizuna satellites each served to connect one township with the wider world.

GNSS capabilities are also used to coordinate search and rescue operations and develop Geographic Information Systems applications for measuring environmental impact and other such goals (e.g., see Doyle and Adams 2015). In general, space systems are going to be primarily utilized in the following circumstances, which constitute basic applications:

- Measuring the stage of a reactor meltdown
- Determining the spread of radioactive particles and measuring radiations in different points within the same geographic area
- Thermal and infrared surveillance of the area in the long-term
- Ensuring the integrity and orderliness of evacuations zones
- Estimating the spread of fallout based on air currents
- Measuring water characteristics including such issues as marine life and plant life in the water and on the ground
- Support for investigations into the causes of the accident
- Assessing collateral damages in neighboring areas of the natural disaster that triggered the nuclear crisis
- Cheap and accessible long-term surveillance of the affected areas.

In addition to regional organizations such as Sentinel Asia, SERVIR and the 2012 European Program "Copernicus Emergency Management Services", all of whom are meant to provide affected governments with relevant satellite data for handling a natural disaster, there is also an "International Charter for Space and Major Disasters", formed in 2000 and numbering 14 members today, among which we find the European Space Agency, Centre National d'études Spatiales[1] and the Canadian Space Agency. The Charter provides the UN and affected states with satellite data at no cost, for the purpose of managing disaster situations.

The presentation of the applications of space capabilities in the Nuclear Sector illustrates uncertainty regarding the critical status of space systems. It is obvious that the criticality fluctuates based on circumstances, becoming especially high in crisis situations. In all other cases, there are already models and instruments for performing some of the functions space systems offer without a dependence on them, even though CSI offer convenience and regularity in surveillance. As long as their functions are easily substituted, without too much of a loss of efficiency, then the criticality of

[1]The French Space Agency: https://cnes.fr/en.

the CSI remains low in the case of the Nuclear Sector, with the aforementioned exceptions.

However, factors underlying the rapidly increasing dependence of terrestrial critical infrastructure systems on critical space infrastructures (CSI) also work in the case of the Nuclear Industry. The Nuclear Industry also has its own series of factors which will serve to increase dependence on CSI:

- Worsening public perception of the nuclear industry, which will demand increasingly sophisticated and responsive security systems that are integrated with CSI, to govern normal activity, but especially the crisis situations. Paradoxically, as nuclear energy has become ever safer statistically, the mass media has put more emphasis on the few incidents which do exist, even when they are irrelevant to other nuclear actors. The 'Fukushima effects' has been to worsen public and political perception of the nuclear industry, even in countries which are safe from the cumulative natural disasters which were themselves a statistical rarity in Japan
- The rapid development, through direct demand, but also independently, of new services with space components which may serve the needs of the nuclear industry. New instruments are devised and launched, their accuracy is increased, their output finds new applications or new interpretations etc. Again, the Fukushima disaster has increased the stake for regulatory authorities and for disaster management groups, which has triggered a new emphasis on CSI
- The West has slowed its adoption of nuclear energy in favor of renewable energy and, in the short term, more fossil fuels, even as these options sometimes prove more expensive and less environmentally sound, but also less prone to triggering public hysteria through the perception of an immediate existential threat. However, emerging countries such as China, Saudi Arabia, India, Russia, along with advanced nations such as South Korea, have increased their emphasis on nuclear energy, preferring their low operating costs, significant reliability and high productivity per installed capacity. Of 60 new reactors planned or under construction, 23 are in China, which is also positioning itself as a civilian nuclear technology exporter
- The threat of nuclear proliferation, not just among "rogue states", but also the potentially more dangerous non-state actors like terrorist organization, requires the development not just of a more functioning anti-proliferation framework, but also for instruments to detect nuclear materials and nuclear tests, to verify the safety of existing stockpiles of nuclear materials and the compliance of various states with norms and directives regarding the use of nuclear technology and the limiting of its spread. Space systems offer such instruments, which are already used in conjunction with other systems and offer concrete advantages
- The growing dependence of other terrestrial CI on CSI will also translate into greater indirect effect on the Nuclear Industry, even if its own consumption of space services were to remain the same, which is unlikely. In particular, the Nuclear Industry registers significant interdependencies with the energy transmission sector, with transport infrastructures and with the financial and banking systems,

all of which are classified as critical infrastructures in every framework devised, including the American and European ones.

Over time, we will see an increase in dependence on space systems, driven, in part, by the risk aversion of the public and the demands for efficiency in the handling of crisis situations, with minimal loss of life, mitigation of damage and rapid resumption of normal activity. In short, the well-being of modern society demands resilient systems. Space systems are a key factor in ensuring resilience of society. Establishing such resilience might was well depend on techniques, tools, methods, models, and frameworks governing such systems.

Final Remarks

The critical dependence of the Nuclear Industry on space systems is mostly limited to its important role in effective crisis and emergency situation management. However, there is potential for space systems to be used as apparatus to affect energy (including nuclear) and the surrounding systems. Emerging issues involve aspects of (1) mining, refining and transport of nuclear material, (2) disposal of nuclear waste, (3) proliferation of nuclear technology and its development along military lines, and (4) handling of public perception of security governance in the field (including all aspects associated with humanity and environmental protection). As leading nations use space as a new playground for achieving parity in battle network-guided weapons warfare and developing of counter-network capabilities, there remains optimism that there will rules for this playground will be set before, if not already done, this space 'ground' is use for nuclear purposes that can have lasting impact on systems in earth.

References

Doyle, B. C., & Adams, M. R. (2015). Statistical evaluation of shoreline change: A case study from Seabrook Island, South Carolina. *Environmental and Engineering Geoscience, 21*(3), 165–180. https://doi.org/10.2113/gseegeosci.21.3.165.

Haeme, T. (2003). *Use of satellite and airborne remote sensing in the safeguards of a nuclear waste repository site* (No. STUK-YTO-TR–199). Espoo: Finland: VTT Information Technology. Retrieved from http://inis.iaea.org/Search/search.aspx?orig_q=RN:34068266.

Jasani, B. (2002). *Chapter 12. Role of satellites in strengthening nuclear security*. King's College London, London.

Komljenovic, D. (2018). Prise de décision en gestion des actifs industriels en tenant compte des risques (PhD). Université du Québec à Trois-Rivières, [Trois-Rivières]. Retrieved from http://depot-e.uqtr.ca/8458/

Pandey, M. D., Lu, D., & Komljenovic, D. (2010). The impact of probabilistic modeling in life-cycle management of nuclear piping systems. *Journal of Engineering for Gas Turbines and Power, 133*(1), 012901-012901–012907. https://doi.org/10.1115/1.4000897.

Chapter 10
Governance by Emerging Technologies—The Case for Sand and Blockchain Technology

Abstract Emerging technologies can also be used for governance. This is the case for sand. Sand is a key ingredient for many industries, including concrete, glass, and electronics. In this chapter, sand governance framework is suggested through the applying blockchain technology with the aim of regulating sand extraction and trade. In this case, blockchain is the technology that can be used for distributed concurrency monitoring. Agent-Based Modeling and Simulation (ABMS) is applied to demonstrate the application of the model.

10.1 The Need for Sand Governance

While space is quietly becoming the next arena of conflict for the world's major powers, it can also be used sustainability driven purposes. Strangely as it might sound, space can also be useful in sand regulating key resources on earth—specifically, sand extraction and trading. First, it is important to note that there is a growing demand for sand and gravel, particularly in the developing countries where rapid economic development influenced growth in the construction industry (de Leeuw et al. 2010; Pour et al. 2018). With the increase in construction activities, the demand for river sand has increased exponentially, causing the depletion and exploitation of natural sand resources thus resulting in adverse effects on the environment, such as sliding of river shores and lowering water table (Agrawal et al. 2017). 1.6 billion tons of cement production annually accounts for consuming 10–11 billion tons of sand, gravel and crushed rock per year (Mehta 2001). Subsequently, environmental impacts concerns have been raised in a number of nations, including China, Ghana, and India.

Sand and gravel are now the most-extracted materials in the world, even exceeding fossil fuels and biomass in terms of weight (Torres et al. 2017). Sand is a key ingredient for concrete, roads, glass, and electronics. Massive amounts of sand are mined for land reclamation projects, shale gas extraction, and beach re-nourishment programs (Torres et al. 2017). There is a growing demand for sand and gravel in the rapid economic development (Leeuw et al. 2010). While scientists are making a great effort to quantify how infrastructure systems such as roads and buildings affect the habitats

© Springer Nature Switzerland AG 2019

A. Georgescu et al., *Critical Space Infrastructures*,
Topics in Safety, Risk, Reliability and Quality 36,
https://doi.org/10.1007/978-3-030-12604-9_10

that surround them, the impacts of extracting construction minerals such as sand to build those structures have been overlooked. Most studies focused on extraction and impacts on specific sites, not in a wider context. As a major ingredient for a variety of products and an integrated part of the supply chain of numerous industries, sand has been decomposed into many sectors based on the intended applications (Mansour 2015). The apparent consumption of industrial sand and gravel of the United States, the world's leading producer and consumer of industrial sand and gravel, was 89.4 million tons in 2016 (Ober 2017). It is difficult to collect definitive data on silica sand and gravel production in most nations because of the wide range of terminology and specifications found among different countries. The United States remained a major exporter of silica sand and gravel, shipping it to almost every region of the world (Ober 2017). It is expected to experience an increased demand for products that sand forms a major input, as a major component in the supply chain for a variety of industries (Mansour 2015). The supply chain is a complex adaptive system and features a dynamics, uncertainty and partial information sharing (Long and Zhang 2014). To overcome the shortcomings of the analytical methods, simulation has been widely used in supply chain evaluation as a decision-making tool (Long and Zhang 2014). Since supply chains are inclined to be decentralized system, blockchain technology appears to be practical to support the supply chain system.

Sand governance is a very wide concept; the focus of this study is only sand supply management. This study aims to provide a framework, applying the blockchain technology to improve the sand supply and reduce the illegal mining, therefore having less impact on the environment.

10.2 Sand as a Key Resource

There is a growing demand for sand and gravel in the rapid economic development (Leeuw et al. 2010). Based on the study by Thornton et al. (2006) sand mining was one of the main causes of erosion rates in bays' shoreline. Jonah and colleagues (2015) assessed the ecological impacts of beach sand mining on three beaches of Ghana over a four-month period. Lai et al. (2014) demonstrated extensive sand mining, which resulted in the wider and deeper outflow channel of a lake in China, caused the decline in the lake levels. Mascarenhas and Jayakumar (2008) suggest that dunes play a defensive role in the wake of tsunami along the coast and can therefore be used as a rescue mechanism to save livelihoods. In the other hand, de Leeuw et al. (2010) suggests mining lakes for sand increases the risk of flooding as well as decreases flood storage capacity. We draw the following conclusions

- There is a lack of adequate information regarding the negative effects and costs of sand mining
- There is an absence of global monitoring of the sand extraction. This gap has contributed to the gap in awareness and knowledge

- It can also be argued impacts of sand mining can have effects on several systems including the ecosystem, wildlife, marine, landscape, and critical infrastructures.

On these accounts, there is need for approaches to address sand as a key resource and consequent governance of sand. A preliminary response might be found in emerging technologies (i.e., Blockchain).

10.2.1 Blockchain Based Approach

An agent is a computer system situated in a certain kind of environment and is capable of autonomous action to meet its designed objectives. Moreover, a multi-agent system is a loosely coupled network of software agents that interact to solve problems that are beyond the individual capacities or knowledge of each problem solver (Kavak et al. 2018). Agent Based approach is general and powerful because it enables to capture complex structures and dynamics. The other important advantage is that it provides for construction of models in the absence of the knowledge about the global interdependencies (Lynch et al. 2017). The dynamics of the system arises from the interaction of agents whereby the behavior of the agent is determined by its cognitive structure (Schieritz and Grobler 2003). The ABM techniques can simulate the effect of the mining on the land change. ABM facilitates the incorporation of subsets of different ancillary data and discrete variables. In addition, the ubiquity of the agents' results in a larger vision of what the agents represent and offer better assessment based on empirical findings.

NetLogo is used as the ABM tool for this study. NetLogo is a multi-agent programmable modeling software program (Wilensky 1999) that can be used to study the interaction between multiple heterogeneous agents and the phenomena that emerge as a result of their interaction over time (Lu et al. 2016). NetLogo has been applied in several studies in supply chain management for modeling and simulation. Ponte et al. (2017) carried out an experiment via NetLogo to simulate the interaction between inventory models depending on the demand. Long and Zhang (2014) used NetLogo to model and simulate an integrated framework of supply chains.

This study proposes a new framework to monitor the sand resource supply by using the blockchain technology. The system is modeled employing Netlogo and the results of the simulations facilitate the decision-making on the mining operation and monitor the supply chain management.

10.2.2 Blockchain Technology

The basic idea of blockchain technology is allowing the actors in a system which are called nodes, to transact digital assets using a peer-to-peer network that stores these transactions in a distributed way across the network (Ølnes et al. 2017; Katina et al. 2019). The blockchain is a mere data structure with distributed multiversion concurrency control (Mattila 2016). Blockchain technology enables the development of a new sand governance system with participatory decision-making and decentralized relevant organizations that can operate over a network of computers without any human intervention (Wright and De Filippi 2015). Smart contracts can be applied to set up autonomous digital entities to manage the sand resource and help societies to become more sustainable (Chapron 2017). Certain basic technologies that are required for any governance system are first a way of both recordings a set of rules, second a way for people to interact with the rules, and third a way to enforce the rules. The blockchain is ideal for recording information that can be later verified as authoritative and offers a novel way of accomplishing the three basic functions of governance. Blockchain would enable end users to verify exactly how, where and by whom the product they intend to purchase has been assembled and made, thereby denying a market for illegal and counterfeit products (Apte and Petrovsky 2016). Recently, several studies have focused on the applicability of blockchain in different aspects of the supply chain management. Tian (2016) studies the utilization and development of blockchain technology in agri-food supply chain. The integration of supply chain through the blockchain technology to achieve disruptive transformation in digital supply chain management and networks is explained in the research by Korpela et al. (2017). Kshetri (2018) examines how blockchain is likely to affect key supply chains management objectives such as cost, quality, speed, dependability, risk reduction, sustainability, and flexibility by presenting early evidence linking the use of blockchain in supply chain activities to increase transparency and accountability. However, the application of blockchain technology on enhancing the decision-making process for the supply-demand planning is not considered, hence this study applied blockchain technology for this concern. If there is a critical situation of sand mining in an area, for instance, the information could be transmitted to the environment agency, so the process can be stopped for a specific duration. Experts of different fields can be used to bridge the gap between limited capacity and the aspiration for the governance. The new artificial intelligence technologies are promoted as a tool improving the evidence base and equity of governance in developing countries. Applying blockchain technology provides new opportunities for enhanced transparency and increased engagement in the environmental management. The impacts of extracting construction minerals such as sand to build those structures have been overlooked in most studies on the field. In addition, the focus of the studies is on extraction and impacts on specific sites, not in a wider context. Therefore, a comprehensive monitoring system is necessary to manage the supply planning, mining operation, and the trade of the valuable resource of sand. In the proposed agent-based system, the heterogeneous agents work independently or in a cooperative and interactive manner to make decisions and communicate in a decentralized environment through the blockchain network. The supply demand can be simulated to provide an estimation of the results of the real situation which can help the decision makers. The blockchain technology is used to track the trade through the complex supply chain from mining operation of the sand resource to consumers. Since the transactions would be through the blockchain platform, the cost per transaction decreases and the security is enhanced.

10.3 Methodology: Blockchain-Based Sand Governance

The methodology contains the first phase of analyzing all the information within a blockchain network. All relative parties can provide their analysis and suggestions for decision-making of the sand mining operation for the area under their supervision. The transactions of sand trades are applied through the blockchain platform, hence, would be verified by all the parties and the risk of smuggling and illegal sand mining as the critical problem mentioned can be decreased. The second phase is providing a blockchain governance framework for regulators to take the necessary actions based on the information provided in the previous phase. The proposed framework is depicted in Fig. 10.1.

The agents determined for this study are the sand resources as the supply agent, the government is the other agent which has the authority to prohibit the sand mining operation in the area. However, the government decisions can be subject to unforeseen factors. Thus, in the proposed framework the agent is modified by adding related experts of the field including the environmental expert who can study and analyze the risks of mining on the sand resource, monitor the impacts of mining on the land. The government agent is a network linked to each other to make the decision. The mining operation would be contingent on the government network decision.

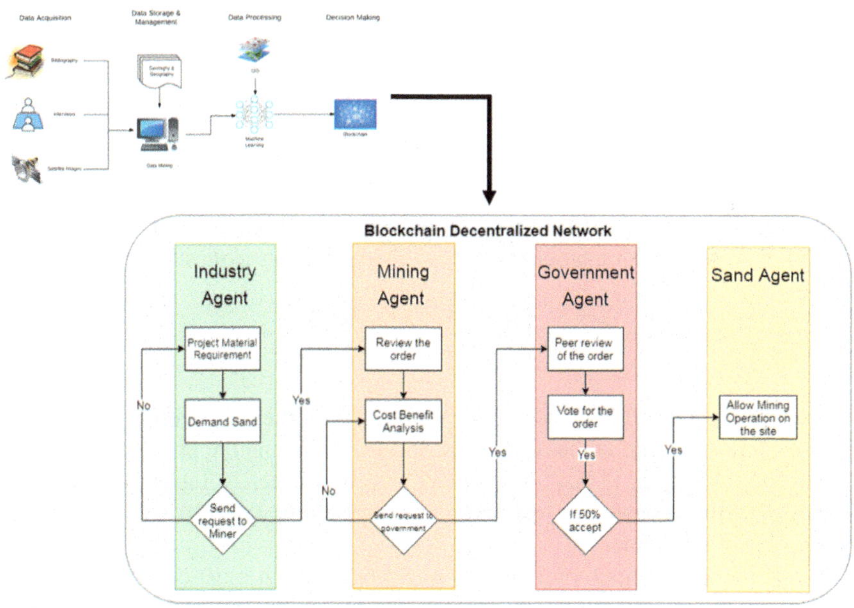

Fig. 10.1 The proposed blockchain-based sand governance model

Table 10.1 Two systems comparison parameters

System	Mining rate	Profit	Fine	Monitoring
Current supply	Based on demand	Miner profit = (Market price – mining cost)	0	Local regulations
Blockchain network	Based on the resource circumstances	Miner profit = Market price – [(mining cost + sand mining price)]	$\left(\delta * \left(\frac{price}{unit} * demand\right)\right)+$ fine deposit	Decision council

The industry agent is defined as the industry or business demand the sand supply. The miner agent tries to make interaction with the sand resource to get the permission for mining the resource, should provide a logical statement of the mining for the network approval. Therefore, the application will only be approved based on environmental circumstances, economic rationale, and the internal policies while all the involved parties within the government decision council accept the trade.

The properties that can help the decision makers to analyze the circumstances are described in Eqs. (10.1) and (10.2). Where (d) is the demand parameter requested by the industry agent, (t) is the duration require by the mining agency to the operation, (a) is the safe availability volume of sand in the resource to be mined on that certain time, and (r) is the condition based on the reason provided by the miner and would be 1 only if is for industrial purpose and zero for other purposes. The permission parameter is a function of the Eq. (10.1):

$$Permission = ((a - (d * t) * r)) \tag{10.1}$$

Equation (10.2) is the decision of the government. If the Eq. (10.1) is a positive number, then the decision would be approving the mining request.

$$Y = \begin{cases} 0, \; otherwise \\ 1, \; if \; permission > 0 \end{cases} \tag{10.2}$$

In addition, in this framework, a fine is allocated for the illegal mining operations. The fine deposit would be defined by the local council and the price of the volume mined would be added to that for the miner and can be seen in Table 10.1. The main properties of the two systems operation are compared in Table 10.1. δ is defined by the local decision council.

In this model, agents represent real world parties that must cooperate to reach desired objectives. Naturally, each agent might want to maximize their utility. However, this must be done without infringing on other agent's utility or goals. In the present model, agents also encompass both monitoring agents and the demand agents.

A key property of agents is decentralization; a key feature offered by the blockchain technology. Every link between demand agents and monitoring agent can be interpreted as a potential supplier-customer relationship. The relationship will be active during the simulation run if there is a trade agreement. Some of the agents' characteristics are predetermined, for instance, agents have given order policies. Other characteristics evolve over the course of the simulation, for example, the volume that is exchanged between two particular agents (Schieritz and Grobler 2003).

The environment defined for this model is a decentralized network which all agents communicate and cooperate through the blockchain platform. All demanding agents advertise their abilities, knowledge in the acquaintance database. The monitoring agents do the transaction by sending messages to the decentralized network. The monitoring agents are autonomous in making decisions by considering all the rules defined for the system through coordination of the decision team within the network.

Through applying blockchain technology, instead of having a central authority that maintains the database and guards its authenticity, a copy of the entire database is distributed to every involved participant to maintain independently. The copyholders follow the predetermined rules and compare their versions together through a continuous process of voting. The most voted version of the network is accepted.

10.4 Results Analysis

In the present study, three different scenarios are defined and simulated for both the current and the proposed systems as indicated in Table 10.2. In the 'current system' the sand is mined by the miners for the contracts with the industries and businesses demanding the sand supply. In the proposed system (i.e., Blockchain-based), the sand is monitored by the government and each demand request needs the decision council approval for the operation. The present model is for illustrative purposes and uses arbitrary numbers. However, the proportions for the values are grounded on the current market condition. Using the defined equations and suggested values (Table 10.2), the results of the comparisons on the simulated cases are shown in Table 10.3. These results indicate that the cost of illegal mining for the miners is higher. Higher than even the profit for 'current' system. Clearly, a monitoring system can reduce the illegal mining as well as providing benefits for the resource in order to implement sustainability programs. A comparative analysis impact of uncontrolled mining operation on sand is provided in Fig. 10.2.

The comparisons of the properties of both situations simulated illustrate the great benefits of having the blockchain technology for governing the sand resources which can have less impact on the environment (Fig. 10.2). Moreover, the sustainability of the sand supply chain can be improved. In the 'current' supply system, there is no monitoring system on the sand mining operation as well as the environmental impact on the surrounded area. The proposed framework contains a government agent that reviews and monitors the operation over the resource and prohibit the mining operation in risky conditions.

Table 10.2 Simulation parameter values

Parameters	Values
Market sand price/unit	\$10x
Sand price/unit for mining	\$5x
Fine deposit	$A = 200x$
δ for the area	$B = 0.5x$
Mining rate/hour	10 unit/day
Safe capacity for mining	$C = 300x$
Sand mining cost/unit	\$$x$

Table 10.3 Results from the simulation

Case	Demand	System	Permission	Profit	Fine	Monitoring
1 (Mining for the cement company demand)	100 units	Current	N/A	\$900x	\$0	No monitoring
		Blockchain supply chain	Yes	\$400x	\$0	Decision council approved the case
2 (Mining for exporting to island A)	200 units	Current	N/A	\$1800x	\$0	No monitoring
		Blockchain supply chain	No	\$−400x	\$1200x	Decision council rejects request with reason
3 (Mining for the glass company demand)	50 units	Current	N/A	\$400x	\$0	No monitoring
		Blockchain supply chain	No	\$−250x	\$450x	Decision council rejected the request due to the concerns about the sand availability and the environmental impacts

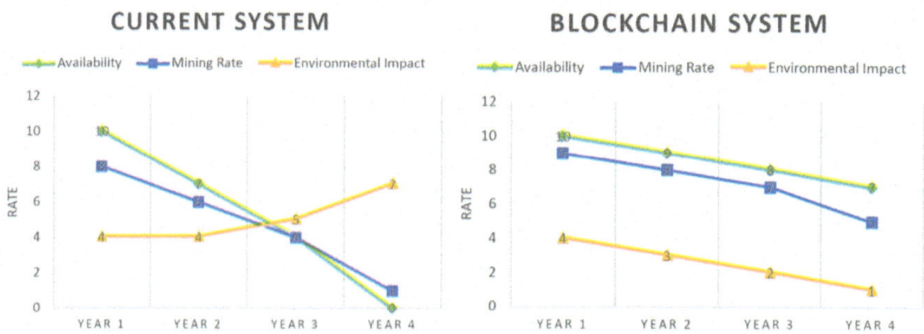

Fig. 10.2 A comparative impact of uncontrolled mining operation on sand

However, there are clear limitations in implementation of this methodology. First, blockchain technology is new and therefore not fully understood. Moreover, implementation of the proposed methodology, requires expertise in blockchain technology. Finally, there is need for regulations given the emerging issues associated with Blockchain and Bitcoin.

Final Remarks

It is time to treat sand like a key resource, on a par with clean air, biodiversity, and other natural endowments that nations seek to manage for the future (Gheorghe et al. 2018). Under this premise, there is need to develop applicable methods, tools, and techniques for addressing risks and vulnerabilities associated with sand. The simulated benefits, clearly, outweigh, continued 'current' approach. Moreover, the proposed framework enables the governance of sand through use of blockchain technology along with use of ABM tools. The purpose of this framework is to reduce (and more ambitiously, eliminate the illegal sand mining. This is can be done by monitoring the sand trades through the blockchain technology and geographic imagery provided by satellites and other technologies including unmanned aerial vehicle (e.g., Doyle and Adams 2015). Moreover, the impacts on the environment can be reduced by taking immediate actions for the critical regions which the applied methodology identifies. In addition, it can arm regulators with policy-making of natural resources treatments. Applying the decentralized blockchain system and using influential parties through the ABMS generates major effects from a discrete decision to the cooperative outcome.

References

Agrawal, U. S., Wanjari, S. P., & Naresh, D. N. (2017). Characteristic study of geopolymer fly ash sand as a replacement to natural river sand. *Construction and Building Materials, 150,* 681–688. https://doi.org/10.1016/j.conbuildmat.2017.06.029.

Apte, S., & Petrovsky, N. (2016). Will blockchain technology revolutionize excipient supply chain management? *Journal of Excipients and Food Chemicals, 7*(3), 910.

Chapron, G. (2017). The environment needs cryptogovernance. *Nature, 545*(7655), 403–405. https://doi.org/10.1038/545403a.

Doyle, B. C., & Adams, M. R. (2015). Statistical evaluation of shoreline change: A case study from Seabrook Island, South Carolina. *Environmental and Engineering Geoscience, 21*(3), 165–180. https://doi.org/10.2113/gseegeosci.21.3.165.

Gheorghe, A. V., Vamanu, D. V., Katina, P. F., & Pulfer, R. (2018). *Critical infrastructures, key resources, and key assets*. Cham, Switzerland: Springer International Publishing. https://doi.org/10.1007/978-3-319-69224-1_1.

Jonah, F. E., Agbo, N. W., Agbeti, W., Adjei-Boateng, D., & Shimba, M. J. (2015). The ecological effects of beach sand mining in Ghana using ghost crabs (Ocypode species) as biological indicators. *Ocean and Coastal Management, 112,* 18–24. https://doi.org/10.1016/j.ocecoaman.2015.05.001.

Katina, P. F., Keating, C. B., Sisti, J. A., and Gheorghe, A. V. (2019). Blockchain Governance, *International Journal of Critical Infrastructures, 15*(2), 121–135.

Kavak, H., Padilla, J. J., Lynch, C. J., & Diallo, S. Y. (2018). Big data, agents, and machine learning: Towards a data-driven agent-based modeling approach. In *Proceedings of the Annual Simulation Symposium* (pp. 12:1–12:12). San Diego, CA, USA: Society for Computer Simulation International. Retrieved from http://dl.acm.org/citation.cfm?id=3213032.3213044.

Korpela, K., Hallikas, J., & Dahlberg, T. (2017). Digital supply chain transformation toward blockchain integration. Presented at the Hawaii International Conference on System Sciences, Waikoloa, HI, HICSS. https://doi.org/10.24251/HICSS.2017.506.

Kshetri, N. (2018). 1 Blockchain's roles in meeting key supply chain management objectives. *International Journal of Information Management, 39,* 80–89. https://doi.org/10.1016/j.ijinfomgt.2017.12.005.

Lai, X., Shankman, D., Huber, C., Yesou, H., Huang, Q., & Jiang, J. (2014). Sand mining and increasing Poyang Lake's discharge ability: A reassessment of causes for lake decline in China. *Journal of Hydrology, 519,* 1698–1706. https://doi.org/10.1016/j.jhydrol.2014.09.058.

de Leeuw, J., Shankman, D., Wu, G., de Boer, W. F., Burnham, J., He, Q., et al. (2010). Strategic assessment of the magnitude and impacts of sand mining in Poyang Lake, China. *Regional Environmental Change, 10*(2), 95–102. https://doi.org/10.1007/s10113-009-0096-6.

Long, Q., & Zhang, W. (2014). An integrated framework for agent based inventory–production—transportation modeling and distributed simulation of supply chains. *Information Sciences, Complete, 277,* 567–581. https://doi.org/10.1016/j.ins.2014.02.147.

Lu, M., Cheung, C. M., Li, H., & Hsu, S.-C. (2016). Understanding the relationship between safety investment and safety performance of construction projects through agent-based modeling. *Accident Analysis and Prevention, 94,* 8–17. https://doi.org/10.1016/j.aap.2016.05.014.

Lynch, C. J., Kavak, H., Gore, R., & Vernon-Bido, D. (2017). Identifying unexpected behaviors of agent-based models through spatial plots and heat maps. In *21st Annual Meeting on Agent Based Modeling & Simulation*. Suffolk, Virginia: Swarmfest 2017.

Mansour, M. (2015). Develop a strategic forecast of silica sand based on supply chain decomposition. *International Journal of Engineering, 9*(1), 9–27.

Mascarenhas, A., & Jayakumar, S. (2008). An environmental perspective of the post-tsunami scenario along the coast of Tamil Nadu, India: Role of sand dunes and forests. *Journal of Environmental Management, 89*(1), 24–34. https://doi.org/10.1016/j.jenvman.2007.01.053.

Mattila, J. (2016). *The blockchain phenomenon: The disruptive potential of distributed consensus architectures* (ETLA Working Papers No. 38). Helsinki, Finland: The Research Institute of the Finnish Economy. Retrieved from https://ideas.repec.org/p/rif/wpaper/38.html.

Mehta, K. P. (2001). Reducing the environmental impact of concrete. *Concrete International, 23*(10), 61–66.

Ober, J. A. (2017). *Mineral commodity summaries 2017* (USGS Unnumbered Series) (p. 202). Reston, VA: U.S. Geological Survey. Retrieved from http://pubs.er.usgs.gov/publication/70180197.

Ølnes, S., Ubacht, J., & Janssen, M. (2017). Blockchain in government: Benefits and implications of distributed ledger technology for information sharing. *Government Information Quarterly, 34,* 355–364. https://doi.org/10.1016/j.giq.2017.09.007.

Ponte, B., Sierra, E., de la Fuente, D., & Lozano, J. (2017). Exploring the interaction of inventory policies across the supply chain. *Computers & Operations Research, 78*(C), 335–348. https://doi.org/10.1016/j.cor.2016.09.020.

Pour, F. S. A., Tatar, U., & Gheorghe, A. (2018). Agent-based model of sand supply governance employing blockchain technology. In *Proceedings of the Annual Simulation Symposium* (pp. 14:1–14:11). San Diego, CA, USA: Society for Computer Simulation International. Retrieved from http://dl.acm.org/citation.cfm?id=3213032.3213046.

Schieritz, N., & GroBler, A. (2003). Emergent structures in supply chains: A study integrating agent-based and system dynamics modeling. In *Proceedings of the 36th Annual Hawaii International Conference on System Sciences, 2003* (9 pp.). https://doi.org/10.1109/HICSS.2003.1174226.

Thornton, E. B., Sallenger, A., Sesto, J. C., Egley, L., McGee, T., & Parsons, R. (2006). Sand mining impacts on long-term dune erosion in southern Monterey Bay. *Marine Geology, 229*(1), 45–58. https://doi.org/10.1016/j.margeo.2006.02.005.

Tian, F. (2016). An agri-food supply chain traceability system for China based on RFID and blockchain technology. In *2016 13th International Conference on Service Systems and Service Management (ICSSSM)* (pp. 1–6). Kunming, China: IEEE. https://doi.org/10.1109/ICSSSM.2016.7538424.

Torres, A., Brandt, J., Lear, K., & Liu, J. (2017). A looming tragedy of the sand commons. *Science, 357*(6355), 970–971. https://doi.org/10.1126/science.aao0503.

Wilensky, U. (1999). *NetLogo*. Evanston, IL: Center for Connected Learning and Computer-Based Modeling, Northwestern University.

Wright, A., & De Filippi, P. (2015). *Decentralized blockchain technology and the rise of lex cryptographia* (SSRN Scholarly Paper No. ID 2580664). Rochester, NY: Social Science Research Network. Retrieved from https://papers.ssrn.com/abstract=2580664.

Chapter 11
Establishing Governance for CSI: An MCDA Approach

Abstract Indeed, the domain of space infrastructure systems has evolved from a benign environment, with a relatively small number of satellites scattered over vast distances, to an increasingly complex, crowded, and hostile 'system-of-systems'. In this chapter, three proposed governance scenarios for the space domain are evaluated efficacy using principles of Complex System Governance. First is a scenario derived from an official document of the United States' National Academies of the Sciences and Engineering. Second is based on an article from the European Institute for Security Studies. A third option is developed based on assumption of accelerating growth in space commercialization and evolution to an Internet-like environment. The three governance scenarios are evaluated using single utility analysis, Fuzzy Sets-based decision analysis, Evidential Reasoning, the Analytic Hierarchy Process, and Logical Decisions for Windows multi-utility software. The results are discussed along with strengths and weaknesses of the tools.

11.1 Basis for Space Governance

Space is quietly becoming the next arena of conflict for the world's major powers. It is not a new arena, having been first explored in the 1950s. It is however, growing in importance. The National Aeronautics and Space Agency (NASA) and European Space Agency (ESA) have both released strategy documents outlining their respective visions for space. Unsurprisingly, they conflict in interest, priorities, and governance. During the writing of this paper, CNN released a documentary "War in Space—The Next Battlefield", describing potential future conflicts.

Complex System Governance may help solve that problem, and Multi-Criteria Decision Analysis (MCDA) could render recommendations from the CSG models developed. The authors used the NASA and ESA reports to evaluate the United States' (US) and European Union's (EU) governance models for 'critical space infrastructure'. Further, they proposed a notional model constructed on the principles of the Viable System Model. Then, they used several MCDA techniques to evaluate the governance models against each other.

© Springer Nature Switzerland AG 2019

A. Georgescu et al., *Critical Space Infrastructures*, Topics in Safety, Risk, Reliability and Quality 36, https://doi.org/10.1007/978-3-030-12604-9_11

Present authors researched both 'space infrastructure' and 'Complex System Governance.' Overall, they found that fragile space-based systems support swathes of modern society, and space continues to become more crowded, both with orbiting assets and conflicting international intent and action. Complex System Governance is a material-agnostic method to construct a governance meta-system, tailored to the functions of the governed system(s). With tailoring, it could be fitted to govern critical space systems.

11.2 Description of the Space Domain

To the average consumer, the threat of a potential satellite attack is seldom considered when compared to other issues regarding our modern way of life. If a satellite were to inexplicably become inoperable, many might assume that a simple loss in navigational software or communications capabilities would be inconsequential and easily restored. However, humankind is far more attached to space-based systems than typical consumers realize.

11.2.1 History

The beginnings of space exploration and satellites dates back to 1957, when the Soviet Union launched humanity's first satellite, named Sputnik, which sparked a race between the world's two major superpowers for nuclear dominance and space superiority. At the time, and for many years to come, space was only seen as useful for national security and defense purposes. In 1967, the United States, Soviet Union, and several other nations signed the UN Outer Space Treaty, which banned the placement of nuclear weapons in outer space but did not provide guidance on other types of weaponry. This treaty that focuses primarily on arms reduction and nation-state international policy remains the only formal document that legally holds nations accountable for their activities in space (Pellegrino and Stang 2016). The remainder of the governance on the space domain is non-binding agreements and voluntary measures, which leaves space relatively unregulated compared to domains like air, land, or the environment.

Another modern concept deeply seated in Cold War tensions is the ability of a nation or other organization to disrupt, dismantle, or otherwise render inoperable the space systems of another nation. The Soviet Union experimented with anti-satellite (ASAT) warfare in the 1960s and eventually declared an operational system in 1973 (CNSSDP 2016). Since then, several nations have developed and tested their capabilities to render satellites inoperable either permanently or for a limited time period. One notable modern test of such a system came in 2007 when China destroyed one of its own satellites with a ground-based ASAT system (Pellegrino and Stang 2016). This capability of nations or other organizations to hinder the satellite operations of another is a crucial force not only in the context of the Cold War, but in our modern defense mindset as well.

After the tensions of the Cold War subsided, space became a new frontier ripe for commercialization and invention. In 1993, the US Department of Defense created the world's first global positioning service (GPS) and marked a major breakthrough in the exploitation and use of space satellites (Mai 2015). While several other nations and entities such as the EU purchase access to the US's GPS system, other states have subsequently developed their own network of positioning, navigating, and timing (PNT) satellites. Examples of such systems are the currently operating Russian GLONASS and the planned Chinese BeiDou and European Galileo systems. In the years following the development of these PNT systems, the link between space systems and the consumer's everyday life increased dramatically.

11.2.2 'Critical Space Infrastructure' (CSI)

Today, the world operates between 1,000 and 1,300 satellites in various orbits around the globe every day (Pellegrino and Stang 2016). Figure 11.1 shows the relative volume of satellites employed by the top ten space powers of the world.

The most notable aspect of this figure comes from the ratio of non-military satellites to all satellites deployed. Approximately 70% of satellites currently in orbit do not provide any military or defense capabilities; they provide commercial and governmental services without which several earth-based systems would collapse.

Recall that the actual term 'critical infrastructure' received increased attention starting in 1997 after the US Clinton administration. However, the concept of a service

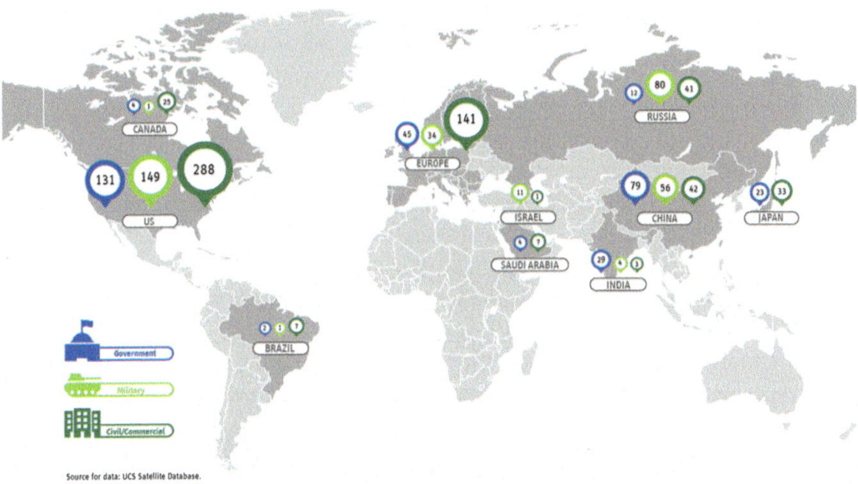

Fig. 11.1 Operational satellites for top 10 global space actors (Pellegrino and Stang 2016)

or key functionality of a nation has been previously defined by other terms such as 'emergency supply' or 'emergency functions.' Since 1997, the modern definition of a CI is an infrastructure (Hesse and Hornung 2015):

> *...that provides an essential support for economic and social well-being, for public safety, and for the functioning of key government responsibilities, including sectors such as information and communication technology, energy production, water and food supply, transport, and nuclear/chemical industry...*

These infrastructures are the key not only to the modern conveniences of everyday living for superpower nations, but also to the most basic government services such as public protection. The loss of any one of these systems would have a grave impact on the socio-economic wellbeing of any nation. Such a loss could be intentionally caused by acts of war or the result of a natural earth or space-based disaster.

Before the 1990s, our CIs were largely earth-bound and unrelated to the activities of satellites, save the national security aspect. Today, systems that utilize one or more components located in outer space (space systems) are intricately linked to nearly all of our CIs. Space systems provide crucial information and guidance through PNT satellites to entire sectors such as power grid management, navigation for air and sea traffic, communications, and even banking operations (Pellegrino and Stang 2016). A total loss of PNT services would cause extreme damage to a given nation's economy and would likely impact the entire world. These CIs are also only a portion of the services that space systems fuel; earth observation satellites are crucial for early warnings of natural disasters and scientific research. As consumer demand continues to drive the creation of new technology, the relationship between space systems and CI will only increase as more satellites are planned and deployed (CNSSDP 2016). This connection between the CIs of a nation and its reliance on space systems has led to the relatively new concept of 'critical space infrastructure' (CSI).

It is important to note that the domain of CSI extends not only to the individual satellites in orbit and the functions that they provide, but also to the infrastructure that launches, maintains, and controls these satellites. Any attack or natural disruption of command and control or maintenance facilities based on earth would have a severe impact on the functionality of the entire satellite constellation. Satellites also do not typically operate alone, but in a system usually referred to as constellation that requires the triangulation or comparison of signals to provide the intended service. In the case of a PNT constellation, the loss of a single satellite could render the entire system unusable if there are no built-in redundancies.

These individual infrastructure components are under the same constant threat of attack, similar to any other CI, with the addition of several new natural and intentional potential disruptions. Since several nations are co-dependent on one another for full access of space services such as the American GPS and space situational awareness tools (SSA), the consequences of such a failure would also likely cascade to affect the livelihoods of other nations and the global economy. Such disruptions should be avoided even at significant cost, but the wide array of threats to space systems will cause countermeasures to be costly.

11.2.3 The Vulnerabilities of CSI

Since the development of the concept of critical infrastructure (CI), nations have slowly come to recognize the criticality of space systems infrastructure (CSI). In 2011, the EU published key strategy documents stating that "space structure is critical infrastructure," and the US has similarly incorporated space infrastructure in their national security plans (Hesse and Hornung 2015). These nations have recognized that while our modern economies and livelihoods depend on space systems, the entire CSI network stands relatively unguarded against potential attacks and disasters. These threats can be grouped into two major categories: intentional and unintentional.

11.2.4 Intentional Attacks

Intentional attacks encompass the wide variety of physical and cyber-based threats that are both specific to space infrastructure and common to all CIs. The main threats to CSI are ASAT weaponry, sabotages to ground stations, signal jamming, and cyber-attacks.

ASAT weaponry further decomposes into kinetic, nuclear, and energy-based (laser) attacks directed either from the ground or from co-orbit with the target satellite. As discussed in the previous section, the capability of one nation to destroy the satellites of another has a history nearly as long as space exploration itself. In today's theatre, several nations such as the US, Russia, and China have all developed ASAT systems capable of destroying target satellites. Kinetic attacks were the first type of ASAT system initially developed, but today's arsenal now includes both nuclear attacks delivering a strong electromagnetic pulse and high-energy laser weapons to destroy targets (CNSSDP 2016). These systems are typically reserved for acts of war and as should be addressed as such but acts of war by other nation-states are not the only threat to a country's CSI.

Non-state actors such as such as terrorist and even organized crime groups can execute attacks on CSI. Moreover, satellites and space-based components do not need to be damaged for a nation's CSI to fail. Ground control stations can be degraded or otherwise sabotaged, and control signals can be jammed using relatively simple means. The vulnerability of CSI to attacks from stateless organizations is a major risk that cannot be overlooked in any plan for CSI protection.

The final threat to be discussed is that of a cyber-attack on the computer systems of either the ground-based control stations or the orbital platforms themselves. In 2008, a cyber-attacker gained full control of a NASA satellite by hacking of its ground station. While the attacker issued no commands to the satellite, he or she did complete all of the steps required to achieve complete control (Pellegrino and Stang 2016). The threat of such a system compromise is severe, but it is consistent with the grave nature of earth-based cyber threats. Both American and European national defense strategies have established cyber security as a priority, and the cyber-protection of national CSI can easily adapt to existing plans. Granted, the potential for a cyber-attack to affect any piece of CI is always present, but the strategies to protect our CSI need not vary significantly from earth-based protection.

11.2.5 Unintentional Hazards

While intentional attacks may seem to be of the most pressing in nature, the list of unintentional threats dwarfs those with malicious intent. Unintentional threats include space debris, solar flares, cosmic radiation, weather interference, human error, and loss of communications or control due to a natural earth-based disruption (Pellegrino and Stang 2016). Some of these threats have been known to exist for several years while others are new.

The threat of space debris is new to the list of unintentional space threats. Space debris is a general term that includes micro meteors, dust, fragments from spent rocket boosters, debris from damaged satellites, and any other form of man-made material. Due to the limited number of usable orbits and increasing satellite congestion, the likelihood of collisions between satellites and debris has increased significantly since the days of early space exploration. Over 30,000 objects greater than 10 cm in diameter are currently catalogued, and the over 1.5 million objects down to one millimeter in diameter are still extremely dangerous. An impact with any one of these objects could cause total satellite disruption and performing repairs in space are extremely expensive (Pellegrino and Stang 2016).

In 2009, the Iridium-33 satellite collided with the Russian Kosmos-2251 and scattered over 2,000 long-term fragments of debris. Collisions like these also fuel concerns about Kessler Syndrome, a condition in which space debris is such a widespread threat that no spacecraft or satellite will be able to enter or pass through orbit undamaged (Pellegrino and Stang 2016).

Space weather also poses significant threats to anything in orbit. These threats include solar flares, coronal mass ejections, and cosmic radiation. Not only can such spikes in energy damage sensitive electronic components in satellites, but the corresponding electromagnetic interference can disrupt communications and control signals. In 2003, the crew of the International Space Station was forced to take shelter in radiation-shielded areas of the station to protect themselves as solar damaged five satellites beyond recovery (Pellegrino and Stang 2016). Such powerful solar storms are unfortunately common in our solar system. The most powerful recorded solar storm occurred in 1859, long before humanity had launched any satellites into space. If a storm this massive were to appear again, it is estimated it would cause trillions of US dollars of economic damage in the first year alone with recovery lasting nearly a decade (Pellegrino and Stang 2016).

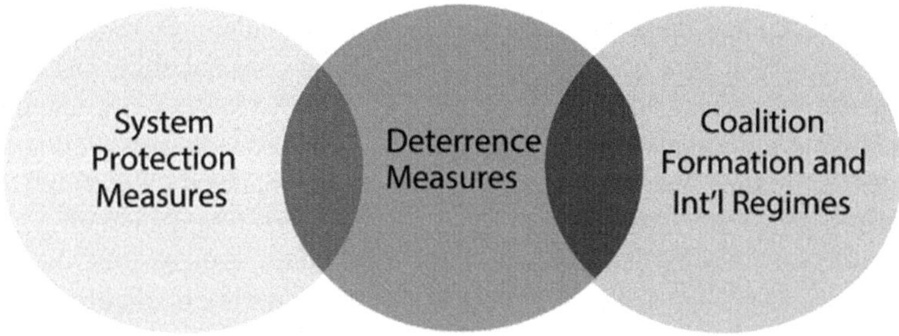

Fig. 11.2 Aspects of US Space Asset Defense

Any failure of the ground-based CSI system can cause widespread orbital drifting and cascading damage for future operations. Space weather can and does affect terrestrial weather, indirectly impacting communication channels from control stations to satellites. Unintentional interference from other man-made sources can also cause degradation and even total disruptions in vital control signals. The combination of these threats further increases the typical challenges faced by engineers and policy makers alike. The issue of CSI protection is beginning to receive some attention, but much work is still required to develop a system that is truly resilient in the face of disasters.

11.2.6 Current Practices and Protections

To begin the monumental challenge of protecting our nation's CSI, US and European agencies have prepared preliminary strategies. The US CSI protection plan involves three related lines of effort, as noted in Fig. 11.2 (CNSSDP 2016). These lines of effort integrate on multiple levels to provide the optimal resistance to space and ground-based threats to CSI.

11.2.7 System Protection Measures

The descriptor 'system protection measures' includes a broad range of engineering and policy decisions designed to maximize the resilience of space systems. Engineering decisions from design, configuration, component redundancy and even entire satellite redundancy can help mitigate risks associated with space weather or damage. To prevent damage from collisions with space debris or other satellites, astronautical engineers employ a concept known as space situational awareness (SSA). Space situational awareness is defined as "the capability to assess activities in space and, in particular, monitor hazards to space infrastructure" (Pellegrino and Stang 2016).

Currently, the United States owns and operates the world's most widely used SSA and tracking system. However, nearly all global space systems utilize this American tracking and collision avoidance system. In 2014, the US SSA system generated over 8,000 collision warnings, which resulted in 121 avoidance maneuvers (Pellegrino and Stang 2016). Without the American SSA system, many nations' CSIs would quickly succumb to collisions with debris or other satellites, creating a cascading chain of orbital collisions. The EU nations and many others have realized their dependence on the American SSA system, and they are currently developing their own systems to compartmentalize damage in the case of American SSA loss.

The development of an entire SSA and tracking system is a large undertaking even for a resource-rich nation. The US produced its own system at great expense, but several other European countries do not have the resources to build an entire system of their own. Instead, the EU is planning a new SSA system that will network together individual system components, each owned and operated by their respective funding nations. Together, these components will act as a functioning SSA system, but each nation retains control over the operation and funding of their specific components. The EU will not contribute any sensors directly, as this will be a project funded and owned by individual member nations (Pellegrino and Stang 2016). Such a model of cooperation is the centerpiece of the European model of CSG.

11.2.7.1 Deterrence

Establishing redundant systems greatly diminishes the effect of an attack or accident on the meta-system. The US CSI defense strategy employs this resilience itself as a means of deterrence. Deterrence is the process of increasing the cost of aggression and denying the benefits of an attack for an adversary (CNSSDP 2016). Several deterrence strategies function in the context of CSI. Deterrence via denial, in the context is a space system, occurs when the defender convinces a potential attacker that destruction of the target satellite would fail to realize the attacker's goal, due to either system redundancy or attack failure. Thus, CSI protection through deterrence is heavily linked to CSI design and system resilience. That is, as a system's resilience, real or perceived, rises, the probability an adversary will launch an attack will fail.

Deterrence through retaliation is another functional strategy, which was "mutually assured destruction" during the Cold War era of nuclear brinkmanship. The concept still holds value for the theoretical scenario in which one nation attacks another's CSI. Currently, most major spacefaring nations have the ASAT weaponry required to damage or destroy the CSI of another, through space-based, ground-based, or cyber-warfare methods. Even if a first strike on the CSI of the target nation is successful, other nations retain the resources necessary to stage an effective retaliation and cause equal or greater damage (CNSSDP 2016). This effectively deters nations from attacking one another, but it does not provide significant deterrence to adversaries without their own CSI to defend. The threat of mutually assured CSI damage is potentially significant enough for nations to develop non-aggression treaties and rules governing the future use of space.

11.2.7.2 Building Coalitions and Alliances

Such treaties and agreements describe the third of the US's CSI protection efforts. International treaties regarding space activities are still limited to the 1967 UN Outer Space treaty, but several other non-binding agreements have been signed that pave the way for future cooperation. Voluntary measures have become a new norm in the governance of space activities since they are flexible and useful for promoting good will among competent space actors (Pellegrino and Stang 2016). Over time, specific practices contained within these agreements came to be acknowledged as customary law for the purposes of UN negotiations, easing the process of future formal treaties. Coalitions form a key portion of US CSI defense strategy by increasing the impact of retaliatory deterrence and decreasing pool of potential attackers. However, until formal treaties are signed, space remains a frontier that is held in place largely with non-binding agreements that have relatively insignificant consequences if broken.

The US three-prong approach to CSI protection offers a comprehensive plan to mitigate the risks posed by natural and man-made threats to CSI. However, no plan can fully mitigate these risks. Cyberspace is too diffuse and dynamic to be fully known and defended. Space debris continues to increase, and it will continue further without concerted efforts to control it. Rogue nations or non-state actors may ignore national-level deterrence efforts and assault on CSIs.

If nations are not prepared for losses to CSI, a single disaster has to potential to damage not just a satellite or a constellation, but all of the economies that depended on it. The vast resources required to reach orbit nullifies the possibility of repair, and the construction of new satellites on Earth is immensely costly by itself. Because our economies and modern lives are so dependent on space systems, CSI protection must remain a priority for all nations, lest the world economy collapse like a house of cards.

11.3 Complex System Governance

Governance should not be confused with the term government. Rosenau (2003) defined governance "sustained by rule systems that serve as steering mechanisms through which leaders and collectivities frame and move toward their goals". Lum (2016) disagreed, sub-dividing governance into four main attributes namely (1) articulating society-focused priorities and goals; (2) assuring goal coherency; (3) steering society towards those goals; and (4) ensuring individual's action accountability.

11.3.1 Definition of CSG

Understanding Complex System Governance requires an understanding of the subjects 'complex systems' and 'governance' before understanding the intersection of

Fig. 11.3 Characteristics of complex systems, adapted from Keating et al. (2015)

the two. The Newtonian paradigm coined 'complex system' as 'unity in multiplicity'. Snowden (2002) and Walters (2016) characterized 'complex system' as that which is "indefinable by a single rule". However, complex systems can be characterized by the set of characteristics that tend to define them (see Fig. 11.3). Understanding both complex systems and governance separately allows us to move towards a definition of Complex System Governance "Design, execution and evolution of the metasystem functions necessary to provide control, communication and integration of a complex system" (Keating et al. 2015).

The earliest research and development into CSG had been application- or content-specific, and narrowly focused on the domains of Information Technology, Risk, and Corporate and Common-pool resources (Calida 2016). However, the subsequent theory-crafting has been focused on building a foundation for further growth (Keating et al. 2015, p. 277):

> *From a systems theoretic conceptual foundation, a set of nine interrelated functions is enacted through mechanisms. These mechanisms invoke metasystem governance to produce the communication, control, coordination and integration essential to ensure continued system viability.*

A graphical representation of the CSG framework is summarized in Fig. 11.4.

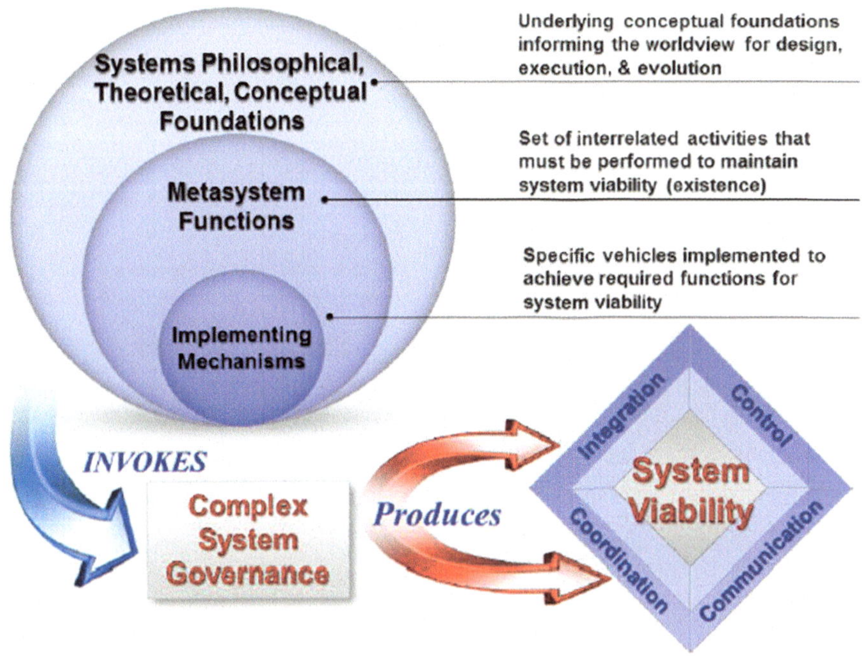

Fig. 11.4 CSG paradigm (Keating et al. 2015)

11.3.2 Emergence and Current State of CSG

The concept and framework described above are based on the disciplines of Governance, Systems Theory (axioms and propositions defining system functionality) and Management Cybernetics (the theory of effective organization) with the intent of developing a practical approach to effectively tame complex systems and problems (Keating et al. 2015).

Though each field offers independent solutions and applications to complex system problems, CSG's evaluation and solution approach is highly dependent upon the roles of external elements including environment, context, systems, and metasystems (a mechanism-based set of nine functions for system viability governance (Keating et al. 2015). Though each element has its own significance to CSG, metasystems constitute the largest portion of CSGs' development. Important considerations involving CSG and metasystems include (1) metasystems operation at logical levels beyond the level of element integration (2) metasystems conceptual foundation with systems theory and management cybernetics (3) metasystems definition of interrelated functionality with achievement goals for system viability but limitation to the procedural development (4) metasystem functionality and operational basis on viability but not performance and (5) metasystem design ranging from purposeful design, execution and maintenance to self-sustaining systems (Keating et al. 2015).

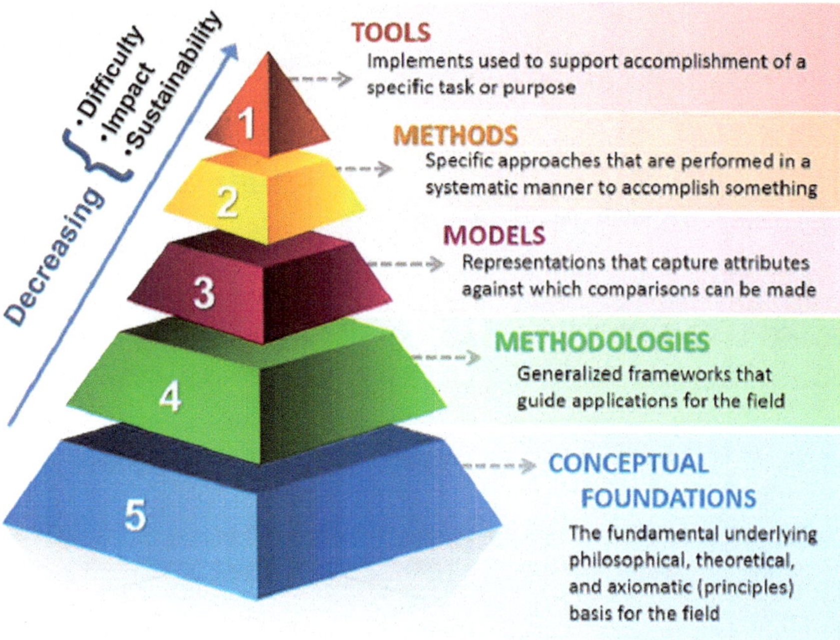

Fig. 11.5 CSG development and implementation stages (Keating and Ireland 2016)

At present, most of the work in CSG is focused on the conceptual foundation of the field. Near future work will extend that foundation into methodologies (e.g., Keating and Katina's 2016 paper *Complex System Governance Development: A First-Generation Methodology*), which will likely be followed by models, methods, and mature tools (see Fig. 11.5) (Keating and Ireland 2016).

To continue enhancing the field's foundations, thinkers need to proceed in a systematic approach: taking the existing theory, applying it to a problem, advancing it via application and practice, and returning lessons learned to developmental stage as suggested in Fig. 11.6 (Keating and Ireland 2016).

11.3.3 Benefits and Advantages of CSG

For all its newness and cyclical development, CSG can still offer several advantages over other forms of governance. CSG adopts a holistic approach, considering the entire system as a dynamic entity, which helps the user choose the correct problem(s) to solve. However, its goals are simple and narrowly defined: control, communications, coordination, and integration (see Sect. 11.4 for definitions). To achieve them, CSG adopts the metasystem functions from the Viable System Model, but tailors them to the unique context as identified in the holistic view. Further, rather than

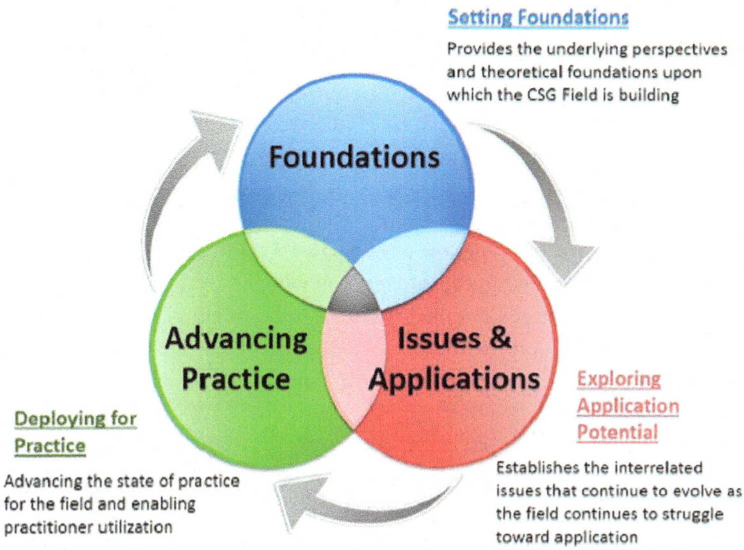

Fig. 11.6 Current advancing in CSG field (Keating and Ireland 2016)

attempt to control all outside inputs, CSG theory robustly builds the metasystem to maintain desired performance levels while tolerating those negative inputs. To maintain existence and keep all nine metasystem functions healthy, CSG theory demands that system development be a continuous process. Thus, it permits developers to actively match governance solutions to emerging problems, thereby compensating for internal or external fluctuations that the metasystem cannot tolerate. Finally, CSG development is scalable, from the individual to the enterprise level; the benefits increase directly as the practitioner's scope of application. (Keating and Ireland 2016).

CSG offered scenario models can be constituted among the detailed with insights on the structural relationship, context, and systemic deficiencies and providing the 'big picture' view along with associated factor prioritization for any complex system or problem. With such detailed evaluation and hierarchical distinctions, better evaluation is anticipated for all prevailing scenarios and hence better, more coherent decision making (Keating et al. 2015).

CSGs' strong emphasis on monitoring and regulating governance and development in areas of design, analysis and evolution allows organizations and states with in-depth monitoring and analysis of complex systems and their by-products with efficient maintenance possibilities for system viability (Keating et al. 2015).

With CSG objectivity to expand technology centric infusion, practitioners shall be engaged with the development of tools and model for in-depth modern complex system analysis, hence enabling organizations with extraordinary leads on complex systems and their derivative problems (Keating et al. 2015).

11.3.4 Challenges and Recommendations with CSG

While CSG's parent fields are mature, with a corresponding wealth of literature, CSG is not, and its immaturity brings various challenges. The theory is subject to some tension, based on its heritage as the child of three parent fields (management cybernetics, systems engineering, and governance). Worldview is "the system of values and beliefs that allow us to process events" and assign meaning correctly (Keating and Ireland 2016). With three parent worldviews, CSG practitioners find strengths where those three worldviews agree and tension where they do not. Further, management cybernetics and systems engineering tend to be practiced at the operational or technical levels of a problem, while governance uses a broader, strategic approach. Again, differences in the parent fields create tension in CSG (Keating and Ireland 2016).

In addition to internal tensions, CSG theory is still finding its place. Practitioners of one or more parent or related disciplines are wary of adopting CSG, because those fields are littered with dead-end approaches and failed theories. Only time and its cyclical development will allow CSG to grow into acceptance. Further, CSG is open to (and may actually require) injection of material from a fourth or fifth field, which would expand CSG's scope and applicability (Keating and Ireland 2016).

11.4 Objective Analyses

Multi-Criteria Decision Analysis (MCDA) tools can be use evaluate proposals for governing 'critical space infrastructure' via the tenets of Critical System Governance theory. Both military and civilian entities use space infrastructure for myriad functions, including earth observation and environmental monitoring, early warning, reconnaissance, navigation and communication. Governing the orbital environment and its contents would require controlling a titanic array of nation-states, their civil and military branches, corporations, industrial groups, and regulatory bodies, all with different ethical frameworks, competing priorities and varied governance modes.

Some organizations prioritize their own interests first, in the manner of a nation-state, while others advocate cooperative control for the greatest common good, as a federation might. Finding common ground between even the major players and constructing a governance model for CSI are excessively complex tasks, which might be eased via the CSG model. However, the inherent complexity of the problem requires that even evaluation of competing models be robust; thus, the project will use several MCDA models to shape a recommendation.

In the current synthesis, the objectives are: (1) to conduct literary research into the backgrounds of both CSI and CSG, (2) to construct governance models based on that research, (3) to evaluate those models using MCDA tools, and (4) to recommend a preferred course of action based on the analyses.

11.4.1 Identification of Decision Attributes

Control

Control over the space domain and related systems became an issue with the Soviet Union's Sputnik launch in 1957. Since then, several countries have conducted space launches to further military, scientific and civil endeavors. This issue became more complex when the US Department of Defense launched Global Positioning System (GPS), and the European Union countered with their Galileo program, which sparked debates who ought to be control information from space systems, and how that control ought to be exercised (David-Braunschvig and Jeremy 2003). Neither side disputed ownership of their respective space-borne assets, but international debates have continued regarding the merits of independent control over such systems and the information they provide.

Discussions of 'space control' and 'space superiority' first surfaced in the US during the Strategic Space and Defense Conference in 2006, with many officials formally raising concerns over the importance of safety and control of space assets (Scott, 2006).

Definition: Control

Keating and Katina (2015) define control by specifying what it must deliver. "The primary function of control by the metasystem in CSG is to provide the minimal constraint necessary to ensure continued system performance and behavior, while maximizing autonomy of governed entities." Both in this context and in the context of 'critical space infrastructure', this maximized autonomy allows member states greatest possible freedom of action, internally and externally, as long as the governed infrastructure is performing as required.

Criteria: Control

Thus, in addition to the effectiveness of the model's holistic degree of control, the analysis must assess the model's control over hazards. The uncontrollable elements in space are mostly natural hazards, but a CSG model functioning at the national or international level can impose some controls on the hazards that stem from human activity. Thus, a measure of the model's control effectiveness is the degree to which it defends against known threats.

However, the National Security Space Defense and Protection report (CNSSDP 2016) on the US's space strategy points out that addressing only known threats to CSI is purely reactive. However, taking steps to address and mitigate unknown or evolving threats manages those risks and opens response options even before the risk events have occurred. The last evaluation attribute of the control criteria is the control metasystem's ability to respond to future or unknown threats.

Communication

Definition: Communication

Communication in CSG is "the flow and processing of information within and external to the system that provides for consistency in decisions, actions and interpretations made with respect to the system" (Keating and Katina 2016).

Criteria: Communication

If any system, effective communication must be timely and accurate. Other elements are vital to communication, like sender, receiver, message, and medium. However, this analysis will avoid such in-depth interrogation of the models by assessing all factors overall with a third criterion, effectiveness.

Coordination

Due to the international fabric of interdependence on critical infrastructures, various nations and organizations are exploring the benefits of cooperation and coordination. The European Space Agency (ESA) and Canada who describe their partnership as "… opening a new chapter that will offer opportunities for Canadian contributions to ESAs programmes and ESA contributions to Canadian programmes" (Mumford 2011).

Definition: Coordination

Coordination for complex system governance is defined as "providing for interactions (relationships) between constituent entities within the system, and between the system and external entities, such that unnecessary instabilities are avoided" (Keating and Katina 2016). For 'critical space infrastructure', coordination can be further divided into two categories: (1) internal, pertaining to actions completely internal to a nation, state, corporation, or other body; and (2) external, which is any coordination action involving more than one nation, state, corporation, other body, or the governance organization itself.

Criteria: Coordination

Successful coordination is that which facilitates open, de-conflicted usage and defense of space assets, providing safety and security information to participants. In addition, a successful coordination function will be scoped such that it reaches both all participating organizations and all assets in any of their systems. Far from being excessive monitoring, this scope of coordination is required to avoid unforeseen conflicts between previously unknown or unseen assets.

Integration

Integration, the third pillar of Complex System Governance, does not require complete coupling of constituent systems. It is instead a balance between independence and dependence.

Definition: Integration

In their seminal article, "Complex System Governance Development: A First-Generation Methodology", Keating and Katina define integration as "…continuous maintenance of system integrity. This requires a dynamic balance between autonomy of constituent entities and the integration of those entities to form a coherent whole" (Keating and Katina 2016). For the space domain, "system" must be defined as the system of systems existing in space, their ground-based control systems, and the many users that depend on or add to the data provided by those space-based systems. A holistic view is required due to the probability of collisions and increasing competition in space.

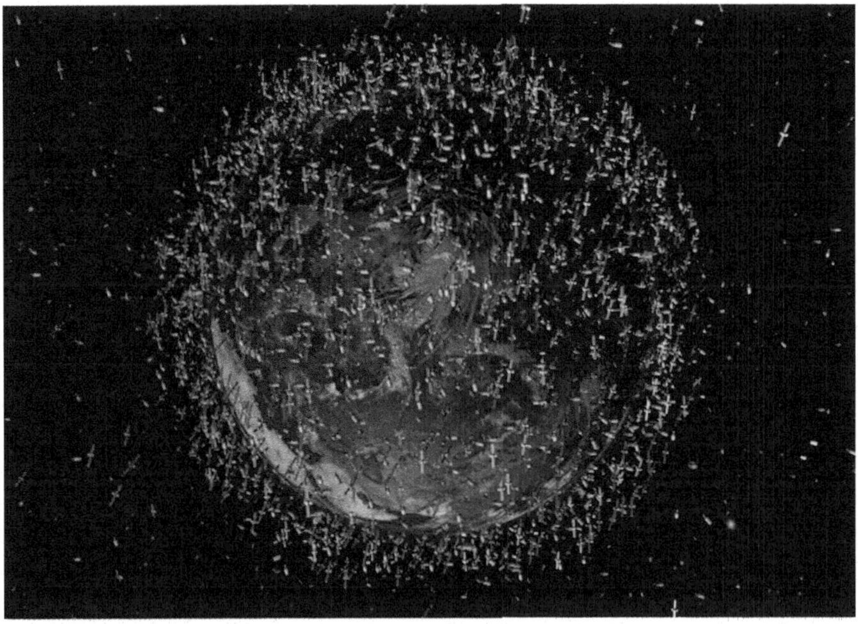

Fig. 11.7 Existing manmade objects orbiting Earth (O'Callaghan 2014)

Criteria: Integration
Integration is difficult to measure, since it may be physical or organizational. To avoid overly constraining the system, it should be the minimum necessary to maintain viability. If a solution does not provide some level of integration for all constituent systems, it should be considered deficient. Further, integration should include a continuous, maintenance-oriented component, which adapts the integration functions as the component, governed systems evolve.

Monitoring
The potential for disastrous interaction among space-based systems elevates monitoring to a critical attribute for space infrastructure governance. While still a rare occurrence (O'Callaghan 2014), the potential for collision is sure to grow (see Fig. 11.7).

Definition: Monitoring
Keating and Katina identify system monitoring as a critical metasystem, providing "…oversight of the system performance indicators at a strategic level, identifying system level performance that meets, exceeds, or fails to meet established performance expectations" (Keating and Katina 2016).

Criteria: Monitoring
Since a complete view must include both space objects and the systems of which they are a part, the degree to which governance can provide that view is the criteria. Governance scenarios that provide only a partial picture would be less viable and

therefore must be rated lower. The monitoring system's ability to continuously update is also crucial. Given the velocities at which space-based assets move, the monitoring system must update information (or accurately predict it) at sufficient frequency to avoid negative impact to the control function.

Intelligence
While intelligence is not an established attribute of complex system governance, its inclusion in space infrastructure governance seems appropriate. Access to space is not and, in all likelihood, will not be confined to cooperative states. North Korea launched a satellite in 2016 that malfunctioned and took on an unstable orbit (Ap 2016).

Definition: Intelligence
Intelligence is an extension of system monitoring, capturing the status of uncooperative states' efforts at launching satellites, ASAT rockets, or deploying ground-based weapons. Its object is to gather information before a launch or an incident occurs, giving governance the information needs to take action proactively rather than reactively.

Criteria: Intelligence
The degree to which governance can provide intelligence that can be shared with cooperative systems is the criteria. Governance scenarios that provide only partial information would be less viable and therefore must be rated lower. Further, all participants in the governance model have a vested interest in the rogue actors who do not participate; the intelligence function ought to track those actors in some fashion.

11.4.2 Hierarchy of Attributes

Figure 11.8 illustrates the overall goal, protection of space infrastructure, and the attributes required to meet the overall goal. Now, let's pace focus on scenario development.

11.5 Scenario 1: The US-Centric Model

The United States has maintained its technology superiority in numerous domain the outer space. The US presently enjoys the dominance in satellite navigation especially by using Global Positioning System (GPS) as the ultimate tool to dominate and conquer the outer space. The GPS is a satellite-based infrastructure developed by the US Department of Defense. It provides the unique globally available signal for navigation, an essential feature to the operations of US and Allied military forces and to a growing number of civilian users (David-Braunschvig and Jeremy 2003).

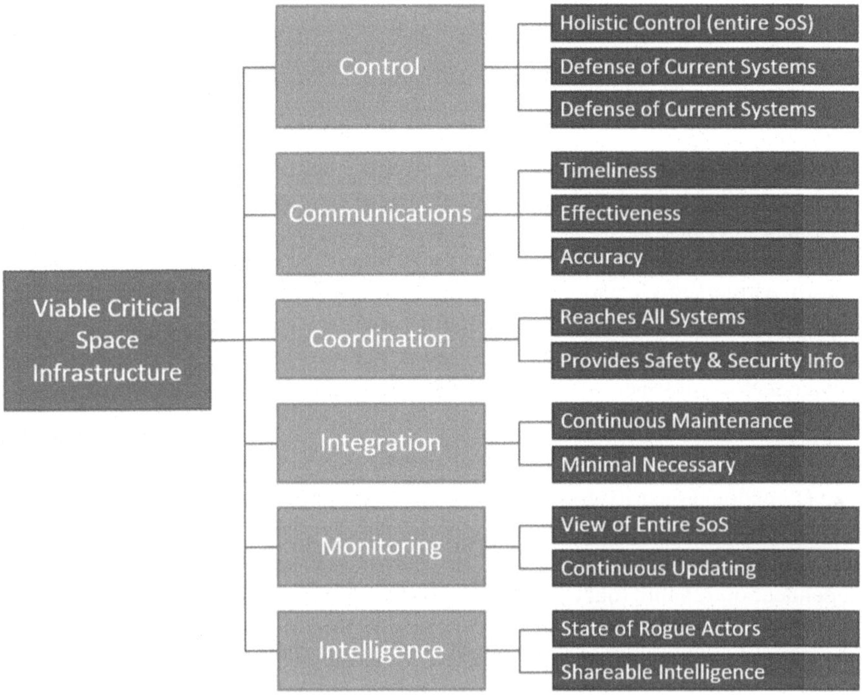

Fig. 11.8 Attribute hierarchy for viable 'critical space infrastructure'

In the US, the National Aeronautics and Space Administration (NASA) was viewed as the sole dominant in the space politics. That view is changing because the old assumption that of state as the unique actor in space politics has been relaxed.

11.5.1 Future of Current US Activities

The space is there to be exploited. Decades ago, two power nations were at front-center in space exploration: The United States and the Soviet Union. The dynamic has changed, and other emerging countries are especially from the European Union and China are making significant gains in the space sector. Americans are determined to develop new, high, and final frontiers to lead all those nations. This is a big challenge because 60 countries now have satellites orbiting the earth which makes the space a congested, contested, and a competitive place. The United States has the obligation to adopt a new space strategy open to adapt to this reality. Recently, commercial parties in space governance have increased in number and significance.

11.5.2 US Space Governance

The United States is still the ultimate master in the outer space governance with respect to the traditional state-centric security and military aspects of the space. Nowadays, private partners are gaining stronger and wider responsibilities in the development and the management of space programs, and the public accountability is at stakes because of the widening of security in space policy.

The US government recognizes that government and non-government actors need to form partnerships to address the increasingly complex task of managing space and space systems. During the Reagan administration, an emerging idea suggested that one of the basic goals of American space policy was to "expand the United States private sector investment and involvement in civil space and space related activities" (POTUS 1990).

Control

The US national security depends on the space dominance. The US national security has enjoyed the reliance on the space dominance over the last six decades, and it has grown tremendously. The US military and other national security institutions are dependent on reliable functions of space systems in peacetime, crisis, and conflicts (CNSSDP 2016). As common to all human-designed infrastructures, space systems are vulnerable from natural causes impacts, from intended or intended human activities, and from international attacks.

The United States work around the clock to preserve its space dominance against increasing threats. To address the later issues, the United states has taken the following measures: (1) the US has developed and will continue to develop the means to deny anyone the ability to use space systems to support hostile actions against the United States; (2) maintain the ability to use space assets for national security purposes in peacetime, crisis, and conflicts; and (3) ideally or idealistically, be assured that space remains a benign operating environment for all civil and commercial activities (CNSSDP 2016).

The US has the potential to disable adversary space systems in time of crisis and conflict even though the US is not the only prayer in the space race. This is why the US has to take into account the of what other countries, adversaries or friends will decide to do. In the NSSDP, the US government outlines its priorities that will allow the US stay ahead in the space sector as follow: (1) develop a space clear vision of what the US wants the future in space to be, (2) understand the extent to which the US can shape that future, (3) identify and develop prudent methods to counter existing, evolving, and emerging threats to US interests in space, and (4) assess those methods in terms of how they affect the future in space and the ability and the commitment of the United States to shape that future.

The US has worked in the past to deny adversary space assets in time of crisis. The Ford administration (1974–1977) shows that the US was involved in effort to deny the Soviet Union space related military activities, and both super powers were investing in technologies to each other satellites. By the late 1970, only two power nations, the US and the Soviet Union had the antisatellite (ASAT) capabilities, but today an increasing number of countries have the abilities to build and launch a vehicle capable to reach the orbit besides the US and the Soviet Union.

The outer-space has shifted big time from military objectives into civilian-oriented consumption. The US is taking the lead and other international commercial companies are flooding the market which may cause risk to the US military operations. The US is aware of competing countries especially China and Russia that are developing counter-space capabilities that can be used to disrupt US capabilities and weaken the US global stature in space exploration (CNSSDP 2016). With the decision to suspend the NASA's Space Shuttle program, private investors are entering the US space sector (Dillon 2014) to fill the void left by NASA and future looks brighter than ever.

Communication

Communication is a key in maintaining the safe operation of existing and future satellites and sustain the high value of satellites orbits. The United States heavily depends on space assets to fulfill many civilian and military critical infrastructure function such as communication, earth imaging, navigation, and early warning systems. Communication in the space governance in America is based on expending international cooperation and promoting safe and responsible operation in space. To enhance transparent communication while preserving US national security and freedom of access to outer space, the US is leading other power nations to establish a framework with accepted normal to sustain a responsible space program.

Coordination

The outer space policy is nowadays characterized by strong indicators that management of the outer space domain goes beyond the terrestrial or typically "sovereign" domain. The space politics has remained state-centric in the US, but the after cold war geopolitics has brought the feature of globalization and the private sector involvement in space management. Key factors that are fueling the private sector in the space management are increasing costs of spaces exploration; a series of failed and in some cases disastrous governmental space programs, shrinking agency budgets, shift in political ideology favoring private management models.

Integration

Satellite systems are highly complex with diverse civil, military, and commercial functions. In addition, they are constructed, installed, and overseen by pool of diverse experts. The integration tasks including hardware, software, and malware are enormous. The complexity and potential for both accidental and intentional failure are a big concern to make sure all involved subsystems are working in unison.

Monitoring

Rivalries between traditional and emerging space power complicate the monitoring of the outer space. Monitoring requires transparency but emerging space nation especially China are reluctant to share their activity publicly. The United States does its best to push transparency through international treaties.

Intelligence
In the United States, Critical infrastructures are vital to the country survival so that their loss would have a disastrous consequence on the security and the economic welfare of the country. After the 11 September 2001 terrorist attack on the US soil, the industrial intelligence, environmental monitoring, counter-insurgency, crisis management, and counterterrorism activities brought a great deal interests in using the space in intelligence activities.

11.6 Scenario 2: The Euro-Centric Model

In contrast to the US-centric model, this analysis will also consider a Euro-centric model. Founded on the guiding principles of Europe's planned SSA and tracking system, this model would network multiple space system components together while maintaining individual national control and authority.

11.6.1 Current European Activities

Current space activities of the ESA and EU are focused on releasing Europe from the dependency of American space systems such as GPS and SSA capabilities. The EU is currently working on Galileo, the first European global navigation service system that will utilize a set of EU-owned satellites available for open use by all EU member states. The ESA is also working on a path to non-dependence on the American SSA and tracking system to avoid total system collapse in the event of American service disruption or denial. It is important to note that the EU does not own the SSA project. Rather, the ESA facilitates the networking of various individual system components that are independently owned and operated by their parent member states.

The Galileo PNT constellation is currently scheduled to begin early operations by the end of 2016, with full system operation scheduled for 2020 (European Space Agency 2015). Significant progress has been made towards decreasing dependence on American GPS but plans for an eventual European SSA network have only begun to take shape. Nearly the entire world depends on the US SSA and tracking system and the European solution still has several political and engineering challenges ahead.

11.6.2 European Space Governance

The proposed European CSI protection model takes heavily from the proposed SSA framework. Satellites divided into various functional categories such as PNT, SSA, and earth observation will all be produced or retrofitted by parent nations through an agreed upon plan. The plans will contain design criteria for the construction of new satellites, as well as options for retrofitting any currently operational satellites as necessary. All nations that contribute construction and inspection resources will be entitled to the benefits of such a system and the resiliency it provides. Parent nations will maintain ownership and control over their individual satellites and ground-based components, but together the system will fulfill the world's global needs for CSI.

Such a system will provide significant redundancies and deterrence against attack or damage, and the efficient sharing of peacetime information will benefit all nations involved. While still incomplete, the Euro-centric model will offer a different approach from the US-centric model across several key attributes as discussed below.

Control

Control for such a system in the terms described above, would be high when considered in relation to other models. Several redundancies and resistances to physical and cyber-attacks will be implemented in order to provide maximum system resilience. Where one nation's component fails, another nation's assets can replace until a permanent replacement is identified. Such resistance to disruption will provide significant deterrence for potential adversaries and will offer multiple options for routine maintenance and repairs.

Communications

Communications efficiency in a Euro-centric model will likely be its main weak point. While electronic communications are nearly instant, the interpretation of messages and data across national and lingual barriers may hinder the effectiveness of this model. The accuracy of such data will likely be relatively high as any given data point will have several comparisons for verification purposes, but the interpretations of this data will still be subject to debate before any action is taken.

Coordination

Coordination is one aspect in which the Euro-centric model would excel. The collaboration of each nation's space agencies together with the appropriate resources would allow an open flow of information and education for all participating nations.

Integration

The large amount of systems integration in a Euro-centric model will produce a significant negative impact to the overall costs of such a system. System components will be relatively useless when separated from the system. As such, large amounts of 'system-of-systems' integration must take place for full system functionality to be achieved.

Monitoring

The monitoring functionality of such a Euro-centric system will provide maximum data distribution with relatively few limitations on what data to restrict. With the exception of specifically designated defense or national security data, all pertinent SSA and functional information will be transmitted across the entire system for monitoring and use by all components.

Intelligence
The intelligence sharing capabilities of the Euro-centric model are extremely powerful when used in the proper manner. The fact that individual nations retain the command and control of their satellites gives them the freedom to pursue defense and other earth observation functions at their will. The decision to share the information obtained from such activities will remain the prerogative of the parent nation, but if so chosen, the intelligence can instantly be spread to the entire system for maximum awareness.

11.7 Theorizing a Market-Centric Model

The market governance model for CSI is complementary to the European and American models described before and fulfills several of the CSG requirements. Fundamentally, it develops a set of solutions for governance issues in the context of growing involvement of private entities in space, as owners/operators/administrators of critical space infrastructures. The basic principle of this model is to tie governance priorities to mechanisms incentivizing an emergent positive behavior on the part of stakeholders, with positive security externalities, regardless of whether they intended this. It is no secret in CIP that the profit motive and the uncertainty of loss as opposed to costs lead private actors to downplay investment in security by regarding it as a cost. Such incentive issues and misalignments with competent authorities and decision makers are apparent in space as well.

The CSG elements, such as the association between private actors, cooperation for standards, exchanges of information and dissemination of best practices remain important, but what the market governance model brings to the table is the intrinsic motivation to incentivize such behavior. The basic idea is that there must be an incentive to nudge stakeholders into actions with positive outcomes for the CI and environmental stability and security while discouraging and punishing actions and policies with negative externalities or who are free riding on other market participants.

The most important such elements, in the context of CSI, are:

- The cost of financing space assets;
- The cost of insurance and reinsurance of the asset;
- The liability claims and damages for responsibility in a negative event materializing;
- The cost of damages to the "orbital commons";
- The damages extracted by clients as compensation for losses suffered;
- The cost to intangible value for the actor in question (prestige, public image etc.) since this will diminish the goodwill which is a component, for instance, of stock prices;
- Regulatory costs and other impositions, such as fines for certain practice.

Such a model would work, for instance, in the area of tackling the space debris threat by incentivizing sustainable behavior to limit the creation of new debris, by creating sources of funding for research into clean up, by establishing liability costs for damaging another system with debris created by the actor in question etc.

11.8 Elements of Governance Models of Other Countries

In the currently fragmented global governance domain, we can distinguish a multitude of individual approaches to the issue of critical space infrastructure governance. These "models" are derived from the realities and needs of the country in question and reflect their organizational culture and economic structure (Table 11.1). While the country in question may qualify as a powerful, their models are not necessarily scalable to form a regional or a global model but are still interesting from the perspective of systems and mental modes.

Also included in the table below is a generic "non-spacefaring nation". It seems dubious to discuss CSI governance from the perspective of an actor or actors who

Table 11.1 Elements of other CSI governance models

Russia	• The governance system resembles the American one, in which CSI must ensure security and the power of the country, in a military sense • Projects such as the GLONASS GNSS are aimed at reducing reliance on American CSI, for the purpose of redundancy, but also in order to secure the military aspect of CSI use (guided munitions and so on) • Continued developing ASAT weaponry, as another area for deterrence • Its space industry is heavily centralized, but is open to cooperation with the West and the various commercial opportunities it brings in order to surmount the financial difficulties affecting the industry since the fall of the USSR or even earlier. This policy was on display when Russia became the only transport provider for astronauts travelling to and from the ISS • Russia has plans to expand its space activities on its own, also for prestige purposes, but is also looking to acquire dominance in niches of the global system
China	• Ascendant superpower, in military, economic and geopolitical terms, with space ambitions and the second largest inventory of space system after the United States • The space industry is subordinated to the state, without Russia's financially-motivated drive to seek out international cooperation • Space is a source of prestige and of military capabilities and deterrence • Cooperation with the West is made more difficult by the latter's fears of unauthorized technology transfers • It is building up a complete array of basic CSI capabilities (the Beidou/Compass GNSS is an example) • Is undergoing a process of rapid catch-up growth to the US and Russia, with serial accomplishments that enhance its prestige and certify the reaching of certain technological milestones in terms of homegrown capabilities • The specific CSI governance system of China would be rigidly hierarchic, with possible problems with communication. The importance placed on considerations related to national security defined in the widest possible sense would lead to a certain opacity diminishing the communication and coordination with other actors

(continued)

Table 11.1 (continued)

Japan	• We have compared China and Russia to the US on the governance side, but Japan resembles most a country from the European governance model • Multilateralism and good cooperation with other countries are key, including for division of labor and specialization within certain domains • Japan would support collective solutions to space governance issues, focusing on achieving consensus • It would prioritize the provisioning of critical space services from multiple sources, while acting, where possible, to achieve some degree of independence. An example in this sense is the Zenit system, which boosts GPS capabilities in Japan • The experience of the Fukushima-Daiichi disaster, described in more detail in Chap. 9, Japan started not only to invest in the space capabilities it requires and can afford, but also to explicitly aim for the resilience of national CSI through measures such as redundancy • The Fukushima disaster also highlights Japan's capacity to utilize membership in certain groups or in multilateral initiatives to compensate for a failure in critical space services provisioning. Examples in this regard are the International Charter on Space and Major Disasters and the Sentinel-Asia initiative. Membership allows Japan to have fallback capacities should crisis and emergency situations take place that outstrip its critical space services provisioning capabilities
Non-spacefaring nations (Georgescu et al. 2016)	• The general dependence on the provisioning of critical space services (and someday also goods) creates the responsibility on the part of all countries to involve themselves in governance activities • A non-spacefaring nation may be critically dependent on space systems, while having no satellites of its own or other systems • Such a country has the following CSI governance priorities (indirect governance is mostly what can be attempted) or, rather, we can term them as strategies to manage its critical dependencies: – To satisfy the trilemma of access to a critical resource – the accessibility, affordability and sustainability of critical space services provisioning – To ensure its minimum required critical space services supply (for the normal functioning of society) not just in peacetime, but especially during wars or crises – To tailor national investment priorities in order to increase resilience to the disruption of the supply of critical space services and to reduce its vulnerability to such crises • In addition to the risks stemming from dependencies that all countries register from a certain level of development technological sophistication onward, the non-spacefaring nation is also confronted with political risk of access to its required critical space services • The CSI governance of a non-spacefaring nations will rely on: – Active participation in all multilateral or global CSI governance initiatives such as UN-COPUOS – If possible, a presence to the level and extent it can afford and manage in ESA is a powerful equalizer of access to space – Using specially designed contracts or diplomatic means to reduce uncertainty of supply during shocks to critical space services provisioning – Assess the possibility of strategically investing in national space capabilities or in a project with a group of other countries, given a reduction in barriers of access to space – Promoting the interests of non-spacefaring nations in international or global for a, arguing against discrimination in access to space services, against the militarization of space and in favor CSI resilience and protection strategies, from mandatory technical standards to space conduct codes

lack some or all of the prerequisites of being a space player, but we must take into account a couple of issues:

- The global nature of the CSI system-of-systems and the requirements of running an advanced, globalized, higher functioning and interdependent economy turn every country into a CSI stakeholder;
- Through participation in international organizations (the European Space Agency), multilateral agreements (International Disaster Charter), global bodies (the UN with COPUOS) and through bilateral partnerships, the non-spacefaring stakeholders can, individually or collectively, advance their own vision for governance on a point-by-point basis;
- True global governance must be legitimized through the consent of all recognized national stakeholders. Therefore, the development of a global governance framework for CSI will, in some way or another, factor in the articulated interests of non-spacefaring stakeholders, to the extent that they can lobby for them;
- Development will increase the reliance on space systems, as was argued elsewhere in this book, thereby increasing the motivation to influence governance to their relative advantage.

Final Remarks

At this point in this reading, one should realize that authors have taken liberty to offer a different view on the space governance issue. One might as well suggest that we have attempted, knowingly or not, followed thoughts commonly attributed to Albert Einstein that: *we cannot solve our problems with the same thinking we used when we created them.* In the present text, we have attempted to offer a viewpoint, perhaps a misguided one, that could be used to address emerging issues in critical space infrastructures, key resources, and key assets.

To the theorized American and European Models for Space Governance, which are summarized in the Table 11.2, we have also tentatively added a market-centric model of governance reliant on incentives (negative and positive) and emergent phenomena to promote good governance.

The final chapter enhances previously articulated ideas in the context of ***Complex System Governance***, as a forward-looking approach to the issues related to critical space infrastructure governance.

Table 11.2 Summary of theorized governance models

Attribute	European governance model	American governance model	Market governance model
Control	Decentralized and diffuse; resilient but difficult to coordinate; features internal redundancies	Its most developed attribute, thanks to the fusion between the civilian and military sides when it comes to technology and dual use systems	Decentralized; reliant on voluntary adoption of norms and standards
Communication	A weak spot for the transmission of information; the making of decision and their reevaluation become more difficult because of culture, language, jurisdiction, long communication circuit etc.	This is the key for the usefulness and the sustainability of the most valuable orbital real estate. The US depends on partnerships and international organizations to establish the communication lines which would lead to the implementation of common standards	A weak link; communication must be open, using industry bodies or intermediary organizations; communication also performed through price signals in stock markets, betting markets and capital markets, aggregating, in theory, information
Coordination	Due to its innate diversity and pluralism, it will have the toolbox, the organizational models and the mental models for effective coordination at macro level which would include other stakeholders	Developed constructively through the routine cooperation of space dominant institutions within the state, while respecting NASA authority. The system functioned quite well as long as space was dominated by just two main actors. The model is encountering difficulties not just with the rise of new state actors, but also as globalization and commercialization have changed the operating parameters of space	Coordination is very difficult to achieve. One solution is to have governments coordinate in setting the legislation which companies in their jurisdiction must then follow; another model is the voluntary adoption of protocols and practices which become standard through mass adoption and standardization phenomena. Industry bodies and other associations could also play a part. When everything comes together, it becomes a self-regulating, emergent body

(continued)

Table 11.2 (continued)

Attribute	European governance model	American governance model	Market governance model
Integration	The cost of the system will be significant, because of disparate construction, replication of efforts and decentralization. At the same time, there will be difficulties in the informational and operational integration. The nature of multinational system-of-systems "condemns countries to cooperation"	The US has an innate advantage in the integrated functioning of the CI system-of-systems with a space component and has developed significant competencies in the field, helped by economies of scale, common standards and the unity of political and regulatory authority. An important role in cooperation as well, through the division of labor	Companies pursue opportunities and respond to incentives. Integration will take place through cross-service agreements, corporate takeovers or mergers and acquisitions to increase efficiencies and partnerships/alliances
Monitoring	With minor exception for security and cost distributions issues, the access to the data streams inside the infrastructure system will be as free as possible for all members to benefit, and collective decision making will tend to lead to more openness and fewer restrictions, also with the possibility of extending to outsiders (ex: free access to Copernicus and unrestricted to Galileo)	Monitoring is a problem in the American system, because it may lead to the development of rivalries among nations with a wide spectrum of capabilities, lowering transparency. Countries like the US and China act unilaterally to test anti-satellite weaponry and, sometimes, they avoid coordination with actors affected by their actions, such as when the destruction of satellites leads to the creation of debris	Monitoring is each actor's responsibility, utilizing its own resources or acquiring services from specialized providers. Companies may pool resources to reduce monitoring costs since they share an "orbital commons", or they may lobby for a form of Public-Private Partnership to address the issue or to provide actual SSA services. Actors are myopic, in that they have a limited field of vision for risks and strategic planning, however good their performances are inside that field

(continued)

Table 11.2 (continued)

Attribute	European governance model	American governance model	Market governance model
Intelligence	The European model has an innate advantage at multinational level compared to the American one, despite the latter prioritizing coalitions in its CNSSDP (2016). The maintenance of national control over own systems, but also the institutionalized capacity to quickly share information presents advantages from a military and intelligence perspective, because it allows for the collection of own data as well as its deliberate dissemination	It is a priority for the US and its 17 intelligence agencies, as well as the other security actors, to work together and develop new information sources. Data gathering through satellites was one of those capabilities. In the long-term, US intelligence needs will guarantee investment in new space systems and in research	Actors actively compete to obtain intelligence and utilize it to improve market position. For-profit entities provide intelligence services, often as an extension of terrestrial business intelligence (ratings, business assessments etc.). Intelligence is limited by a rule of relevancy and immediacy, so actors may develop blindspots which only authority figures (states, international organizations) may be able to rectify

References

Ap, T. (2016, February 9). North Korea satellite 'tumbling in orbit,' U.S. official says. Retrieved November 24, 2016, from http://www.cnn.com/2016/02/08/asia/north-korea-rocket-launch/.
Calida, B. Y. (2016). Complex system governance: moving diverse theory to practice. *International Journal of System of Systems Engineering, 7*(1/2/3), 22–42. https://doi.org/10.1504/IJSSE.2016.076125.
CNSSDP. (2016). *National security space defense and protection: Public report*, Committee on National Security Space Defense and Protection; Division on Engineering and Physical Sciences; National Academies of Science, Engineering, and Medicine. Washington, DC: The National Academies Press.
David-Braunschvig, R. L. G., & Jeremy, C. M. (2003). Space diplomacy. *Foreign Affairs, 82*(4), 156–164.
Dillon, R. M. (2014, May 29). SpaceX's Elon Musk unveils spacecraft designed to fly astronauts to space station. *US News & World Report*. Retrieved December 4, 2016, from http://www.usnews.com/news/science/news/articles/2014/05/29/elon-musk-unveils-spacecraft-to-ferry-astronauts.
European Space Agency. (2015, December 18). *What is Galileo?* Retrieved December 6, 2016, from http://www.esa.int/Our_Activities/Navigation/Galileo/.
Georgescu, A., Botezatu, U. E., Popa, A. D., Arseni, Ş. C., & Sava, A. C. (2016). Romanian National Security: A critical space infrastructure perspective. *MTA Review*, publicat de Academia Tehnică Militară, *XXVI*(1), 53–64, martie 2016. ISSN 1843-3391.
Hesse, M., & Hornung, M. (2015). *Space as a critical infrastructure*. New York: Springer Science & Business Media.
Keating, C. B., & Ireland, V. (2016). Editorial. *International Journal of System of Systems Engineering IJSSE, 7*(1/2/3), 1–21. https://doi.org/10.1504/ijsse.2016.07.
Keating, C. B., & Katina, P. F. (2015). Editorial: Foundational perspectives for the emerging complex system governance field. *International Journal of System of Systems Engineering, 6*(1/2), 1–14.
Keating, C. B., & Katina, P. F. (2016). Complex system governance development: A first generation methodology. *International Journal of System of Systems Engineering IJSSE, 7*(1/2/3), 43–74. https://doi.org/10.1504/ijsse.2016.076127.
Keating, C., Katina, P., & Bradley, J. (2015). Challenges for developing complex system governance. In *IIE Annual Conference Proceedings* (pp. 2943–2952).
Lum, R. A. (2016). A futures perspective on constitutional governance. *International Journal of System of Systems Engineering, 7*(1/2/3), 207. https://doi.org/10.1504/ijsse.2016.076123.
Mai, T. (2015, July 30). *Global Positioning System history*. Retrieved December 6, 2016, from http://www.nasa.gov/directorates/heo/scan/communications/policy.
Mumford, R. (2011). ESA and Canada renew space partnership. *Microwave Journal, 54*(2), 46.
O'Callaghan, J. (2014). *Why don't objects collide often in Earth orbit?* All About Space. Retrieved December 6, 2016 from https://www.spaceanswers.com/space-exploration/why-dont-objects-collide-often-in-earth-orbit/.
Pellegrino, M., & Stang, G. (2016). *Space security for Europe* (Rep. No. 29). Paris, France.
POTUS. (1990). *National Security Directive 42: National Policy for the Security of National Security Telecommunications and Information Systems* [Press release]. *Intelligence Resource Program*. Retrieved November 30, 2016, from https://fas.org/irp/offdocs/nsd/nsd42.pdf.
Rosenau, J. N. (2003). *Distant proximities: Dynamics beyond globalization*. Princeton, NJ: Princeton University Press.
Scott, W. (2006). Space control redux. *Aviation Week & Space Technology, 165*(18), 54.
Snowden, D. (2002). Complex acts of knowing: Paradox and descriptive self-awareness. *Journal of Knowledge Management, 6*(2), 100–111. https://doi.org/10.1108/13673270210424639.
Walters, D. (2016). Leadership issues in governance of complex systems. *International Journal of System of Systems Engineering IJSSE, 7*(1/2/3), 130–142. https://doi.org/10.1504/ijsse.2016.076121.

Chapter 12
CSI—A Complex System Governance Approach

Abstract Given the inherent complexity of space systems, a Complex System Governance (CSG) model may be appropriate. Keating and Ireland (2016) call CSG 'transdisciplinary,' stating that it does not belong in any particular industry or commercial sector, but instead reaches across traditional disciplines and domains. They also acknowledge that by nature, CSG inherits the strengths and weaknesses of all its parent fields, not all of which are subsumed in the integration. CSG is a new field, so it may not be mature enough to address the problem of governing 'critical space infrastructure.'

12.1 CSG Blueprint

However, Keating and Bradley (2015) built a model for CSG, which blueprints the major concerns a CSI governance construct must address. They itemize the contributions to CSG from management cybernetics, systems theory, and the VSM theory's meta-system principles. Finally, they created a notional CSG model (see Fig. 12.1) whose metasystem functions roughly correspond to their analogues in the VSM model.

A metasystem is a "set of related functions which only specify 'what' must be achieved for continuing system viability (existence), not specifying 'how' those functions are to be achieved" (Keating and Katina 2016). Keating and Bradley (2015) describe each of the functions in detail, enumerating specific responsibilities for metasystem function (see Table 12.1).

12.2 CSG Development Process

However, this CSG model is too generalized to adopt directly. Therefore, Keating and Katina (2016) developed a preliminary process to develop CSG metasystems for a given system (see Fig. 12.2). Stage 1, Initialization, both frames the system

© Springer Nature Switzerland AG 2019

A. Georgescu et al., *Critical Space Infrastructures*,
Topics in Safety, Risk, Reliability and Quality 36,
https://doi.org/10.1007/978-3-030-12604-9_12

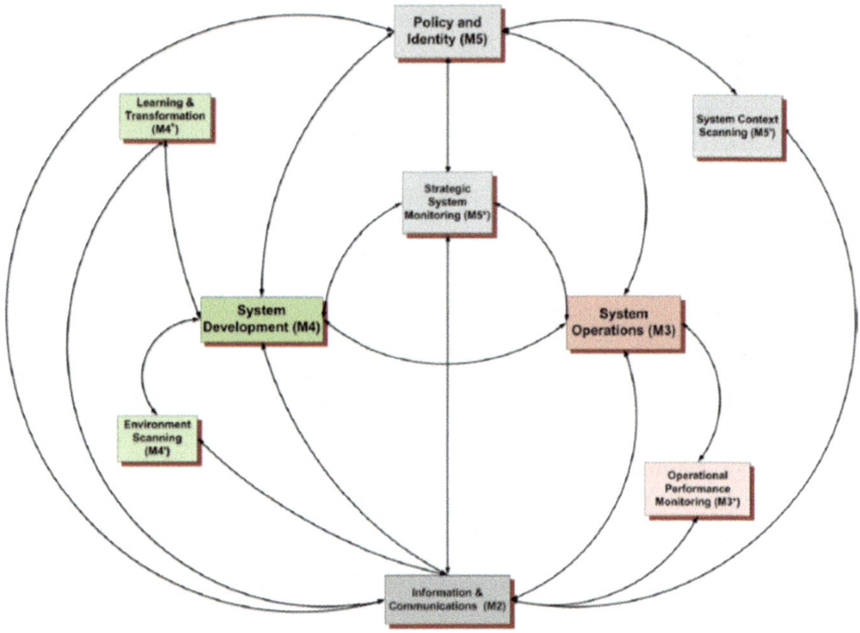

Fig. 12.1 The Notional CSG model, adapted from Keating and Bradley (2015)

Table 12.1 CSG metasystem functions

Metasystem name	Primary role
(M5)—policy and identity	Provide direction, oversight, accountability, and evolution of the system. Foci include policy, mission, vision, strategic direction, performance, and accountability for the system such that: the system maintains viability, identity is preserved and maintained, and the system is effectively projected both internally and externally
(M5*)—system context	Monitor the system context (the circumstances, factors, conditions, or patterns that enable and constrain the system). Maintains system context
(M5′)—strategic system monitoring	Monitor measures for strategic system performance and identify variance requiring metasystem level response, emphasizing variability that may impact future system viability
(M4)—system development	Provide for the analysis and interpretation of the implications and potential impacts of trends, patterns, and precipitating events in the environment. Develop future scenarios, design alternatives, and future focused planning to position the system for future viability

(continued)

Table 12.1 (continued)

Metasystem name	Primary role
(M4*)—learning and transformation	Provide for identification and analysis of metasystem design errors (second order learning) and suggest design modifications and transformation planning for the system
(M4′)—environmental scanning	Provide the design and execution of scanning for the system environment. Focus is on identification of circumstances, patterns, trends, threats, events, and opportunities for the system
(M3)—system operations	Maintain operational performance control through the implementation of policy, resource allocation, and design for accountability
(M3*)—operational performance	Monitor measures for operational performance and identify variance in system performance requiring system level response, emphasizing variability and performance trends that may impact system viability
(M2)—information and communications	Enables system stability by designing and implementing the architecture for information flow, coordination, transduction and communications within the metasystem and between the metasystem, the environment and the governed system

and investigates its environment. Once that study is complete, the developer begins Governance Readiness Level Assessment, which uses the information from the previous stage to describe or infer the state of the system's existing governance. In the second half of Stage 2, the developer addresses found deficiencies by choosing lines of effort for CSG development, based on both likelihood of success and efficacy addressing deficiencies. In Stage 3, Governance Development, the developer identifies broad, possibly overlapping efforts that improve the total effectiveness of the CSG metasystems.

However, Carter (2015) cautions that complex system governance functions and responsibilities can be viewed differently within System of Systems (SoS) and SoS Engineering (SoSE) domains, and Space Infrastructure is the archetype of a SoS/SoSE domain. Thus, the CSG development for 'critical space infrastructure' must account for the unique needs dictated by SoSE principles.

The organizations M2, M3, and M3* are the CSG metasystems that execute Communications, Operations, and Operational Performance, respectively. In a SoS like the global space enterprise, in which the participants answer to no higher authority but assume the shared roles of command and control, the management or governance tasks of these three metasystems should primarily support and mediate conflicts. Their activities should be based, as much as possible, on a single unified standard (or

Fig. 12.2 Model of CSG development process, adapted from Keating and Bradley (2015)

set), thereby reducing confusion and bureaucracy and simplifying both coordination and audits (Carter 2015).

Regarding metasystems S4 and S5 (System Development and Policy/Identity), Carter advocates for the use of accurate modeling and simulation (M&S) to help the system correctly adapt to a dynamic environment. Further, the metasystems ought to use maximize the utility of their limited resources via a decision process to rank those adaptation actions in preference order (Carter 2015).

Development of a robust CSG model for CSI is outside the scope of this paper. Therefore, the model developed will be painted in generalizations, which may be extended and specified in future work.

12.2.1 CSG Development Framing

The US-centric and Euro-centric governance models, as discussed in the previous sections, place a high premium on control. The US-centric Model emphasizes a militarized view of space control. Its proposal of a defensive space triad even echoes the nuclear defensive triad of the United States, which consists of bombers, land-based missiles, and submarine-based missiles (see Fig. 12.3). Seeing space from a military perspective will perpetuate and, in all likelihood, accelerate efforts by other countries to develop anti-satellite weapons. The Euro-centric Model stresses cooperation between governments, while admitting that gaining political concurrence, even

Fig. 12.3 The United States nuclear defensive triad

within the EU, is a difficult and time-consuming process. It should be noted that some countries have still not ratified the Outer Space Treaty of 1967, while the Moon Treaty of 1979 failed and was never ratified[1]. With no obvious solution, examining a new governance model, based on a different, modern example, is warranted.

Future of the Space Domain

Context is needed to develop a different model of complex system governance. This leads to the question: what is, or could be, the future of space? Part of the answer is radically reduced cost for space access. According to one pioneer in the space revolution, Elon Musk:

> *If one can figure out how to effectively reuse rockets just like airplanes, the cost of access to space will be reduced by as much as a factor of a hundred. A fully reusable vehicle has never been done before. That really is the fundamental breakthrough needed to revolutionize access to space (Musk* 2016*).*

While hardly a disinterested a party, Mr. Musk has already developed a reusable booster rocket that can return to Earth and land vertically. IEEE-USA has also endorsed a US-led effort to reduce the cost of access: "…there be a major effort to integrate civilian space activity into the National Aerospace System (NAS) and to incentivize greater use of standardization and commercial off-the-shelf components

[1] See the United Nations Office for Disarmament Affairs: http://disarmament.un.org/treaties/t/outer_space.

and vehicles."[2] With much lower access costs, commercial interests of all sizes will have access to space. Economic benefits could be considerable.

The Internet Model
A technological revolution in space access could lead to an explosion in commercial activity and a corresponding dramatic increase in prosperity. The Washington Post summarizes the results of Internet deregulation as follows: "The Net *has been* open since it was privatized by the Clinton administration. It proliferated globally as it migrated farther away from government control—bringing freedom and prosperity to billions. It grew from a mere 88,000 users in the late 1980s, to more than 3 billion today. Cisco estimates that the exploding "Internet of Everything" (that is, machines talking to one another online, such as your car and your tablet) will generate more than $14 trillion in global economic growth by 2022. In short, the Internet is the greatest deregulatory success story of all time—a simple fact that vexes those seeking new and unnecessary rules" (McDowell 2014).

Internet Governance
How is the Internet governed? With the exception of a few authoritarian governments that attempt to control content (China is the best-known example), the Internet is directed by commercial interest and regulated via near-universal access and transparency. While control by any one country, corporation, or individual is minimal, participants exert sufficient control to maintain viability. Tremendous redundancy in Internet infrastructure means that, according to David Clark of MIT, it cannot be destroyed: '*A vast behemoth that can route around outages and self-heal, the Internet has grown physically invulnerable to destruction by bombs, fires or natural disasters—within countries, at least*' (Wolchover 2012). Other elements of complex system governance (communication, coordination, integration, and monitoring) are strongly supported by the Internet system of systems.

12.2.2 CSG Development: Readiness Level Assessment

The second phase of developing a CSG model is the Governance Readiness Level (GRL) Assessment, which tries to assemble knowledge from the Framing phase to understand both the state of the system's governance and future actions to develop it (Keating and Katina 2016). Specifically, the expert, in-depth analysis will allow the selfsame experts to classify the system, then identify the improvement lines of effort with greatest potential. Those lines of effort most likely to solve governance problems are ranked by the experts in terms of their impact on significant problems and their probability of success (Keating and Katina 2016).

---

[2]See IEEE Position Statement by the Board of Directors: https://ieeeusa.org/wp-content/uploads/2017/02/SpaceAccess0214.pdf.

Two of the major deficiencies in this project are a lack of subject matter experts and a shortage of time in which to engage them on all things 'critical space infrastructure', Governance, and Decision-making. The authors 'expertise' on the first two subjects is derived solely from research. Thus, to avoid overreach, the CSG model developed in the next section is painted with exceedingly wide brush strokes, based on the Framing section.

12.2.3 CSG Development: Space Governance

If commercial uses predominate and proliferate in space, governance should be designed to reflect that reality. The attributes of CSG would be applied as follows under a "WikiSpace" scenario:

Control
Control would be exerted and governed by user self-interest, with a very large number of users helping to maintain viability.

Communications
Communications between all parties must be instantaneous and constantly maintained. This could be fostered by government efforts, as for example, how the Defense Advanced Research Projects Agency (DARPA) did with the Internet.

Coordination
Commercial organizations must take the lead in coordination, motivated by mutual benefit. A transition to commercial dominance and diminishment of military activities could foster better coordination.

Integration
Integration would be handled by commercial standards organizations. Investments by developed countries could stimulate this activity.

Monitoring
Monitoring would be the responsibility of all parties. Monitoring could initially be provided by the United States as a common operating picture of space (COP) freely available to all users. A COP would provide transparency if disabling or destroying a rogue satellite becomes necessary. With a COP, there would be no questions at to intentions and much less suspicion.

Integration
Integration would be handled by commercial standards organizations. Investments by developed countries could stimulate this activity.

Intelligence
Intelligence on the space activities of rogue nations should become the major avenue of investment by governments. With a reduced emphasis on development of satellites and ASAT capabilities (no longer needed if highly redundant commercial capabilities are used by governments), intelligence could be gathered and provided to the commercial space community.

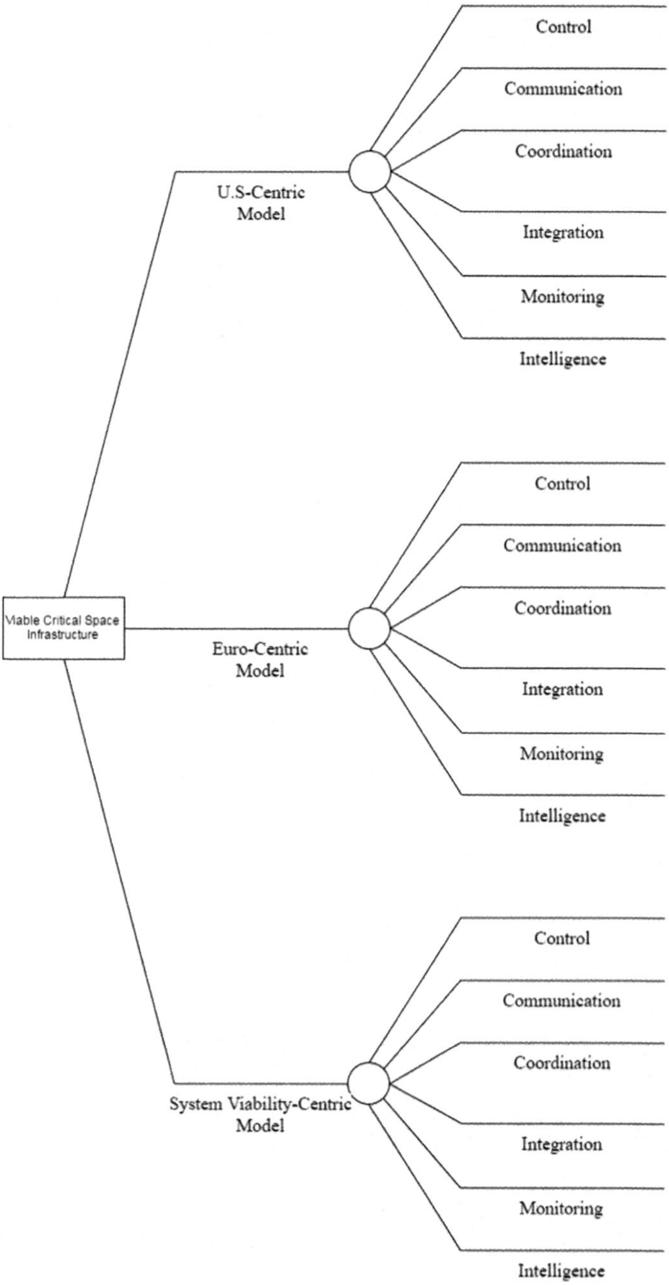

Fig. 12.4 A decision tree model

12.3 Analysis via Multiple Decision Methods

12.3.1 Party Problem

The three models for 'critical space infrastructure' governance will be evaluated with the 'party problem' approach, which ranks all prospects in order from best to worst and uses their relative utility to choose the best alternative. Again, and for the sake of simplicity, authors compressed the 14 criteria into their parent attributes and evaluated all attribute-model combinations. The associated decision tree is shown in Fig. 12.4.

Order Rule

Based on the same data set as the ER model, the authors converted the ER labels to numeric equivalents, as shown in Table 12.2. Best and worst matches became 1 and 0 respectively, and the rest of the labels were assigned numbers in increments of 0.25. That data was averaged for each of the decision tree branches and sorted in descending order (Fig. 12.5).

Equivalence Rule

According to the Equivalence rule, all prospects' preferences can be equated to a chance at the best versus worst outcomes. For any given prospect, the rational decision-maker will be indifferent to receiving either the prospect or a best/worst chance (Howard and Abbas 2015). Figure 12.6 enumerates the preference probabilities for all prospects. In situations that contain measurable probability, the decision-maker can decompose some nodes in the tree into probability trees. However, this evaluation of theory contains no measurable probabilities, so that step will be omitted.

Substitution Rule

Having assigned preference probabilities to every prospect, the 'party problem' method allows the decision-maker to substitute equivalent deals into the decision tree (Howard and Abbas 2015). Figure 12.7 depicts the third of the original decision tree relevant to the US-centric model, appending the relevant preference probabilities to each of the criteria. The equations below the model compute the utility of the US-centric model as a whole by combining the preference probabilities of all six attributes (equally weighted).

Table 12.2 ER data set conversion table

ER label	Assigned
Best	1
Good	0.75
Avg.	0.50
Poor	0.25
Worst	0

Fig. 12.5 Order rule for all possible prospects

B: Best

CSG centric model - Intelligence

CSG centric model - Integration

CSG centric model - Control

CSG centric model - Communication

U.S centric model - Coordination

U.S centric model - Monitoring

U.S centric model - Intelligence

Euro centric model - Coordination

Euro centric model - Monitoring

Euro centric model - Intelligence

CSG centric model - Monitoring

U.S centric model - Control

U.S centric model - Communication

Euro centric model - Communication

Euro centric model - Control

Euro centric model - Integration

CSG centric model - Coordination

W: Worst

U.S centric model - Integration

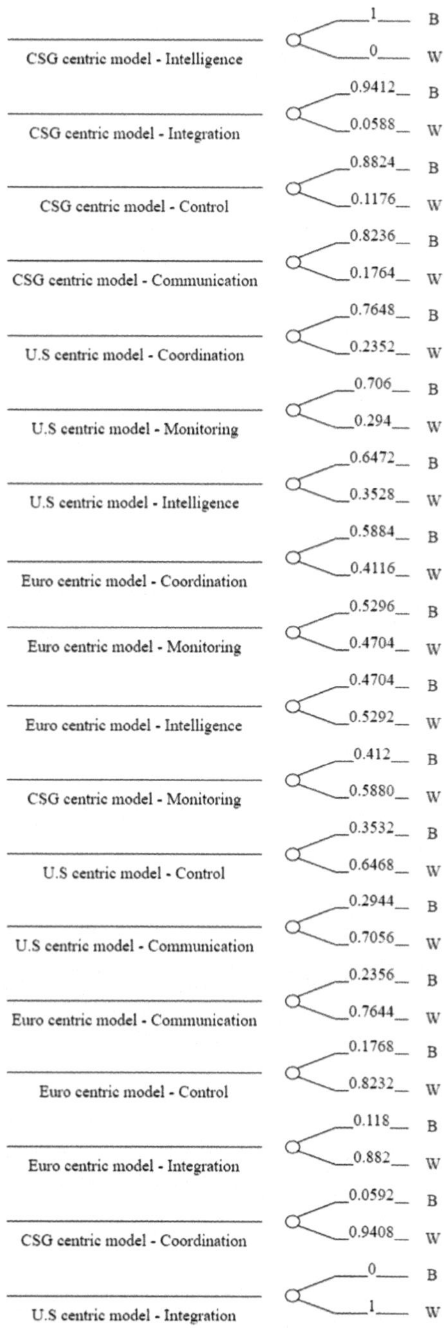

Fig. 12.6 Equivalence rule for all possible prospects

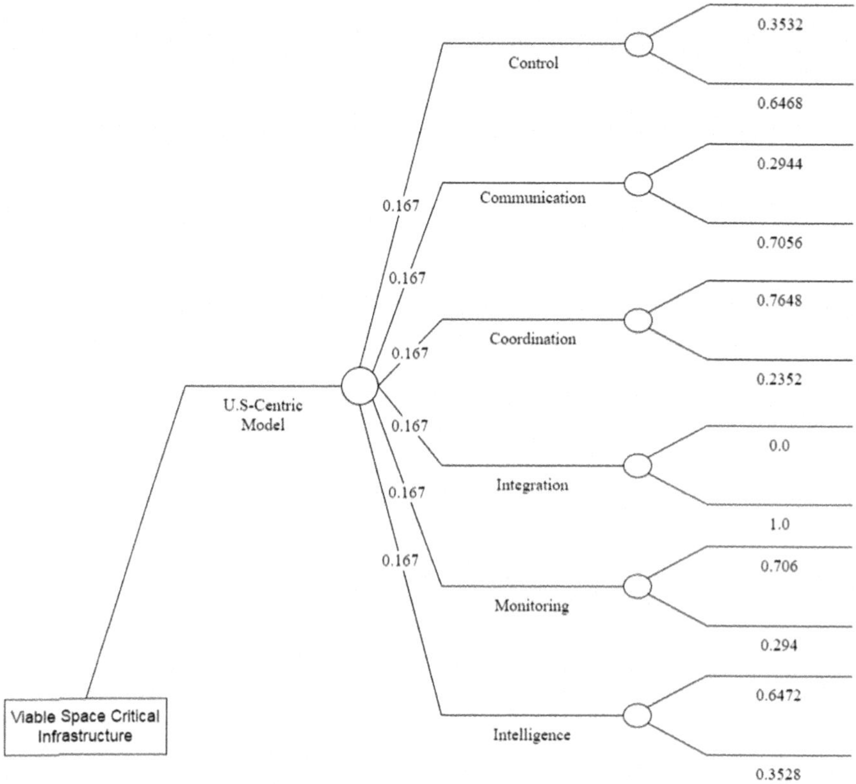

Fig. 12.7 Substitution diagram for US-centric model

Therefore, the preference/utility for US-centric model is:

$$P_{US} = 0.1667 * (0.3532 + 0.2944 + 0.7648 + 0 + 0.706 + 0.6472) = \mathbf{0.4610}$$

Since this is a binary choice, the corresponding chance at the worst-case outcome is:

$$1 - P_{US} = 1 - 0.4610 = \mathbf{0.5390}$$

Figure 12.8 depicts the substitution Diagram for Euro-centric Model, where:

$$P_{Euro} = 0.1667 \times (0.1768 + 0.2356 + 0.5884 + 0.118 + 0.5296 + 0.4708) = \mathbf{0.3533}$$
$$1 - P_{Euro-centric} = 1 - 0.3533 = \mathbf{0.6467}$$

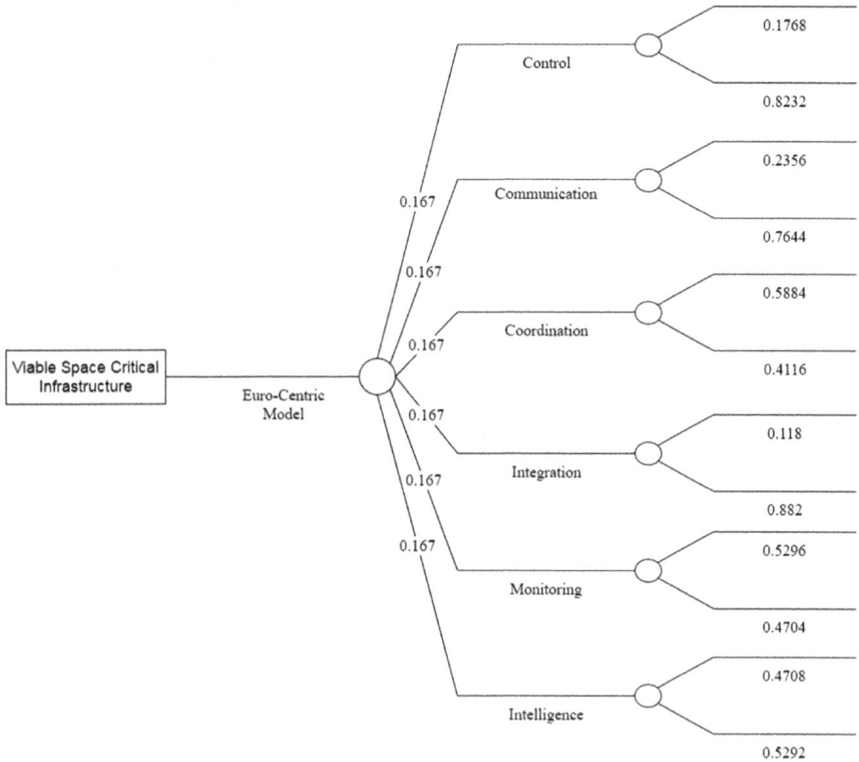

Fig. 12.8 Substitution diagram for Euro-centric model

Figure 12.9 is the substitution Diagram for CSG/Viability-centric Model, where:

$$P_{VSM-CSG} = 0.1667 * (0.8824 + 0.8236 + 0.0592 + 0.9412 + 0.412 + 1.0) = \mathbf{0.6865}$$
$$1 - P_{VSM-CSG} = 1 - 0.6865 = \mathbf{0.3135}$$

Finally, the Preference Probabilities for the three models are presented in Fig. 12.10.

Choice Rule

The choice rule is the party problem's mandate that a rational decision-maker must choose the alternative with the greatest preference probability. Thus, according to Table 12.3, the 'party problem' method supports the CSG/Viability model for governing CSI.

The model's strength lies in its strong performance across the Control, Communication, and Integration criteria. The Euro-centric model performed well in the Coordination, Integration and Monitoring criteria, which allowed it to edge out the US-centric model, whose main strength is Monitoring.

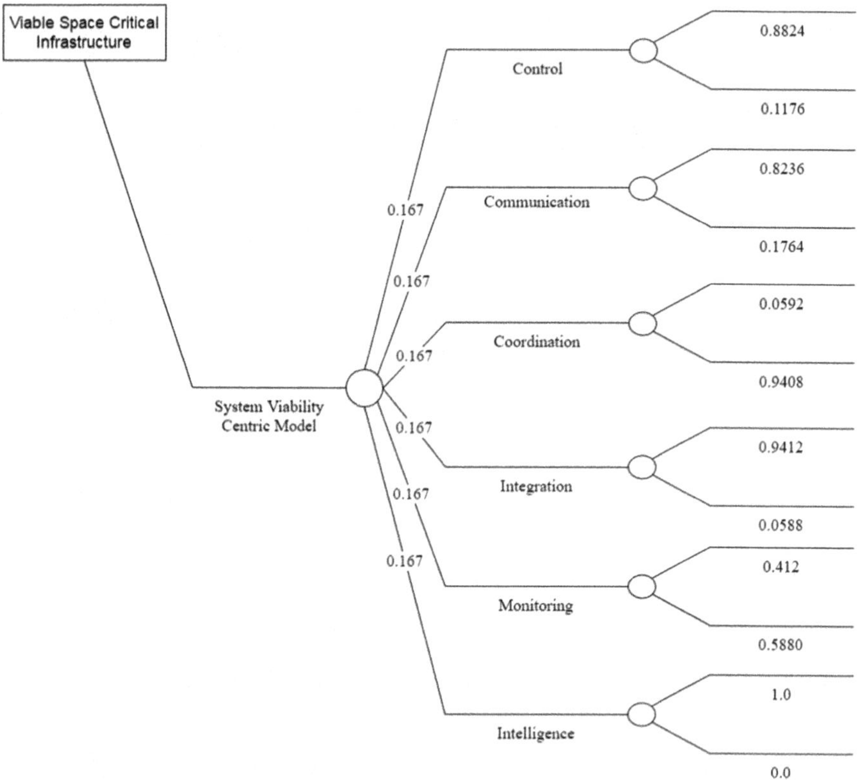

Fig. 12.9 Substitution diagram for CSG/Viability-centric model

12.3.2 Fuzzy Sets

Fuzzy Set Methodology

Fuzzy set methodology uses merit and projections of future behavior to evaluate the merit of different alternatives (Katina and Unal 2015). One of the strengths of the Fuzzy Set methodology is the ability to accept linguistically imprecise words as input and convert them to more precise data before computation. The method assigns grades of membership to possible system states, based on the probability that the system will assume that state in the future, then each alternative is rated on its performance in each state. The theory constructs utility functions for each criterion of merit and selects the alternative with the greatest total utility.

Choosing System States

In order to apply fuzzy set methodology to this problem, we must first identify the different possible states in which the system might exist during its life cycle. Unlike decision alternatives or criteria of merit, these states will represent different levels of functionality or configuration of the given system. The chosen states for

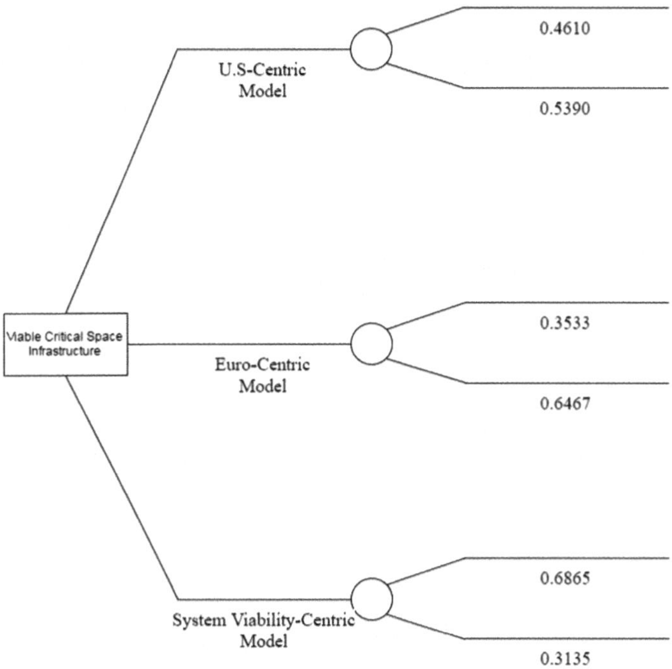

Fig. 12.10 Preference probabilities for the three models

Table 12.3 Model ranking by equivalent preference probability

Model	Preference probability
CSG/Viability-centric model	0.6865
US-centric model	0.4610
Euro-centric model	0.3533

the problem represent system functionality based on overall percentage of required system components (e.g., satellites, ground controls). Each alternative is evaluated either numerically or linguistically for each system state for a given criterion of merit. Table 12.4 provides a linguistic evaluation. It is important to note that, experts are consulted in the gathering of the related information.

System State 1 will refer to a system that is significantly degraded because an unintentional or intentional event has caused system damage. This state, given a membership grade of 0.1 is relatively unlikely to occur, but must still be considered. System State 2 corresponds to 75% component functionality and will represent significant but non-critical system degradation. The system will still act in order to maintain critical services but will little or no redundant capacity.

System State 3 corresponds to 100% component functionality with minimal redundancy. In this state, all operations are normal, and the system does not have to take unplanned actions for maintenance or routine component outages. System State 4

Table 12.4 Linguistic evaluations

Functioning components	50%		75%	100%	125%
System state	1		2	3	4
Holistic control (entire SoS)	US	Very Poor	Poor	Average	Good
	EU	Poor	Average	Good	Good
	CSG	Average	Good	Good	Best
Defense of current systems	US	Average	Good	Good	Best
	EU	Average	Average	Average	Average
	CSG	Poor	Poor	Poor	Poor
Defense of current systems	US	Poor	Average	Good	Good
	EU	Poor	Average	Average	Average
	CSG	Average	Good	Good	Best
Timeliness	US	Poor	Average	Average	Good
	EU	Very Poor	Poor	Average	Average
	CSG	Average	Good	Good	Good
Effectiveness	US	Poor	Poor	Average	Average
	EU	Average	Good	Good	Best
	CSG	Good	Good	Good	Best
Accuracy	US	Poor	Average	Average	Good
	EU	Poor	Good	Good	Good
	CSG	Average	Good	Good	Best
Reaches all systems	US	Very Poor	Poor	Average	Average
	EU	Poor	Average	Average	Good
	CSG	Average	Average	Good	Good
Provides safety and security information	US	Average	Good	Good	Best
	EU	Average	Good	Good	Good
	CSG	Very Poor	Poor	Average	Average
Continuous maintenance	US	Poor	Poor	Average	Good
	EU	Poor	Average	Good	Good
	CSG	Average	Good	Good	Best
Minimal necessary (integration)	US	Poor	Average	Average	Average
	EU	Average	Average	Average	Average
	CSG	Average	Average	Average	Good

(continued)

Table 12.4 (continued)

Functioning components	50%		75%	100%	125%
View of entire SoS	US	Poor	Average	Average	Good
	EU	Average	Average	Good	Good
	CSG	Average	Average	Good	Best
Continuous updating	US	Poor	Average	Good	Best
	EU	Poor	Average	Average	Good
	CSG	Average	Average	Average	Average
State of rogue actors	US	Good	Good	Good	Best
	EU	Poor	Average	Average	Good
	CSG	Average	Poor	Poor	Best
Sharable intelligence	US	Average	Average	Average	Average
	EU	Good	Good	Good	Good
	CSG	Good	Good	Best	Best

Table 12.5 Fuzzy membership function for chosen system states

System state	1	2	3	4
Description	50% functionality	75% functionality	100% functionality	125% functionality
Membership	0.1	0.3	0.8	0.65

corresponds to the ideal state of the system, in which operations are nominal and it possesses 25% or more redundancy at the component level. This state will allow the system to respond to unplanned events without any impact to critical services.

Sample Calculations and Results

To begin a sample calculation, we must first establish the membership grade function associated with the system states. Table 12.5 provides the chosen membership function for the different states the system can occupy.

Next, each criteria of merit are based on a numerical or linguistic evaluator. Since quantifiable data is rare in this situation, linguistic evaluators will be used throughout the exercise. Table 12.6 describes the possible linguistic evaluators assigned to the given criteria and their associated membership for each system state. The decomposed attributes described in previous sections will serve as the problem's criteria of merit.

Table 12.7 shows the calculation of merit for the criteria Shareable Intelligence, including the linguistic assessments of each alternative in each system state for the criterion of shareable intelligence. Each criterion of merit will be computed similarly to identify the optimal alternative for each criterion.

Once we have established our definitions and completed evaluations of each alternative for the criterion of shareable intelligence, we can now create the fuzzy utility set.

Table 12.6 Chosen definitions of fuzzy linguistic evaluators

Description	State 1	State 2	State 3	State 4
Very poor	0.04	0.09	0.25	0.36
Poor	0.2	0.3	0.5	0.6
Average	0.25	0.4	0.6	0.7
Good	0.3	0.5	0.7	0.8
Best	0.4	0.6	0.8	0.9

Table 12.7 Linguistic evaluations of alternatives

Shareable intelligence	State 1	State 2	State 3	State 4
US-centric	Average	Average	Average	Average
EU-centric	Good	Good	Good	Good
CSG/Viability-centric	Good	Good	Best	Best

$$U^f_{US} = \{(0.1, \text{Average}), (0.3, \text{Average}), (0.8, \text{Average}), (0.65, \text{Average})\}$$
$$U^f_{EU} = \{(0.1, \text{Good}), (0.3, \text{Good}), (0.8, \text{Good}), (0.65, \text{Good})\}$$
$$U^f_{CSG} = \{(0.1, \text{Good}), (0.3, \text{Good}), (0.8, \text{Best}), (0.65, \text{Best})\}$$

The US-centric model's fuzzy utility set decomposes into:

$$\begin{aligned} U^f_{US} = \{&(0.1, \{(0.25, 1), (0.4, 2), (0.6, 3), (0.7, 4)\}, \\ &(0.3, \{(0.25, 1), (0.4, 2), (0.6, 3), (0.7, 4)\}, \\ &(0.8, \{(0.25, 1), (0.4, 2), (0.6, 3), (0.7, 4)\}), \\ &(0.65, \{(0.25, 1), (0.4, 2), (0.6, 3), (0.7, 4)\})\} \end{aligned}$$

Taking the minimum of the two grades of membership when distributing the function membership allows us to create one large fuzzy set containing all membership grades with associated system states.

$$\begin{aligned} U^f_{US} = \{&(0.1, 1), (0.1, 2), (0.1, 3), (0.1, 4), (0.25, 1), (0.3, 2), (0.3, 3), (0.3, 4), (0.25, 1), \\ &(0.4, 2), (0.6, 3), (0.7, 4), (0.25, 1), (0.4, 2), (0.6, 3), (0.65, 4)\} \end{aligned}$$

We can then perform the union operation on memberships sharing the same system state and produce the consolidated utility function immediately below. The same method is used to produce the utility sets of the remaining alternatives.

$$U^f_{US} = \{(0.344, 1), (0.743, 2), (0.946, 3), (0.937, 4)\}$$
$$U^f_{EU} = \{(0.344, 1), (0.760, 2), (0.979, 3), (0.957, 4)\}$$

Table 12.8 Fuzzy alternative results

Criteria of merit	A^f			A^f_N		
	US	EU	CSG	US	EU	CSG
Holistic control (entire S-of-S)	0.957	0.979	0.979	0.978	1.000	1.000
Defense of current systems	0.979	0.946	0.888	1.000	0.966	0.907
Defense of future systems	0.979	0.946	0.979	1.000	0.966	1.000
Timeliness	0.957	0.946	0.979	0.978	0.966	1.000
Effectiveness	0.946	0.979	0.979	0.966	1.000	1.000
Accuracy	0.957	0.979	0.979	0.978	1.000	1.000
Reaches all systems	0.946	0.957	0.979	0.966	0.978	1.000
Provides safety and security info	0.979	0.979	0.946	1.000	1.000	0.966
Continuous maintenance	0.957	0.979	0.979	0.978	1.000	1.000
Minimal necessary (integration)	0.946	0.946	0.957	0.988	0.988	1.000
View of entire S-of-S	0.957	0.979	0.979	0.978	1.000	1.000
Continuous updating	0.979	0.957	0.946	1.000	0.978	0.966
State of rogue actors	0.979	0.957	0.979	1.000	0.978	1.000
Shareable intelligence	0.946	0.979	0.990	0.955	0.988	1.000

$$U^f_{CSG} = \{(0.344, 1), (0.760, 3), (0.990, 3), (0.971, 4)\}$$

Once the fuzzy utility sets of each alternative are complete, the system state with maximum utility is selected for the alternative function. The alternative function is then produced from the highest utility of each alternative.

$$A^f = \{(0.946, a_{US}), (0.979, a_{EU}), (0.990, a_{CSG})\}$$

When comparing these alternatives, it becomes clear that the CSG based model is the best alternative because it provides the highest utility; however, the alternative function can be normalized to evaluate the relative utility of each alternative.

$$A^f_N == \{(0.955, a_{US}), (0.988, a_{EU}), (1, a_{CSG})\}$$

This process is then repeated for all criteria of merit. The results have been placed in Table 12.8. Both the actual and normalized attribute sets are important for later use.

One final step is required to determine the best alternative across all criteria of merit. For each criterion, the highest utility produced is selected for inclusion in the fuzzy decision set. If two alternatives share the highest utility for a given criteria, both are included in the decision set. Once the decision set is created, utilities for similar alternatives are consolidated using the union operation. In this instance, the

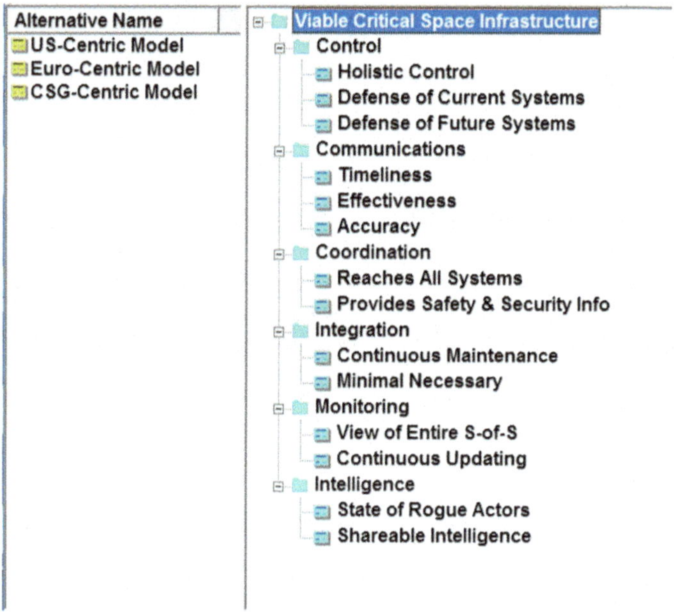

Fig. 12.11 ER attribute hierarchy

values for the decision set are extremely close to one and are therefore displayed in scientific notation. The final fuzzy decision set for this problem is:

$$D^f = \left\{ \left(1 - 4.08 * 10^{-9}, a_{US}\right), \left(1 - 8.58 * 10^{-11}, a_{EU}\right), (1, a_{CSG}) \right\}$$

Overall, the fuzzy set methodology supports the CSG/Viability model for protecting 'critical space infrastructure'.

12.3.3 Evidential Reasoning (Intelligent Decision Systems)

Using the decision attributes developed and defined in Sect. 11.4 (Chap. 11) and *Intelligent Decision System for Multiple Criteria Assessment* software, the authors analyzed the scenarios developed in Sect. 11.5 (Chap. 11), Sect. 11.6 (Chap. 11) and Sect. 3, 4, 5 of this chapter.

Evidential Reasoning Hierarchy

The following Evidential Reasoning hierarchy (see Fig. 12.11) was developed using IDS software. The software supports entry of the hierarchy and then provides a number of options depending on the type of attribute (quantitative vs. qualitative) and the value measurement of the attribute.

Table 12.9 ER scenario ranking and degree of belief

CSG attributes	US-centric model	Euro-centric model	CSG/Viability model
Holistic control (entire SoS)	Poor 0.7	Avg. 0.7	Good 0.8
Defense of current systems	Best 0.7	Avg. 0.7	Poor 0.8
Defense of future systems	Avg. 0.7	Avg. 0.7	Best 0.8
Timeliness	Avg. 0.7	Poor 0.7	Good 0.8
Effectiveness	Poor 0.7	Good 0.7	Best 0.8
Accuracy	Best 0.7	Good 0.7	Poor 0.8
Reaches all systems	Poor 0.7	Avg. 0.7	Good 0.8
Provides safety and security info	Best 0.7	Good 0.7	Poor 0.8
Continuous maintenance	Poor 0.7	Avg. 0.7	Best 0.8
Minimal necessary	Poor 0.7	Avg. 0.7	Good 0.8
View of entire SoS	Poor 0.7	Good 0.7	Best 0.8
Continuous updating	Best 0.7	Avg. 0.7	Poor 0.8
State of rogue actors	Good 0.7	Avg. 0.7	Best 0.8
Shareable intelligence	Avg. 0.7	Good 0.7	Best 0.8

Evidential Reasoning: Ranking by Scenario

The three future scenarios for 'critical space infrastructure' were ranked using IDS software's categories of Best-Good-Average-Poor-Worst and a Degree of Belief from 1 to 0. Since the scenarios the authors evaluated are not exhaustive (for example, the alternative 'take no action' is absent), they frequently avoided the Best and Worst ratings. Similarly, Degree of Belief is always less than one, because the scope of this project did not include soliciting feedback from experts in the field, which would have raised the confidence interval. If more information were available (ER works best when there is ample evidence), the authors would have undertaken a more comprehensive analysis. Therefore, this analysis should be considered preliminary, with more evidence needed for a complete view. Uncertainty could have been applied using a probability distribution, but the uncertainty of the bottom-level attributes would have compounded with the probability function. Note that the higher belief associated with the CSG/Viability Model reflects the probability that commercial interests rather than military/government systems will dominate the space domain. Table 12.9 provides ER Scenario ranking and degree of belief.

Overall Results of ER Analysis

Figure 12.12 provides an overall view of analysis results, with the CSG/Viability Model developed in this document ranking highest. Since the model was developed from a holistic, SoS view of the space domain, it is not surprising that the model excelled when compared to the US-centric and Euro-centric scenarios. Note that the

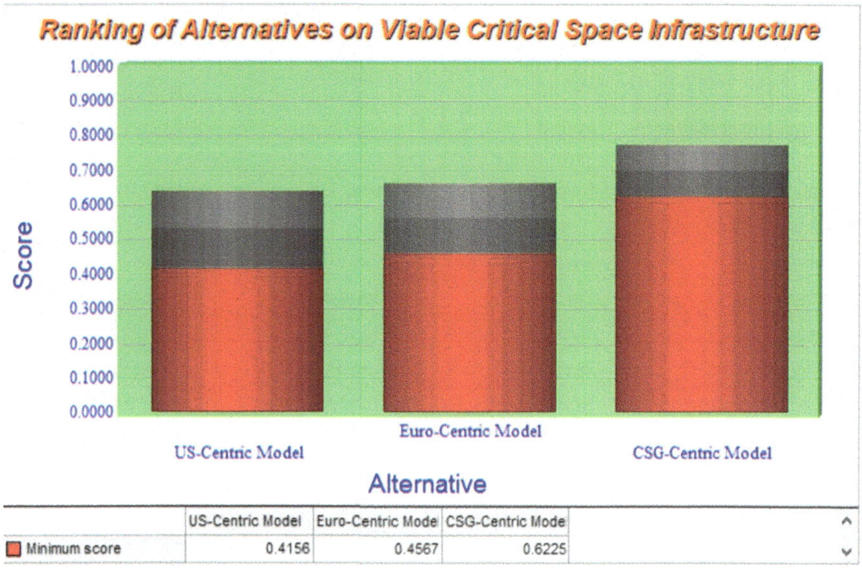

Fig. 12.12 Bar graph of ER analysis results by scenario

gray area of each bar graph reflects uncertainty in the attributes selected and their completeness.

Ranking of Scenarios by Attribute
Figure 12.13 illustrates the ranking of the three examined scenarios overall and by attribute. Note that the CSG/Viability Model (red) ranked higher against most attributes when compared to the US-centric (green) and Euro-centric (blue) models. In most cases the CSG/Viability scenario ranked higher due to the high degree of redundancy in a system dominated by multiple commercial interests and the cooperation (generated by self-interest) across the majority of systems. If commercial interests have a large stake in the domain, governments will follow. Again, this Internet-like model with accelerated growth due to major advances in technology.

Scenario Utility
Figure 12.14 illustrates the utility of each scenario over the range of values used in the IDS ER application (note again that red represents the CSG/Viability scenario, blue the Euro-centric scenario, and green the US-centric scenario). Since the attributes derived from Complex System Governance are considered to be essential to the viability of a system, the authors assigned equal utility across the board. If some of the attributes were not met to any degree, the system would fail.

Analytic Hierarchy Process
Using *Logical Decisions for Windows* (LDW), the authors completed an AHP Analysis of the space governance models. Each alternative was rated against the others for every criterion (called Measures in LDW), and the resulting utility scores combined

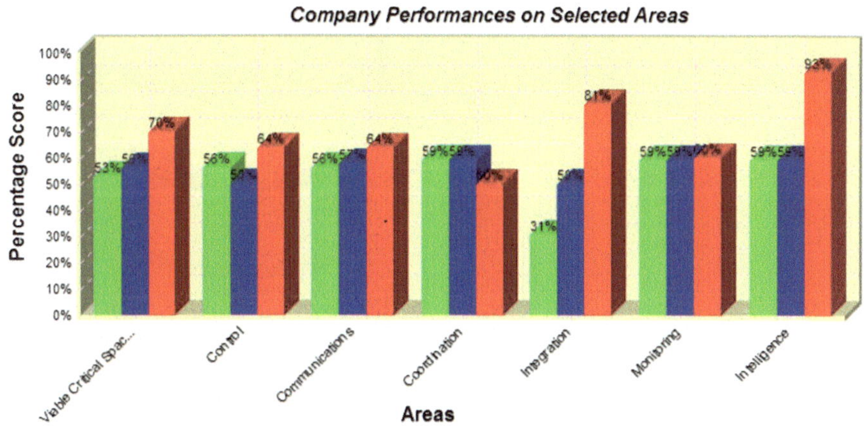

Fig. 12.13 ER ranking of scenarios on an attribute basis

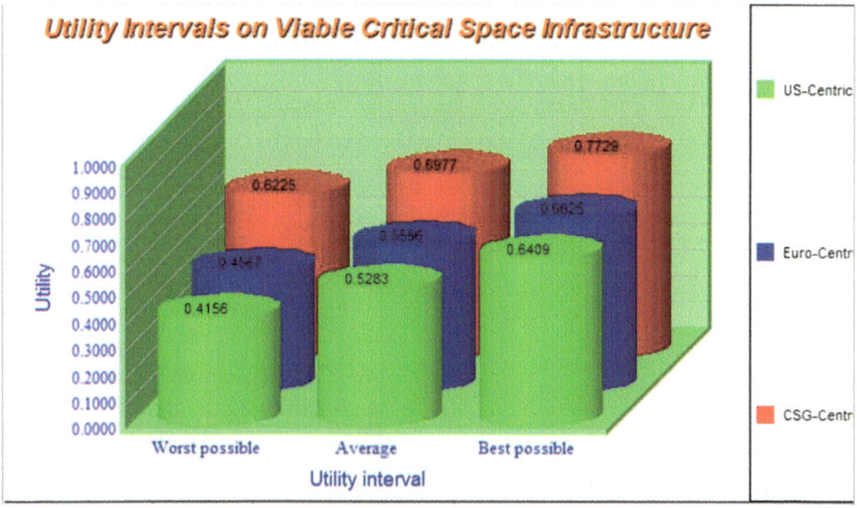

Fig. 12.14 ER scenario utility rankings

using a Multi-variable Utility Function to generate final utility scores and rankings. The LDW software provided primary analysis, but the authors used Microsoft Excel AHP matrices (Landaeta 2015) to handle data at the attribute level.

AHP Structure

The AHP process used the hierarchy of attributes and criteria is shown in Fig. 12.15. This structure is based on underpinnings suggested in Chapter XI.

Assessing Scenarios by Attribute and Criteria

The ER data is not usable for AHP, so the ER scoring matrix was mapped to the AHP preference structure. Recall that ER data is a matrix of alternatives versus grading

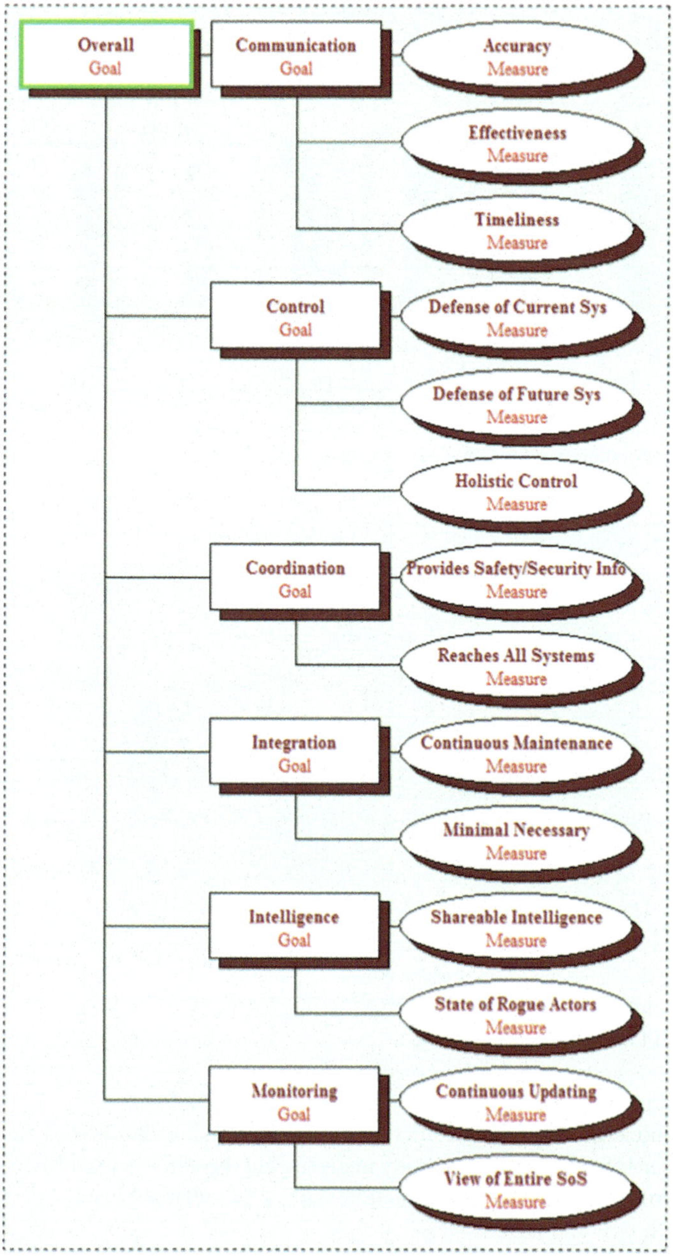

Fig. 12.15 LDW model setup for AHP

Table 12.10 Conversion table for creating AHP preference from ER label

	Worst	Poor	Average	Good	Best
Worst	**1.00**	0.33	0.20	0.14	0.11
Poor	3.00	**1.00**	0.33	0.20	0.14
Average	5.00	3.00	**1.00**	0.33	0.20
Good	7.00	5.00	3.00	**1.00**	0.33
Best	9.00	7.00	5.00	3.00	**1.00**

Table 12.11 Example data from ER table

CSG criteria	US-centric ER label	Euro-centric ER label	CSG/Viability ER label
Holistic control (entire SoS)	Poor, 0.7	Average, 0.7	Good, 0.7

Table 12.12 AHP matrix for criterion: holistic control

Criterion: Holistic Control			
	US-centric	Euro-centric	CSG/Viability
US-centric	**1.00**	0.33	0.20
Euro-centric	3.00	**1.00**	0.33
CSG/Viability	5.00	3.00	**1.00**

criteria, with each matrix cell ranked as Best, Good, Average, Poor, and Worst. (AHP disregards confidence estimates.) However, AHP is a set of matrices, one per assessment criteria, of alternatives rated against alternatives. For each criterion's AHP matrix, the authors compared the ER ratings of each alternative pair (CSG-Euro, CSG-US, and Euro-US) and used the difference to assign an AHP preference in the range 1–9, inclusive. According to the AHP method for describing degrees of preference (Saaty 2008), Table 12.10 was constructed as a reference.

For example, examine the first line of the ER table, partly reproduced below (Table 12.11). In that table, the CSG/Viability model has the ER rating (Good), preferable to the Euro-centric model (Average). Consulting the (row, column) pair (Good, Average) of Table 12.7 returns the preference "3", which in the AHP model signifies "moderate importance… experience and judgment slightly favor one activity over another" (Saaty 2008, p. 86). This preference value, 3, is assigned to the cell (CSG, Euro) in Table 12.7, the Holistic Control AHP matrix. The next comparison, Euro vs US, uses the labels (Average, Poor), so the (Euro, US) cell in Table 12.12 becomes 3. Finally, the (CSG, US) comparison uses the labels (Good, Poor), which is assigned a preference of 5 in Table 12.7. Therefore, the bottom left cell in Table 12.12, (CSG, US), becomes 5. This method was used to fill out a preference matrix for all 14 criteria in the analysis.

Overall Results of AHP

Because this is a Complex System Governance problem, all segments of the governance metasystem are equally necessary for functional governance. The implementing preferences in an AHP and the resulting matrix are provided in the following section.

12.4 Analytic Hierarchy Process and Preferences

The Analytic Hierarchy Process (AHP) approach allows the decision-maker to set up preferences between assessment criteria or groups of them. A grouping the 14 criteria (Tables 12.13 and 12.14) according to their respective parent attributes is indicated.

Table 12.13 AHP criteria preference Matrices

	Holistic Control	Defense Current	Defense Future
Holistic Control	1.00	0.20	3.00
Defense Current	5.00	1.00	0.33
Defense Future	0.33	3.00	1.00

	Timeliness	Effectiveness	Accuracy
Timeliness	1.00	3.00	1.00
Effectiveness	0.33	1.00	0.33
Accuracy	1.00	3.00	1.00

	Global Reach	Safety/Security
Global Reach	1.00	0.20
Safety/Security	5.00	1.00

	Continuous Maint	Minimal Req'd
Continuous Maint	0.50	0.50
Minimal Req'd	0.50	0.50

	View of Entire SoS	Continuous Updating
View of Entire SoS	1.00	1.00
Continuous Updating	1.00	1.00

	State of Rogue Actors	Shareable Intelligence
State of Rogue Actors	1.00	5.00
Shareable Intelligence	0.20	1.00

Table 12.14 AHP attribute preference matrices

	Control	**Comm**	**Coord**	**Integration**	**Monitoring**	**Intel**
Control	**1.00**	1.00	2.00	4.00	3.00	3.00
Comm	1.00	**1.00**	2.00	4.00	3.00	3.00
Coord	0.50	0.50	**1.00**	3.00	2.00	2.00
Integration	0.25	0.25	0.33	**1.00**	0.50	0.50
Monitoring	0.33	0.33	0.50	2.00	**1.00**	1.00
Intel	0.33	0.33	0.50	2.00	1.00	**1.00**

Table 12.15 Preference-based AHP sub-scores

	Without Preferences				With Preferences			
	US	Euro	CSG	Normalized Weight	US	Euro	CSG	Normalized Weight
Holistic Control	0.11	0.26	0.63	0.0556	0.11	0.26	0.63	0.050
Defense Current	0.72	0.19	0.08	0.0556	0.72	0.19	0.08	0.061
Defense Future	0.13	0.25	0.63	0.0556	0.13	0.25	0.63	0.055
AHP Scores	**0.318**	**0.233**	**0.448**		**0.340**	**0.231**	**0.429**	
Timeliness	0.26	0.11	0.63	0.0556	0.26	0.11	0.63	0.071
Effectiveness	0.07	0.28	0.64	0.0556	0.07	0.28	0.64	0.024
Accuracy	0.64	0.28	0.07	0.0556	0.64	0.28	0.07	0.071
AHP Scores	**0.326**	**0.224**	**0.450**		**0.398**	**0.207**	**0.395**	
Global Reach	0.11	0.26	0.63	0.0833	0.11	0.26	0.63	0.028
Safety/Security	0.64	0.28	0.07	0.0833	0.64	0.28	0.07	0.139
AHP Scores	**0.375**	**0.272**	**0.354**		**0.554**	**0.279**	**0.167**	
Continuous Maint	0.08	0.19	0.72	0.0833	0.08	0.19	0.72	0.083
Minimal Req'd	0.11	0.26	0.63	0.0833	0.11	0.26	0.63	0.083
AHP Scores	**0.095**	**0.227**	**0.678**		**0.095**	**0.227**	**0.678**	
View of Entire SoS	0.07	0.28	0.64	0.0833	0.07	0.28	0.64	0.083
Continuous Updating	0.72	0.19	0.08	0.0833	0.72	0.19	0.08	0.083
AHP Scores	**0.399**	**0.238**	**0.363**		**0.399**	**0.238**	**0.363**	
State of Rogue Actors	0.26	0.11	0.63	0.0833	0.26	0.11	0.63	0.139
Shareable Intelligence	0.11	0.26	0.63	0.0833	0.11	0.26	0.63	0.028
AHP Scores	**0.235**	**0.132**	**0.633**		**0.235**	**0.132**	**0.633**	

The most noticeable impact of the preferences was to raise the score of the US-centric model, almost exclusively at the expense of the CSG/Viability model. The tables on the following page detail the changes. However, it must be noted that the preference selection process, might be suspect as it could be influenced by expert background. For example, if one (the expert) spent time spent in the US military, this background could influence their viewpoints. In this particular case, preference-based AHP scores and ranks are shown in Tables 12.15 and 12.16. Table 12.17 provides AHP score change with preferences.

However, to correctly handle the lack of preference in Complex System Governance theory, all criteria were weighted equally within their respective attributes, which were themselves weighted equally across the overall goal.

Note that an AHP matrix without preference rankings reduces to a series of flat multi-dimensional variable function (MDVF). Without weights or any kind, those MDVFs reduce to simple averages. Thus, the authors used a Multiple Utility Function in LDW to average the scores of each alternative model across the attributes. Table 12.18 details the scores of each scenario across all criteria, across each attribute, and the weights used in the MUF.

Table 12.16 Preference-based AHP scores and ranks

	Weight	US	Euro	CSG
Control	0.287	0.340	0.231	0.429
Comm	0.287	0.398	0.207	0.395
Coord	0.170	0.554	0.279	0.167
Integration	0.059	0.095	0.227	0.678
Monitoring	0.098	0.399	0.238	0.363
Intel	0.098	0.235	0.132	0.633
AHP Scores		**0.374**	**0.223**	**0.403**
RANK		**2**	**3**	**1**

Table 12.17 AHP score change with preferences

	No Preference	With Preference	Difference	% of original
CSG/Viability	0.495	0.403	-0.092	(-18.6%)
US-centric	0.284	0.374	+0.090	(+31.7%)
Euro-centric	0.221	0.223	+0.002	(+0.1%)

Table 12.18 Normalized AHP sub-scores, grouped by Attribute

		Weight in Attribute	US	Euro	CSG/Viability	Normalized Weight
Control	Holistic Control	0.33	0.11	0.26	0.63	0.0556
	Defense Current	0.33	0.72	0.19	0.08	0.0556
	Defense Future	0.33	0.13	0.25	0.63	0.0556
	AHP Scores		**0.318**	**0.233**	**0.448**	
Communication	Timeliness	0.33	0.26	0.11	0.63	0.0556
	Effectiveness	0.33	0.07	0.28	0.64	0.0556
	Accuracy	0.33	0.64	0.28	0.07	0.0556
	AHP Scores		**0.326**	**0.224**	**0.450**	
Coordination	Reaches All Systems	0.50	0.11	0.26	0.63	0.0833
	Safety/Security Info	0.50	0.64	0.28	0.07	0.0833
	AHP Scores		**0.375**	**0.272**	**0.354**	
Integration	Continuous Maintenance	0.50	0.08	0.19	0.72	0.0833
	Minimal Required	0.50	0.11	0.26	0.63	0.0833
	AHP Scores		**0.095**	**0.227**	**0.678**	
Monitoring	View of Entire SoS	0.50	0.07	0.28	0.64	0.0833
	Continuous Updating	0.50	0.72	0.19	0.08	0.0833
	AHP Scores		**0.399**	**0.238**	**0.363**	
Intelligence	State of Rogue Actors	0.50	0.26	0.11	0.63	0.0833
	Shareable Intelligence	0.50	0.11	0.26	0.63	0.0833
	AHP Scores		**0.183**	**0.183**	**0.633**	

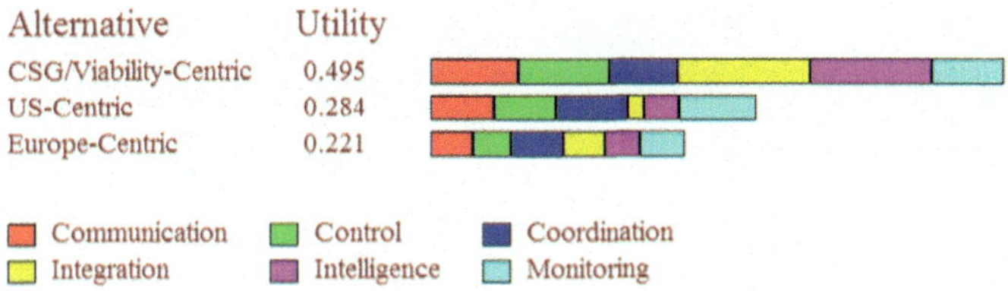

Fig. 12.16 Scenario ranking, via MUF average of AHPs

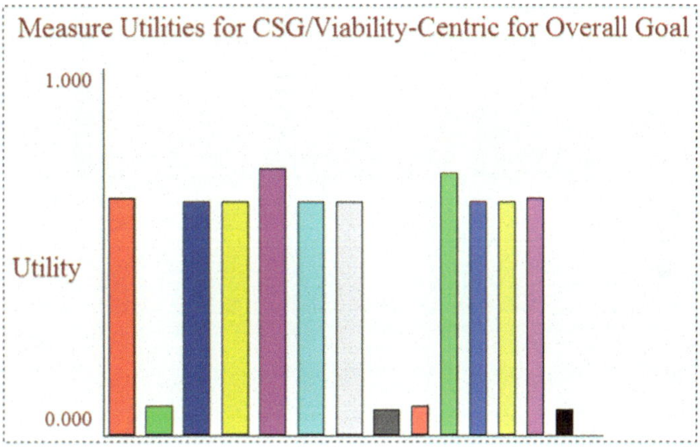

Fig. 12.17 Utility of CSG/Viability by attribute

The AHP analysis selected the CSG/Viability course of action. As shown in both Table 12.9 and Fig. 12.16, the CSG/Viability model of space governance scored highest in Control, Communication, Integration, and Intelligence, delivering especially high sub-scores in the last two attributes. With an overall utility score of 0.495, it nearly doubles the utility of its nearest competitor.

12.4.1 Individual Scenario Results

The dominance of the CSG/Viability governance model is even more plainly evident in Figs. 12.17, 12.18, 12.19 and 12.20, which group the individual AHP matrix scores from each scenario. Figure 12.17 is the legend for all three. Of the 14 separate criteria assessments, the CSG/Viability model led all but four: Continuous Updating, Safety/Security Information, Defense of Current Systems, and Accuracy. In the other 10, its focus on collaboration, information-sharing, mutual support, and industrial interest propelled it to the top of the preference stack, The US-centric model, with its focus on defense, accuracy, punctuality, and command/control, topped the other four AHP matrices. The Euro-centric model was never the preferred alternative.

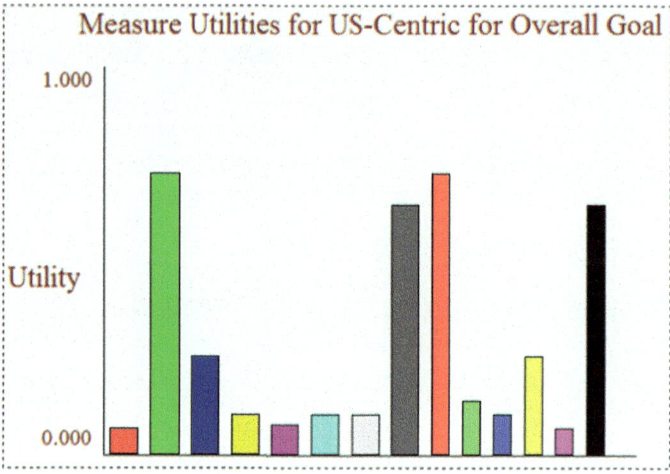

Fig. 12.18 Utility of US-centric by attribute

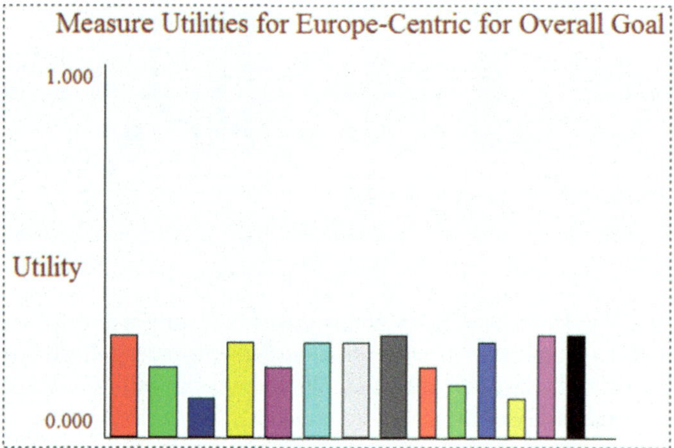

Fig. 12.19 Utility of Euro-centric by attribute

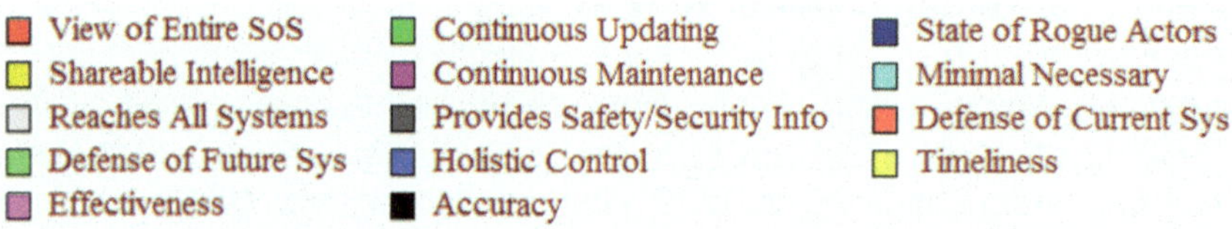

Fig. 12.20 Utility category key

12.4.2 Logical Decisions for Windows

Logical Decision for Windows (LDW) lets the user evaluate choices by considering many variables at once, separating facts from value judgments, and explaining choices to the audience. LDW uses techniques from the field of decision analysis to help make better and logical decisions.

12.4.3 Decision Alternatives

Three measure alternatives are examined: The CSG/Viability-centric, the Europe-centric, and the US-centric, and 6 attributes were exanimated. The creation and manipulation of utility functions for those attributes is automated. Another advantage of using the computer application is also used in this analysis is that the application of *Logical Decisions for Windows* allows rapid changes in values, with results that can be immediately viewed and analyzed.

12.4.4 Expansion and Clarification of Decision Attributes

Table 12.19 shows the expansion and clarification of Decision Attributes.

Table 12.19 Decision attribute expansion

Communication	Accuracy
	Effectiveness
	Timeliness
Control	Defense of current system
	Defense of future system
	Holistic control
Coordination	Provide safety/security info
	Reaches the whole system
Integration	Continuous maintenance
	Minimal necessary
Intelligence	Shared intelligence
	State of rogue actors
Monitoring	Continuous updating
	View of entire SoS

	A	B	C	D	E	F	G	H
1	Ranking for OVERALL Goal using nominal weights (effects of interactions excluded)							
2								
3	Alternative	Utility	Control	Communic	Coordinatic	Integration	Monitoring	Intelligence
4								
5	US-Centric	4.50000	0.83333	0.5	0.833333	0.833333	0.666667	0.833333
6	Europe-Cei	2.66700	0.50000	0.833333	0.333333	0.333333	0.333333	0.333333
7	CSG/Viabili	1.50000	0.33333	0.166667	0.166667	0.5	0.166667	0.166667
8								
9	Ranking for OVERALL Goal using effective weights (corresponds to ranking results bar lengths)							
10								
11	Alternative	Utility	Control	Communic	Coordinatic	Integration	Intelligence	Monitoring
12								
13	US-Centric	4.50000	0.17241	0.103448	0.172414	0.172414	0.172414	0.137931
14	Europe-Cei	2.66700	0.10345	0.172414	0.0689655	0.0689655	0.0689655	0.0689655
15	CSG/Viabili	1.50000	0.06897	0.0344828	0.0344828	0.103448	0.0344828	0.0344828

Fig. 12.21 Ranking for overall goal weights

Matrix: OVERALL Goal × | Goals Hierarchy for OVERALL Goal | Brainstorming window | Ranking for OVERALL Goal | Assess Weight

	Communication	Control	Coordination	Integration	Intelligence	Monitoring
CSG/Viability-Centric	1	2	1	3	1	1
Europe-Centric	5	3	2	2	2	2
US-Centric	3	5	5	5	5	4

Fig. 12.22 Overall goal matrix in LDW

12.4.5 Weight for Alternative Measures

The weight score varies from Zero to 5. Zero is the lowest and 5 the highest. The US-centric model has the highest weight score. Figure 12.21 shows the weight distribution.

The US-centric Model has the best weight for all attributes which means that the US-centric Model is the most credible and transparent (Fig. 12.22). The more weight means the most important attribute.

Overall, the US-centric was ranked high due to the fact that the US is the pioneer in space exploration and in addition the US is open to share and to collaborate with other space power nations for a peaceful space program.

12.4.6 Initial Assessment and Results of Analysis

Overall, the US-centric was ranked high due to the fact that the US is the pioneer in space exploration and in addition the US is open to share and to collaborate with other space power nations for a peaceful space program.

Percentage Weight for Preference Set NEW PREF. SET

Measure	Percentage Weight	Effective Weight
Control	16.7	14.286
Communication	16.7	19.048
Coordination	16.7	19.048
Integration	16.7	14.286
Monitoring	16.7	14.286
Intelligence	16.7	19.048

Fig. 12.23 Percentage weight for preference set

Fig. 12.24 Measure of equivalent ranking: communication

Equivalent Levels of Communication (new units)

Alternative	Utility	Equivalent
CSG/Viability-Centric	1.500	4
Europe-Centric	2.667	11
US-Centric	4.500	22

Preference Set = NEW PREF. SET

According to the analysis (Fig. 12.23), communication, coordination, and intelligence share the same effective weight, whereas control, integration, and monitoring are evenly split for a well-managed space program model.

12.4.7 *Measures of Equivalent Rankings*

The US-centric Model is more open than the rest of the models (Fig. 12.24). Accuracy, effectiveness and timeliness of the US-centric model are higher than other space models discussed in the present chapter.

Also, the US-centric Model is more controlled than the rest of the models (Fig. 12.25). Defense of current system, defense of future system and the control of the system as a whole (holistic control) are well applied to the US-centric model.

US-centric Model is more coordinated than the rest of the models (Fig. 12.26). Safety and security are evenly provided to the whole system.

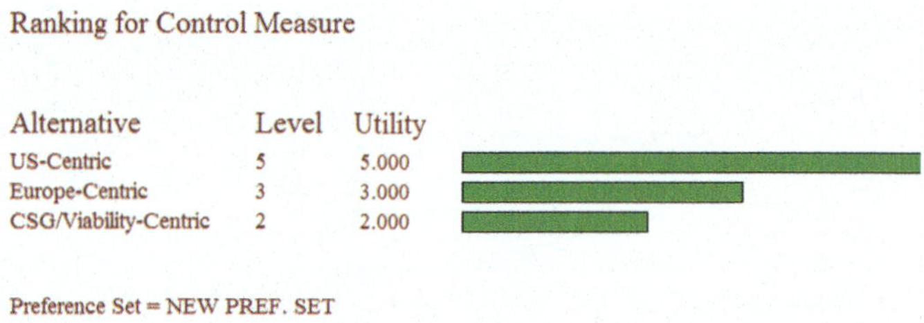

Fig. 12.25 Measure of equivalent ranking: control

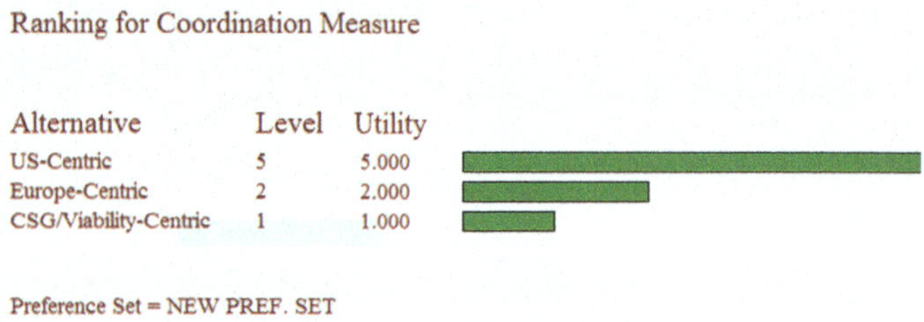

Fig. 12.26 Measure of equivalent ranking: coordination

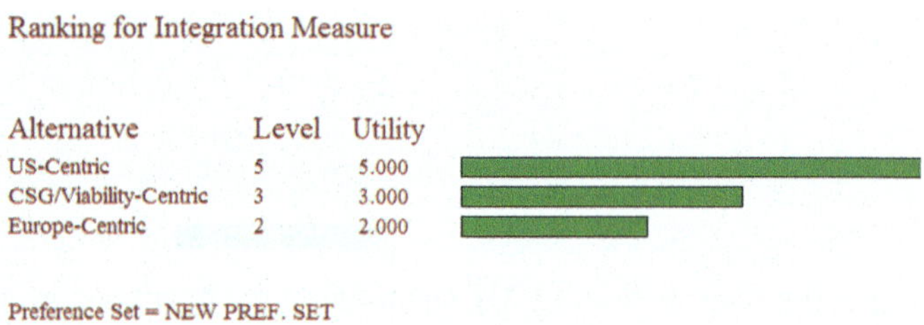

Fig. 12.27 Measure of equivalent ranking: integration

US-centric Model is more integrated than the rest of the models (Fig. 12.27). Continuous maintenance is provided which make the US-centric model the first among other competitors: Europe-Centric and the CSG/Viability-centric.

Also, in the US-centric Model, the way the intelligence is shared is better than other models (Fig. 12.28). The existence of US-centric Model is not about to harm other nations but it's to protect the US nation security and US interest around the globe while acting as a deterrence force to potential rogue nations with evil plans.

Finally, the US-centric Model is better monitored with specific goals and objectives as indicated by Fig. 12.29 The monitoring is continuous, and it is applied to the whole model as System of Systems.

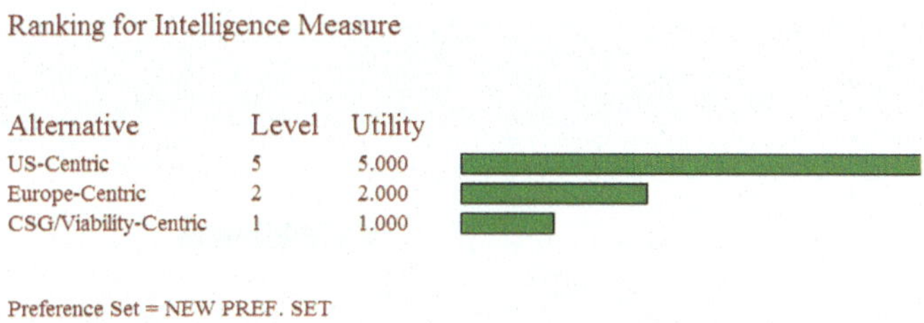

Fig. 12.28 Measure of equivalent ranking: intelligence

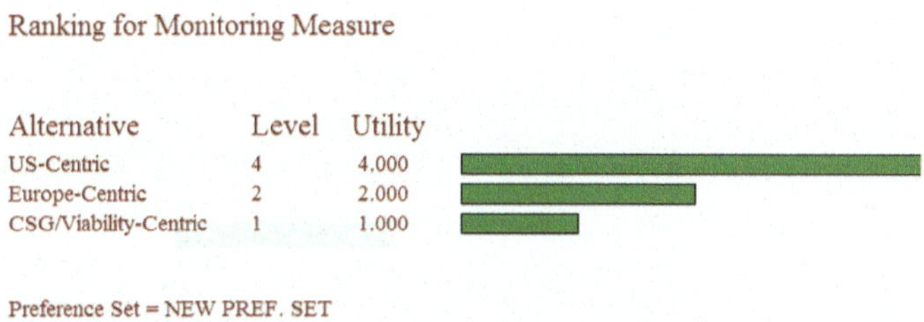

Fig. 12.29 Measure of equivalent ranking: monitoring

12.4.8 Overall Goals

The below (Fig. 12.30) graph was generated with the help of the LDW software tool to provide overall goal hierarchy.

The overall ranking reaffirm that the US-centric model is the best space model program with all attribute's alternatives taken into consideration (Fig. 12.31).

The overall goal utility for the US-centric model ranked better than the Europe-centric in all alternatives except communication alternative (see Fig. 12.32). This is due to reason that space exploration is not an open book. Even though the use shares its space exploits than any space power nation, national security has to be preserved. The European partners are protected under the NATO umbrella, and most of their space exploits are commercial or scientific research oriented.

12.5 Analysis of Results

The '*Party Problem*' approach ranked the prospects across their attributes and explicitly used the rules of Actional Thought (Howard and Abbas 2015) to select a prospect. By a clear margin, (33%), the PP method selected the CSG/Viability model, largely because the attributes Intelligence, Integration, Control, and Communication occupied the top 4 slots in the Order Rule hierarchy.

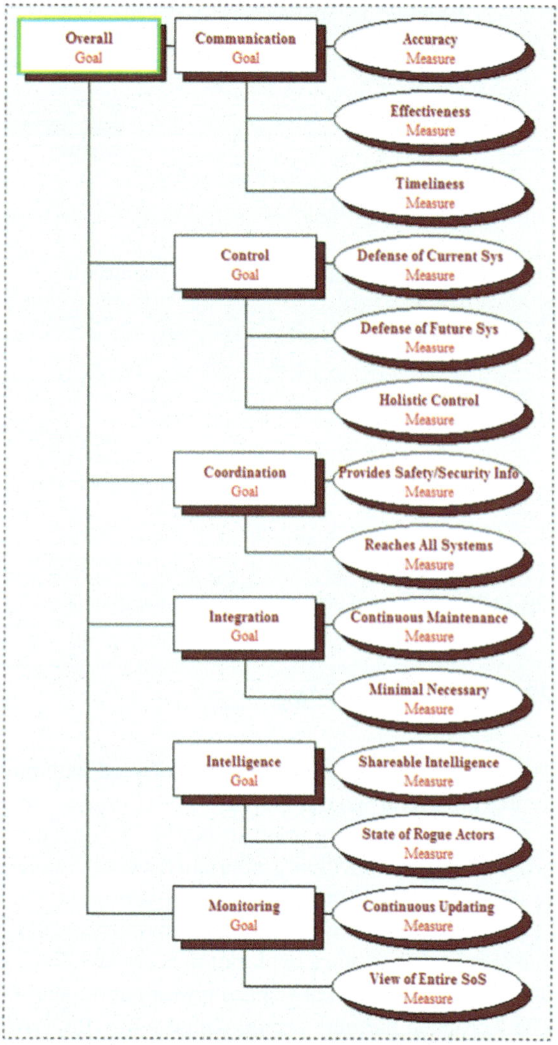

Fig. 12.30 Overall goal hierarchy

The Fuzzy Set method defined possible system states, assessed the probability that each alternative-criterion pair might be in each state, and used fuzzily-defined linguistic labels, all to establish a large set of utility scores for each alternative-criterion pair. After several set operations and normalization, the FS method selected the CSG/Viability model as the most preferred. However, the maximization functions in the selection process produced a margin between CSG/Viability and the next closest model of approximately 4×10^{-9}, which is the narrowest in the project by several orders of magnitude.

The Evidentiary Reasoning method also used linguistic labels, but it coupled them with degrees of belief. The IDS software performed all calculations, selecting the

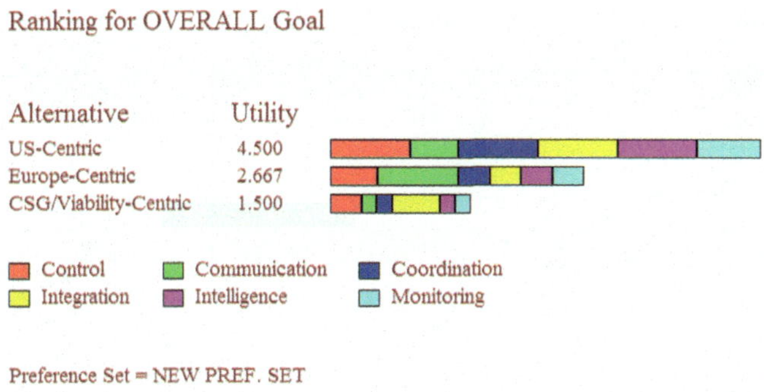

Fig. 12.31 Rankings for overall goal

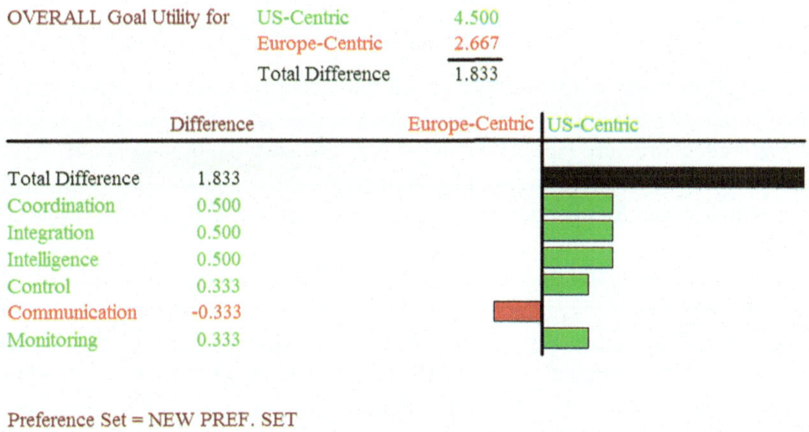

Fig. 12.32 Comparison of US-centric and Euro-centric alternatives

CSG/Viability model with a score about 26% higher than the next closest alternative. This method highlighted Integration and Intelligence Attributes as CSG/Viability's unique strengths.

The AHP method's unique preference matrices were filled based on the performance of each alternative relative to the others, creating a matrix for every criterion. Those matrix scores fed into an evenly weighted multi-attribute utility function to generate overall utility scores. As might be expected, the four highest attribute scores were the CSG/Viability model's Integration, Intelligence, Control and Coordination, driving it to an overall score almost 45% higher than the next nearest competitor.

Finally, using the LDW software, the models' attributes were evaluated via single-variable utility functions, which fed into an evenly-weighted multi-attribute utility function for each alternative. The US-centric model dominated using this method, based on its credibility, transparency, international diplomacy and its maturity as

Table 12.20 Collected MCDA results, normalized

	Normalized Scores			
MCDA Method	CSG-Via	US	Euro	Winner
Party Problem	1	0.672	0.515	CSG-Via
Fuzzy Sets	1	0.9999999959200	0.9999999999142	CSG-Via
Evidentiary Reasoning	1	0.668	0.734	CSG-Via
Analytic Hierarchy Process	1	0.574	0.446	CSG-Via
Logical Decisions for Windows	0.333	1	0.593	US

a government-commercial partnership. Overall, the US model outscored the next strongest alternative by just over 40%.

This analysis proved that multiple MCDA models together still may not entirely agree. Present authors worked with five experts to develop three models against fourteen criteria. It was generally agreed that the CSG/Viability-centric model would best meet the requirements they set forth for governing Complex Space Infrastructure (see Table 12.20).

However, in at the AHP analysis, an individual author's attempt to create a preference structure both exposed personal value biases and invalidated the results of that preference structure (see Appendix A). It also reasonable to conclude that the authors' assorted perspectives strongly influenced the LDW analysis. Certainly, there is a need to account for abstract qualities like transparency and tendency to promote peaceful diplomacy. Finally, there is a call for case applications (see Appendix B for serious gaming model that could also be used in the presented scenarios). Such case applications can only 'improve' our understanding of weaknesses and strengths associated with this emerging research. Case in point is the anti-satellite weapons (ASAT) model that designed to incapacitate or destroy satellites for strategic military purposes (Gheorghe and Vamanu 2007). Destroying might as well be considered a form of governance. Several nations possess operational ASAT systems, with others in development or design. While no ASAT system has yet been utilized in warfare, several nations have shot down their own (defunct) satellites to demonstrate their ASAT capabilities.

Final Remarks

The disparity in normalized answers (Table 12.20) and alternative selection became noteworthy. Due to the fundamental mechanical differences in the models, the authors expect that this disparity may inform us more about our processes than about the CSI prospects. Authors must submit that present research might have been impacted by the time constraint and the lack of robust number of subject matter experts. The authors attribute the disparities in answer and recommendation mostly to these two factors, though we did not attempt to quantify that attribution. We expect that a similar future study that addresses both of these flaws would find itself on much more sure footing, both logically and numerically. The Evidentiary Reasoning assessment, which was the only method in this project that calculated uncertainty, would greatly benefit from expertise that reduced its uncertainty from 20 to 30%.

The authors also suggest that a cross-functional team, versed in Complex System Governance, Multi-Criteria Decision Analysis, space infrastructure, and national/international space policy, re-engage the development of CSG for 'critical space infrastructure'. As an emerging system-of-systems field with no current governance structure wider than the European Union, it is excellent ground for CSG development. In a methodology shift from this project, we suggest that MCDA be used as needed to build CSG models, maximizing their potential utility during the construction phase, before selecting the most effective from among them.

Nonetheless, we had the audacity to offer a different approach and perhaps capitalize on a thought commonly attributed to Albert Einstein: *we cannot solve our problems with the same thinking we used when we created them.* What we have attempted to offer in this chapter a not-so-new approach to an emerging situation. More importantly, however, is the need to apply the presented ideas and models to advance the stated fields and push the boundaries of critical infrastructures into the open fields of space.

References

Carter, B. (2015). A metasystem perspective and implications for governance. *International Journal of System of Systems Engineering, 6*(1/2), 90. https://doi.org/10.1504/ijsse.2015.068807.

Gheorghe, A. V., & Vamanu, D. V. (2007). Risk and vulnerability games. The anti-satellite weaponry (ASAT). *International Journal of Critical Infrastructures, 3*(3/4), 457–470. https://doi.org/10.1504/IJCIS.2007.014120.

Howard, R. A., & Abbas, A. E. (2015). *Foundations of decision analysis*. Upper Saddle River, NJ: Pearson Education.

Katina, P. F., & Unal, R. (2015). Application of fuzzy sets in decision analysis for prioritising critical energy infrastructures. *International Journal of Decision Sciences, Risk and Management, 6*(1), 1–15. https://doi.org/10.1504/IJDSRM.2015.072762.

Keating, C. B., & Bradley, J. M. (2015). Complex system governance reference model. *International Journal of System of Systems Engineering, 6*(1/2), 33. https://doi.org/10.1504/ijsse.2015.068811.

Keating, C. B., & Ireland, V. (2016). Editorial. *International Journal of System of Systems Engineering IJSSE, 7*(1/2/3), 1–21. https://doi.org/10.1504/ijsse.2016.07.

Keating, C. B., & Katina, P. F. (2016). Complex system governance development: A first generation methodology. *International Journal of System of Systems Engineering IJSSE, 7*(1/2/3), 43–74. https://doi.org/10.1504/ijsse.2016.076127.

Landaeta, R. (2015, January 26). *Numeric project selection models*. Lecture presented at Old Dominion University, Norfolk, Virginia.

McDowell, R. M. (2014). *This is why the government should never control the Internet*. The Washington Post. Retrieved November 22, 2016, from https://www.washingtonpost.com/posteverything/wp/2014/07/14/this-is-why-the-government-should-never-control-the-internet/?utm_term=.faafa83e428a.

Musk, E. (2016). *Reusability: The key to making human life multi-planetary*. SpaceX. Retrieved November 22, 2016 from http://www.spacex.com/news/2013/03/31/reusability-key-making-human-life-multi-planetary.
Saaty, T. L. (2008). Decision making with the analytic hierarchy process. *International Journal of Services Sciences, 1*(1), 83–98. https://doi.org/10.1504/ijssci.2008.017590.
Wolchover, N. (2012). *Could the Internet ever be destroyed?* LIVESCIENCE. Retrieved November 22, 2016, from http://www.livescience.com/18030-internet-destroyed.html.

Chapter 13
CSG: Towards CSI Research

Abstract This chapter is developed with an inside out and forward-looking view of governance for the current topic: 'critical space infrastructure'. The need and utility of CSG field are provided. The nature and definition of CSG are explored. Applicability of CSG to the design, execution, and development of governance functions is explored. As part of this exploration advantages, limitations, and challenges brought by CSG to practitioners and the practices for governing complex systems are suggested. Implications of CSG development to enhance practice are examined. The chapter closes with some concluding thoughts and several exercises that serve to underscore central concepts tenets of present research. Explorations in this chapter includes several vignettes and scenarios that serve to demonstrate the utility and contributions offered by CSG.

13.1 Complex System Governance: Re-visited

The problems facing practitioners in modern systems appear to be intractable given the apparent ineffectiveness of the responses provided to address them. These problems continue to proliferate into all aspects of human endeavor and the systems designed to orchestrate those endeavors. They are not the privilege, or curse, of any particular field or sector (energy, utilities, healthcare, transportation, commerce, defense, security, services), as none are immune to the effects of this problem domain. Problems stemming from this domain do not have a precise cause-effect relationship that would make understanding and resolution easy. In fact, they are more likely products of a 'circular causality', where the precise singular determination of cause is doubtful (von Foerster et al. 1953). Instead, these problems are consistent with the notion of Ackoff's (1997) 'messes' (interrelated sets of problems that are not well formulated, understood, or easily resolved) and Rittel and Webber's (1973) 'wicked problems' (problems that are intractable with current levels of thinking, decision, action, and interpretation). This problem domain is likely to continue, and perhaps accelerate, as we continue to grapple with 21st century complex systems and their problems (Casti 2012).

© Springer Nature Switzerland AG 2019

A. Georgescu et al., *Critical Space Infrastructures*,
Topics in Safety, Risk, Reliability and Quality 36,
https://doi.org/10.1007/978-3-030-12604-9_13

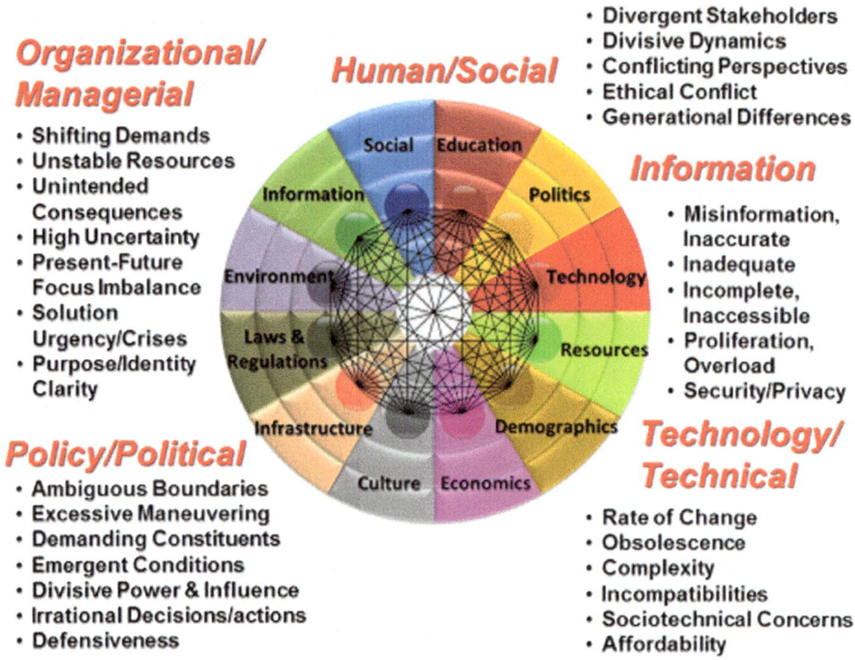

Fig. 13.1 Challenges for practitioners in the complex system problem domain

Arguably, complex systems and their associated problems have been in existence long before the 21st century. However, the landscape for modern systems has changed appreciably into a much more 'complex problem space'. This problem space (Fig. 13.1) is marked by difficulties encountered across the holistic range of technical, organizational, managerial, human, social, information, political, and policy issues. The different aspects of this 'new normal' complex problem space has been previously established (Keating and Katina 2011, 2012; Keating 2014) as being characterized by conditions identified in Fig. 13.1.

While this listing is not presented as exhaustive, it illustrates two important points. First, the issues emanating from this domain continue without consistent resolution methods. Thus, there is certainly room for new thinking and derivative approaches to address this domain. Second, the conditions identified are not likely to recede in the future. In essence, this domain represents the 'new normal' for the practitioners dealing with complex systems.

The problems emanating from this domain appear to be intractable. In any cursory look at present system and their problems it is easily concluded that we have not, and are not, mounting an effective solution. Given the current level of ineffectiveness in dealing with complex system problems that have proliferated into all aspects of human endeavor, CSG has been introduced. It is in the domain presented above that CSG is being postured to impact practitioner capabilities to more effectively address growing concerns. CSG is primarily based in Systems Theory (Adams et al. 2014; Skyttner 2005; von Bertalanffy 1968) and Management Cybernetics (Beer 1979) and has been built upon their philosophical, theoretical, and methodological underpinnings. At this point in our development, we introduce CSG as it has been previously defined as the design, execution, and evolution of the metasystem functions necessary to provide control, communication, coordination, and integration of a complex system (Keating et al. 2014a, b).

In many cases, our systems, including space systems, that is the governance of those systems, have developed over time through processes of accretion or self-organization. Accretion is a process whereby elements are added in a piecemeal fashion until the whole system appears fragmented and no longer makes sense. Self-organization involves letting system structure and resulting behavior develop with minimal design oversight. This can produce results that may or may not be consistent with expectations or desirable performance. The result of either of these system development processes, accretion or self-organization, can and often do result in systems that fail to meet performance expectations. In effect, development is not purposeful, resulting in a condition we refer to as 'system drift'. Just as a powerless ship drifts along its intended course subject to uncontrollable currents, so too can our systems experience drift resulting from development by accretion or self-organization. System drift symbolizes a system that is subject to the unintended consequences that accrue in the absence of a purposefully executed design. In the end, system drift describes a condition all too familiar to practitioners who must navigate systems through the increasingly complex environment, while confronting seemingly intractable issues on a daily basis. CSG is a coherent response to system drift.

CSG is one of many systems-based approaches designed to better deal with complexity and what we referred to earlier as 'system drift'. System drift denotes systems that, irrespective of the noblest intentions, have either never been properly designed or whose execution continually fails to meet desired performance expectations. In short, these 'drifting' systems fall short of delivering minimal value expected, much less producing high performance. We do not need to look far to see examples of drifting systems. In fact, it would be a rare day that we would not be impacted by systems in drift. Consider the following examples: (1) launching of a new Enterprise Resource Planning initiative that collapses due to emergent incompatibilities with existing systems, (2) a costly crisis from discovery of noncompliance to a regulatory requirement that has been in existence for several years but never identified, or (3) introduction of a new purchasing policy that achieves intended reductions in supplier costs but increases overall costs due to resulting schedule delays. Unfortunately, the impacts of system drift are not limited to increased costs. These drifting systems have considerable associated human cost. These human costs are borne by those that must suffer through these drifting systems by compensating for their ineffectiveness. CSG supports thinking, decision, and action to proactively and purposefully address system drift. Ultimately, CSG is intended to reduce the high human costs characteristic of these systems in drift.

Systems-based approaches, such as CSG, and the systems thinking upon which they are founded, are certainly not 'new' in trying to address what we described as system drift. In fact, the foundations of systems thinking have been traced as far back as the ancient Chinese work The I Ching (translated as Book of Change dated prior to 400 B.C.) that noted the dynamic nature of changing relationships among elements. Additionally, the central philosophical tenet of systems thinking, holism, can be traced back to the writings of Aristotle, who suggested that 'the whole is more than the sum of its parts'. Thus, approaches based in systems thinking and 'holism' are not new and have historically represented a significant step toward dealing with system drift. However, what is new in bringing CSG applied research to the problem domain is the fusion of Systems Theory and Management Cybernetics to provide practitioners with perspective, supporting methods, and tools to confront drifting systems. This practitioner focused CSG research seeks to increase capabilities for better understanding, decision, and action in dealing with complex systems and their associated problems. In essence, CSG seeks to increase effectiveness in dealing with system drift.

CSG is focused on providing practitioners with perspective, methods, and tools to better understand and deal with complexities they must routinely confront. In essence, CSG helps avoid system drift through purposeful design, similar to a ship changing heading or speed to compensate for the effects of wind or current. Continual treatment of symptomatic conditions contributes to 'system drift' by focusing on temporary correction of deficiencies at a superficial level. Unfortunately, this correction behavior is endemic to modern systems, fostering 'system superheroes.'

In development of CSG, Management Cybernetics brings three important contributions. First, the strong grounding in Cybernetics provides a strong theoretical foundation for CSG. Cybernetics, at a most basic level, is concerned with communication and control—in effect deriving from the Greek notion of 'steering'. This is consistent with the function of governance as providing the direction and monitoring the movement of the system along that trajectory. With respect to control, taking a cybernetic viewpoint allows inclusion of the more expansive perspective of control consistent with providing the highest degree of autonomy within a system, while preserving system performance. Second, the work of Beer (1979) provides a model (Viable System Model) which includes functions (metasystem) consistent with achievement of governance for a system. This reference model, identified by the functions above, provides CSG with an established frame of reference upon which to build. The Management Cybernetics foundation provides strong systemic/cybernetic set of underpinnings, is logically consistent with CSG articulation from a systems perspective, and offers a field which has withstood several decades of scrutiny. Since its development in the 1970s, Management Cybernetics has been successfully applied for over four decades. It has maintained a sustainable footing, even with the arrival and departure of a multitude of other approaches that have ceased to exist in any formidable fashion.

A CSG Vignette—No Way Out of the Crisis Mode

The mantra is frequently heard in the corporate halls, "all we do is continually deal with crises, moving from one fire to the next". One executive, tired of the continual 'firefighting mode', decided to examine their system from a 'governance' perspective. This entailed structured accounting for the design, execution, and development of the system. The discovery from an introspective examination suggested that, while people were working harder, they were masking system deficiencies that created inefficiencies, inconsistencies, and a seeming endless engagement in the 'status quo' operation in continual crisis mode. Engaging in a protracted 'CSG study' of their system, focused on identification and assessment of critical systemic issues (governance) in the design and execution of their system. The result of the 'guided' self-study identified and prioritized multiple design issues based on their impact on system performance. For example, one design issue was the near absence of coordinated efforts with respect to scanning the environment for trends, patterns, and shifts. This absence precluded early identification, analysis, and response planning – prior to the inevitable crises that would eventually erupt upon their 'too late' discovery. The stage was set to establish and integrate appropriate mechanisms for more effective environmental scanning.

13.2 Defining Complex System Governance

In this section we examine the nature and definition of CSG. We begin with introduction and amplification of a definition of CSG. This is followed by examination of the systems-based paradigm that captures the essence of CSG. We close this section by the examination of five fundamental aspects that captures the essence of CSG. There is a growing body of knowledge related to CSG. The essence of CSG lies in the current state of the definition captured as the "*Design, execution, and evolution of the metasystem functions necessary to provide control, communication, coordination, and integration of a complex system*" (Keating 2015, p. 227). There are several points of emphasis for this depiction of CSG. First, *design* accentuates the necessity to purposely and proactively engage in the creation of the governance system. While this seems as though it should be a taken for granted proposition, we suggest that truly purposeful, holistic, and comprehensive design of governing systems represents the rare case. While we might argue the merits of this conclusion, at this point it suffices to say that based on the current level of performance and issues propagating all manner and form of our 'manmade' complex systems, the anecdotal evidence suggests that what we are doing with respect to our systems is not working. Irrespective of purposeful/purposeless design, *execution* embodies the notion that a design without deployment offers little more than good intention. Execution is where a design meets the harsh realities of the 'real world' which is fraught with complexity and emergent conditions that that are sure to test the most thoughtful designs. We should note that the need to adjust a system during execution is not indicative of poor design, but rather recognition that all designs are flawed. They must be flawed because they are abstractions of real-world complexity that can be neither totally captured nor completely understood. The third leg of CSG, *evolution*, recognizes that systems as well as their environments are in constant flux. Therefore, governance must also be able to flex (evolve) in response to internal and external changes impacting the system. *Evolution* by its very nature suggests that the emphasis is on long term sustainability, notwithstanding the need to operate a system in real time. In effect, governance must be capable of absorbing, processing, and responding to external turbulence and internal system flux to ensure the system remains viable (continues to exist) in both the short-term operational sense that delineates current system existence as well as the long-term evolutionary sense that positions the system for the future.

The second aspect of the CSG definition lies in the articulation of the metasystem as the set of functions that produce governance for a complex system. These functions find their basis in and offer an extension of Beer's metasystem concept in the Viable System Model (1979). The metasystem for CSG is the set of 9 interrelated functions that act to provide governance for a complex system. These functions include:

- ***Metasystem Five (M5)—Policy and Identity***—focused on overall steering and trajectory for the system. Maintains identity and balance between current and future focus.
- ***Metasystem Five Star (M5*)—System Context***—focused on the specific context within which the metasystem is embedded. Context is the set of circumstances, factors, conditions, or patterns that enable or constrain execution of the system.
- ***Metasystem Five Prime (M5′)—Strategic System Monitoring***—focused on oversight of the system performance indicators at a strategic level, identifying performance that exceeds or fails to meet established expectations.
- ***Metasystem Four (M4)—System Development***—maintains the models of the current and future system, concentrating on the long-range development of the system to ensure future viability.
- ***Metasystem Four Star (M4*)—Learning and Transformation***—focused on facilitation of learning based on correction of design errors in the metasystem functions and planning for transformation of the metasystem.
- ***Metasystem Four Prime (M4′)—Environmental Scanning***—designs, deploys, and monitors sensing of the environment for trends, patterns, or events with implications for both present and future system viability.
- ***Metasystem Three (M3)—System Operations***—focused on the day to day execution of the metasystem to ensure that the overall system maintains established performance levels.
- ***Metasystem Three Star (M3*)—Operational Performance***—monitors system performance to identify and assess aberrant conditions, exceeded thresholds, or anomalies.

- ***Metasystem Two (M2)—Information and Communications***—designs, establishes, and maintains the flow of information and consistent interpretation of exchanges (communication channels) necessary to execute metasystem functions.

A third primary aspect of the metasystem construct is found in the *communication channels* that provide for the flow of information between system entities as they perform functions. These channels support the flow of information for decision and action as well as produce consistency in interpretation for exchanges within the metasystem and between the metasystem and external entities. The ten communication channels are adapted from the work of Beer (1979) and extensions of Keating and Morin (2001). Table 13.1 provides a concise listing of the communication channels, their primary metasystem function responsibility, and the particular role they play in metasystem execution.

The final part of the definition of CSG is focused on the elements of control, communication, coordination, and integration. These terms and their basis emanate from

Table 13.1 Communication channels of the metasystem for CSG

Communications channel and responsibility	CSG metasystem role
Command **(Metasystem 5)**	• Provides non-negotiable direction to the metasystem and governed systems • Primarily flows from the Metasystem 5 and disseminated throughout the system
Resource bargain/ Accountability **(Metasystem 3)**	• Determines and allocates the resources (manpower, material, money, methods, time, information, support) to governed systems • Defines performance levels (productivity), responsibilities, and accountability for governed systems • Primarily an interface between Metasystem 3 to the governed systems
Operations **(Metasystem 3)**	• Provides for the routine interface concerned with near term operational focus • Concentrated on providing direction for system production of value (products, services, processes, information) consumed external to the system • Primarily an interface between Metasystem 3 and governed systems
Coordination **(Metasystem 2)**	• Provides for metasystem and governed systems balance and stability • Ensures design and achievement (through execution) of design: (1) sharing of information within the system necessary to coordinate activities, and (2) ensures decisions and actions necessary to prevent disturbances are shared within the Metasystem and governed systems • Primarily a channel designed and executed by Metasystem 2

(continued)

Table 13.1 (continued)

Communications channel and responsibility	CSG metasystem role
Audit **(Metasystem 3*)**	• Provides routine and sporadic feedback concerning operational performance • Investigation and reporting on problematic performance issues within the system • Primarily a Metasystem 3* channel for communicating between Metasystem 3, the governed systems, and the metasystem concerning performance issues
Algedonic **(Metasystem 5)**	• Provides a 'bypass' of all channels when the integrity of the system is threatened • Compels instant alert to crisis or potentially catastrophic situations for the system • Directed to Metasystem 5 from anywhere in the metasystem or governed systems
Environmental Scanning **(Metasystem 4′)**	• Provides design for sensing to monitor critical aspects of the external environment • Identifies environmental patterns, activities, or events with system implications • Provided for access throughout the metasystem as well as governed systems by Metasystem 4′
Dialog **(Metasystem 5′)**	• Provides for examination of system decisions, actions, and interpretations for consistency with system purpose and identity • Directed to Metasystem 5′ from anywhere in the metasystem or governed systems
Learning **(Metasystem 4*)**	• Provides detection and correction of error within the metasystem as well as governed systems, focused on system design issues as opposed to execution issues • Directed to Metasystem 4* from anywhere in the metasystem or governed systems
Informing **(Metasystem 2)**	• Provides for flow and access to routine information within the metasystem or between the metasystem and governed systems • Access provided to entire metasystem and governed systems • Primarily designed by Metasystem 2 for utilization by all metasystem functions as well as governed systems

Management Cybernetics (communication, control) and Systems Theory (coordination, integration). Here are the extended perspectives for each of these elements provided by CSG:

- *Control* (constraints necessary to ensure consistent performance and future system trajectory). In our formulation of control, we look to a more informed system view for guidance. This view suggests that control is not a pejorative term, to be scorned as a form of domination over a particular venue, activity, or entity. On the contrary, in the systems view we take, control is essential to ensure that the system stays on the trajectory that will provide future viability in response to changing conditions and circumstances. This is achieved by providing the greatest degree of autonomy (freedom and independence of decision, action, and interpretation) possible while still maintaining the system at desired levels of performance and behavior. In effect, this suggests that over constraint of a system waste resources (constraint is not free), limits system initiative/creativity, and diverts important emphases of a metasystem unnecessarily to lower levels of the system (inefficiency).

- *Communication* (flow and processing of information necessary to support consistent decision, action, and interpretation across the system). Communication is essential to governance and operation of the metasystem. Communications includes not only the exchange of information, but also the interpretative schemas that permeate the system. These interpretative schemas are necessary to provide coherence in making, understanding, and interpreting the myriad of exchanges in a system. Communications may range from formal to informal, explicit to tacit, and patterned to emergent. There is not an optimal configuration for communication in a system and the arrangements are certainly subject to shifts over time and emergent patterns. However, from a Complex System Governance perspective, communications are something that would be better off not left to chance self-organization. Instead, purposeful design and evolution of communications within a system is more likely to produce and maintain desirable results.
- *Coordination* (providing for effective interaction among different entities within the system, and external to the system, to prevent unnecessary oscillations). Certainly, coordination is an essential aspect to ensure that a system provides sufficient interaction among different elements to maintain consistency. Quite possibly the most important aspect of coordination is the damping of unnecessary fluctuations as the system operates. In effect, this implies that there must be sufficient standardization to provide routine interface as well as a sufficiently robust design to absorb emergent conditions that could not have been known in advance. While original work in management cybernetics focused on coordination as an internal function, we should also consider the necessity for coordination external to the system.
- *Integration* (design for system unity with common goals, accountability, and balance between individual autonomy and system level interests). The primary focus of integration is to ensure that the system achieves desirable levels of performance while: (1) providing the maximum level of autonomy to constituents, (2) invoking the minimal constraint necessary for the system to function as a unity in achieving the intended purpose, and (3) strategically shifting the balance point between autonomy and integration based on changes in contextual factors and system performance levels. Integration is not achieved through serendipity, but rather by active design and continuous evolution.

The definition of CSG is incomplete without recognition of the underlying paradigm within which it is embedded. We now turn our attention examine this paradigm and its importance to the deeper understanding and development of CSG.

13.2.1 The CSG Paradigm

CSG is developing and exist in the early stages of emergence at the intersection of the governance, systems theory, and management cybernetics fields. As such, it has the advantages of being tied to three fields with substantial substance, acceptance, and longevity. On the contrary, this intersection also invokes the criticisms and limitations of the fields as well at the potential for incompatibilities of the fields. Potential incompatibilities might exist across philosophical, theoretical, or methodological lines.

This does not diminish the pursuit of CSG as an integrated field, but rather establishes a set of cautionary considerations in movement forward. In light of this caution, we have produced a succinct paradigm for CSG:

> *…a set of functions is enacted by mechanisms that invoke metasystem governance to produce the communication, control, coordination, and integration essential to continued system viability.*

A second means of system development is self-organization, where the structure and functions of a system are permitted to develop 'on their own' without imposition of external constraints. This approach works great, as long as the system continues to produce expected behavior and desired performance levels. In effect, with self-organized system design 'you get what you get', which may or may not continue to meet expectations given the present and future system realities. System design by self-organization might be great for low stakes endeavors (e.g. a dinner party). However, for high stakes complex systems, such as the maintenance system, example exclusive reliance on self-organization is a recipe for disastrous system performance.

The third means of development is by purposeful design. This development involves the rigorous examination of a system through a set of systemic lenses. Although there are other systems-based approaches for applicability in system development, CSG is offered as a rigorously grounded systems-based approach to see underlying systemic issues and generate potential alternative paths forward.

However, CSG is not a 'silver bullet' or 'magic elixir' promising to cure all ills of modern systems. It requires hard work and commitment, but the payoff can be substantial. What is the payoff? Imagine having to navigate to a destination in the dark, without a map, having questionable directions, and no local knowledge of 'bad spots' to avoid. The result is very likely the 'trip from hell'. CSG provides practitioners with the equivalent of a real-time guide—providing directions, identifying impediments along the way, and tailoring the route to the capabilities of the vehicle (system) and practitioners making the trip. CSG is an invitation to generate a different experience in navigating complex systems and their problems. In effect, a *GPS (Governance Positioning System)* to provide directions to the future via more effective and compatible routes.

CSG has been developed as an alternative to ad hoc or *self-organized* system design, execution, and evolution. The CSG alternative is one of '*purposefully designed*' systems, or p-systems. P-systems are focused on active design, execution, and evolution of governance functions in ways that are consistent with the laws (principles) of systems.

A Complex System Governance Vignette—Where is the Metasystem?

Our systems continually act to disappoint by producing behavior, performance, and outcomes that are inconsistent with our intentions. An exemplar of this can be found in the case of an urban university, seeking to better understand their 'system' for bring new students into the university system. Engaging in examination of the 'system' used for bringing on new students, several discoveries came to the forefront. Among these were the realization that the 'system' for student entry was not designed, executed, maintained, or developed as a system at all. Instead, what was purported to be a system was a loose 'aggregate' of different processes, elements, and components. This aggregate had developed over time, without the benefit of a higher-level view concerning 'how it actually fit together' and made sense to effectively guide student entry. What was discovered was that there were individual units (e.g. finance, admissions, housing, etc.) that individually functioned very well. However, from a metasystem viewpoint the individual units failed to function together. This was evidenced by the many problems that spanned multiple units and required their integrated efforts to address. Unfortunately, lacking a design or understanding as to how the 'metasystem' functioned, each issue that required multiple units for resolution generated inconsistent performance, emergent crises, and high human costs (frustration for administrators as well as students who had to navigate the system). The active examination from the 'lenses of the CSG metasystem helped to identify system design deficiencies and understand their source in systems principles/laws violated). Thus, a different path forward was made visible. A new decision, action, and performance interpretation space was opened to the system designers.

CSG is a systems-based approach to enable practitioners to better deal with complex systems and their problems. In essence, CSG purposefully addresses '*system drift*'. *System drift* denotes systems that, irrespective of the best intentions, have either never been purposefully designed or whose execution continually fails to meet desired performance expectations. In short, these '*drifting*' systems fail to deliver minimal value expected, much less produce high performance. We do not need to look far to see examples of drifting systems. In fact, it would be a rare day that we would not be impacted by systems in drift. Consider the following examples: (1) launching of a new Enterprise Resource Planning initiative that collapses due to emergent incompatibilities with existing systems, (2) a costly crisis from discovery of noncompliance to a regulatory requirement that has been in existence for several years but never identified, or (3) introduction of a new purchasing policy that achieves intended reductions in supplier costs but increases overall costs due to schedule delays. Unfortunately, the impacts of *system drift* are not limited to increased monetary costs. These drifting systems have considerable associated human cost, borne by those destined to suffer as victims of drifting systems. CSG supports thinking, decision, and action to proactively and purposefully address *system drift* and its consequences.

A Complex System Governance Vignette—Where is the Owner's Manual for this System?

We have all been in the situation where we are driving a rental car and can't seem to find where a particular control is located and operates (e.g. heat, windshield wipers, gas tank release, trip mileage control, cruise control, radio, resetting clock, etc.). Since we do not 'own' the vehicle, there is a certain 'acceptance' of the annoyance, unless we breakdown and go to the owner's manual in the glovebox to learn how the function we desire is performed. However, for the vehicle we 'own' we have an owner's manual that provides guidance to make the intricate system adjustments we desire (e.g. Bluetooth settings). We would not purchase a complex vehicle without also receiving the owner's manual that tells us critical things about our car (system) such as maintenance intervals, troubleshooting problems, meaning of indicators, and performance of essential functions. Suggesting that a car is nowhere near the complexity we find in a modern organization, why do we not have an integrated owner's manual that specifies the design, execution, and development for our organization (system)? In many ways we have pieces and parts – for example processes, policies, and procedures provide some indicators of system execution. However, at the 'metasystem' level for governance, it is the rare case that we find an owner's manual equivalent for governance of an enterprise.

CSG can provide value across several levels (Fig. 13.2), including:

- ***Practitioner***—Enhanced capacity of individual practitioners to engage in the level of systems thinking necessary to more effectively deal with the issues related to design, execution, and evolution of complex systems and their problems.
- ***Enterprise***—Provide competency development (knowledge, skills, abilities) for targeted entities (units, staff teams, departments) across the enterprise to better engage complex systems and problems.
- ***Support Infrastructure***—Examination and development of support infrastructure

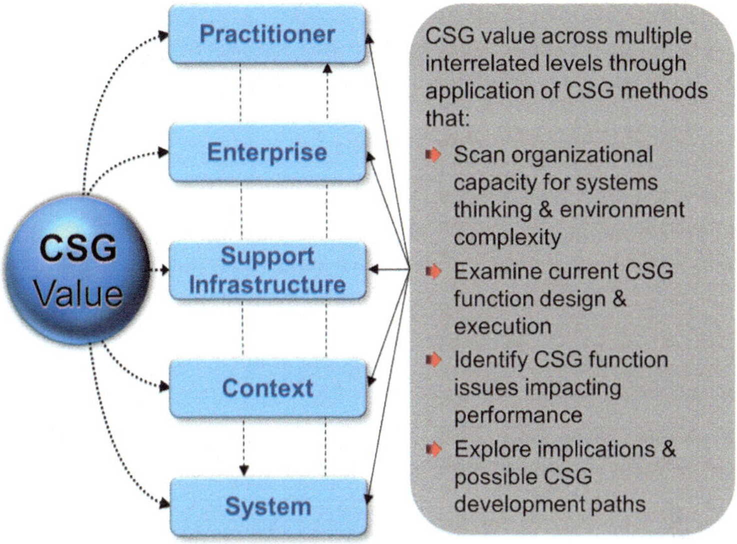

Fig. 13.2 Value added at multiple levels from CSG

(processes, technologies, systems) for compatibility with system governance design, execution, and development.

- ***Context***—Identification and development consideration for unique circumstances, factors, and conditions that influence (constrain/enable) achievement of system governance functions and system performance (e.g. stakeholders, regulatory requirements, staff, leadership style, etc.).
- ***System***—Providing identification of impediments to system performance rooted in specific deficiencies in design, execution, and development of governance functions and corresponding system laws.

The value accrued by CSG stems from: (1) scanning of the capacity an organization (entity) to engage in a level of systems thinking compatible with the complexity demands of the system environment, (2) exploration of the design and execution of essential governance functions, (3) identification and prioritization of system performance constraints tracked to problematic governance functions and violations of systems laws, and (4) establishment of developmental strategies across multiple levels essential to enhancing CSG to improve system performance.

To illuminate the applicability of CSG and potential contributions, we examine three scenarios of application.

13.3 Application Scenario

13.3.1 SCENARIO 1: Workforce Capacity for System Thinking

Situation
A workforce is continually behind in producing innovative thinking to effectively respond to complexity demands of their environment—resulting in crises, surprises, or inefficiencies. The errors continue to mount with increasingly deficient performance, discontent in the workforce, and the seeming inability to effective function in relationship the demands of the complex environment within which the system and practitioners must function.

CSG Perspective Discussion
A critical element of CSG is the dependence on the capacity of the workforce to engage at a level necessary to realize the inherent value in CSG. Without the correct frame of reference (System Thinking Capacity), the results desired from CSG are not likely to be achieved. In essence, if the workforce does not have the necessary systemic thinking skills, then CSG is just another approach that an organization might grasp at for relief. Regardless of how dire the organization circumstances might be, there is not shortcut to having the requisite capacity in individuals to effectively engage any systems-based endeavor. There are two primary drivers for this situation. First, as mentioned is the capacity of the workforce to think systemically. Second, is the degree to which the environment demands with respect to systems thinking capacity. Performance will largely be determined by the degree that there is a sufficient 'match' between the systems thinking capacity that exist in the workforce to that demanded by the environment they must navigate.

CSG Response Discussion
Systems Thinking Capacity (ST-Cap) and Environment Complexity Demand assessment instruments can be used to identify gaps between ST-Cap of the entity (team, department, organization) and the demands of their environment. 'Critical' areas for enhancing ST-Cap are identified. Figure 13.3 depicts this gap along the seven dimensions of systemic thinking.

13.3.2 SCENARIO 2: System Governance Pathologies Identification

Situation
A focal system is experiencing continual failures (e.g. cost overruns, schedule delays, missed performance targets) that are resistant to improvement efforts. The external manifestations of failures are evident in either product/service quality, missing milestones or required customer completion schedules, or conflicts in the adequacy,

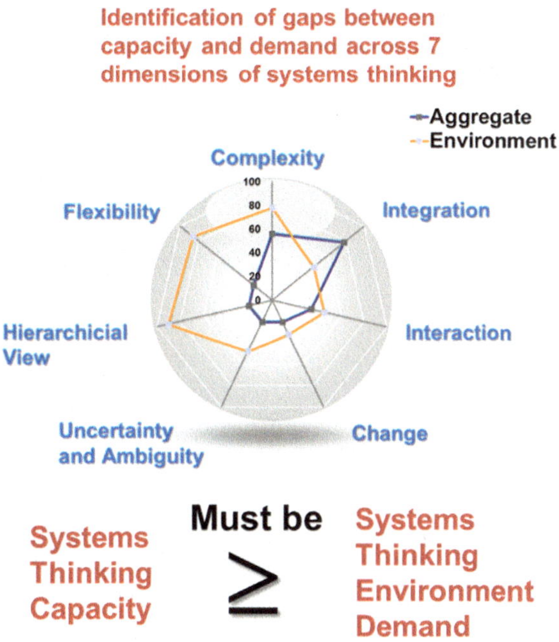

Fig. 13.3 The gaps between systems thinking capacity and environment demand

utilization, or outcomes achieved for resources consumed. There have been several failed attempts to locate the source of the deficiencies, but there does not appear to be a singular root cause to which failures can be attributed. The result is a sagging customer confidence, resource scarcity, and a diminished workforce from the anxiety and frustration being experienced without an apparent path forward or end in sight.

CSG Perspective Discussion

It is quite easy to identify the results for violation of underlying systems principles (evidenced as pathologies). Pathologies are the outward manifestation of underlying system design, execution, or development issues. Being able to properly trace the systemic issue requires a 'deeper dive' into the actual system producing the performance issues. In essence, a system can only produce what it produces, nothing more and nothing less. If the system performance is not consistent with that we desire, we must understand the system that is producing the undesirable behavior/performance. Focusing only on the outward signs of the underlying systemic issues can at best provide a temporary fix. At worst, more damage than good might accrue from superficial treatment of symptoms of underlying system deficiencies (pathologies).

CSG Response Discussion

Focal group completes a System Governance Pathologies assessment instrument (Katina 2015). Deep system pathologies (aberrations from healthy system conditions) across 9 governance functions are identified, mapped, systemically explored, and prioritized for response. This approach provides an opportunity to discover the underlying source of deficiencies in a system. These are not necessarily observable

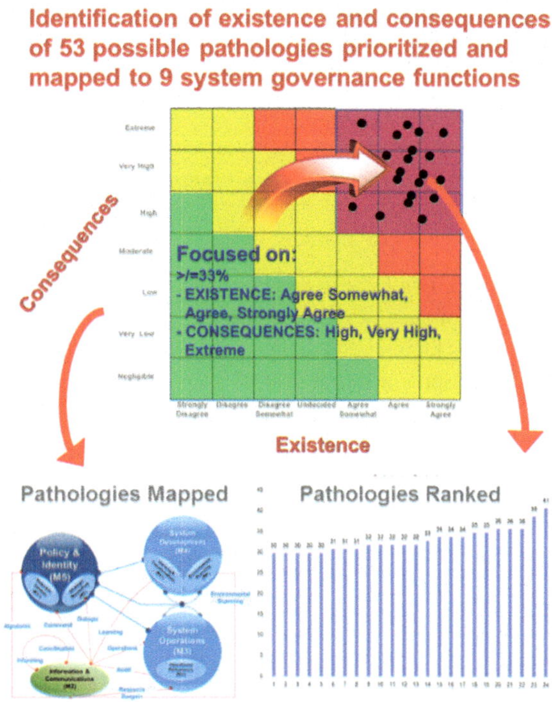

Fig. 13.4 Mapping deep seated system pathologies in an organization

from the inspection of their superficial deficiencies produced. Figure 13.4 shows a mapping of one particular pathology (of 53 different pathologies) in a system. There can be various articulations of system pathology (Davidz 2018; Katina 2016; Troncale 2013).

13.3.3 SCENARIO 3: System Governance Development

Situation

An organization has difficulty in providing a clear, coherent, and accountable system innovation strategy to address persistent criticisms from oversight bodies. External forces are continually challenging the organization to provide information, performance indicators, and reasons for major decisions and strategies being pursued.

CSG Perspective Discussion

Although it is common to receive 'oversight' in performance of organizational mission, care must be taken to understand the degree to which the system is designed, executed, and developed such that oversight is not a burden but rather a welcomed opportunity to 'demonstrate' the system. As most systems are not purposefully designed, it is not uncommon to look at external 'hands' as an annoyance at best and at worst an impediment to performance. The need to constrain a system may in fact stem from inadequacies in the design or execution of a system to be commensurate to that which is demanded. Without a robust design against which to reference external perturbations, it is not likely that a system will generate sufficient resilience to effectively direct external 'meddling' in a system. System development should emphasize development of robustness in the design such that externally imposed 'reaching' can be better understood and responses can question the system design/execution for appropriateness to 'classes' of probing, not just individual cases.

CSG Response Discussion
Mapping of the CSG landscape provides visualization for analysis of the most critical challenges facing CSG development (peaks). Past, on-going, and future planned system development initiatives are mapped against the existing governance landscape, pathologies, and system criticisms. 'Holistic' analysis provides clarity and focus for integrated system development response strategy. Adjacent figures are representative of the current research, including application of a 14 Point CSG Governance Check. Figure 13.5 shows a mapping of a CSG landscape for a system.

13.4 Implications

With respect to development of CSG, there are two interrelated aspects. First, there is the development of active governing systems. This *Governance Development* is focused on identifying and engaging in a set of interrelated activities designed to establish, execute, and evolve the continuing development of the CSG metasystem functions. CSG development is always focused on identification and execution of feasible development activities consistent with initial assessments of the state of governance in an organization (system). Development involves purposeful improvement of the system of interest (context, pathologies, system). Ultimately, the purpose of governance development is to enhance system performance through the process of continual (integrated activities) to move the system to a more desirable, feasible, achievable, and sustainable level of performance. It would be shortsighted not to include the multiple aspects of development for CSG, including practitioners, organization, larger enterprise, support infrastructure, context, and system.

Table 13.2 identifies the details of the five interrelated development activities that can be engaged to further *Governance Development*. These five elements include: (1) *Exploration*—examination of the performance of the metasystem functions with input from the prior two stages, (2) *Innovation*—identification and prioritization of feasible decisions and actions to improve the metasystem functions, (3) *Transformation*—implementation of innovation strategies and initiative deployment planning to improve the metasystem functions, (4) *Evaluation*—continuous monitoring of the impact of strategies and initiatives undertaken to enhance metasystem performance, and (5) *Evolution*—monitoring development of system governance toward more desirable levels of performance and higher states of maturity.

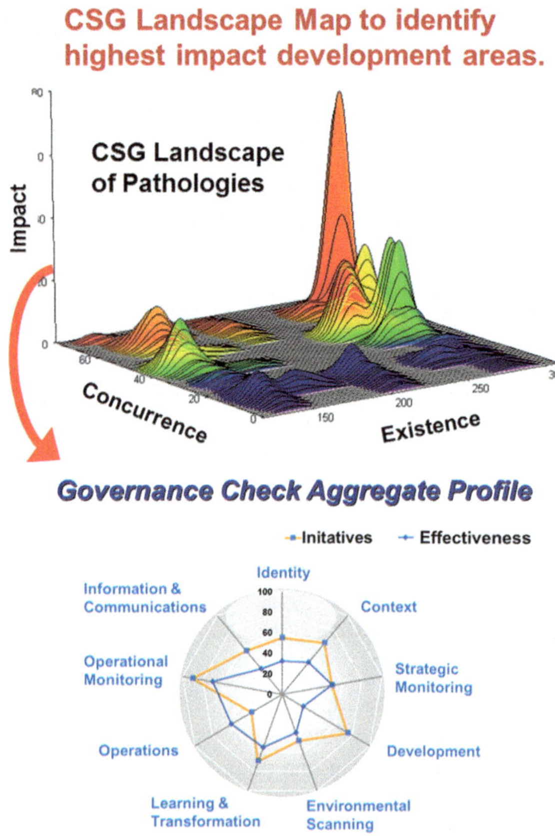

Fig. 13.5 Mapping CSG landscape to identify highest priority development areas

There are three critical points of consideration for implications concerning CSG development. First, while the different governance development activities listed above are presented as separate, they are not independent or linear in execution. In fact, they are considered to be interrelated and overlapping. Therefore, the consideration and performance of the different activities are not mutually exclusive of one another. In essence, they set a frame of reference for a holistic and continuous conversation concerning execution of CSG development. Second, the conversation and actions invoked in CSG development are directed to enhance the overall function of CSG. This is achieved by engaging activities targeted to make improvements in the state of CSG and context for the system of interest. There is an advantage that accrues from the depth of exploration that should be achieved on the 'front end' of CSG development. In particular, engagement in CSG without a workforce commensurate to the engagement offers limited probability of success.

A third critical consideration for CSG development stems from the explorations and mapping of historical, presently existing and future initiatives in relationship to CSG development priorities. This serves as a 'litmus test' to question the relationship

Table 13.2 Interrelated activities for CSG development

Governance development activity	Purpose	Objectives
Exploration	Holistic analysis and synthesis of metasystem context, design, execution, and pathologies	• Systemic investigation and self-study of the metasystem • Identification of completeness of the set of mechanisms performing CSG functions as well as the effectiveness of individual mechanisms • Conduct systemic inquiry to explore multiple perspectives and implications for the metasystem state • Identify, represent, and prioritize systemic meaning and implications of deficiencies, pathologies, and patterns • Define the metasystem current state and trajectory • Examine models of the current system, future system, environment, and context
Innovation	Definition of compatible and feasible metasystem development decisions, actions, priorities, strategies, and initiatives	• Develop the high-level strategy for systemic modifications to the metasystem • Identification, evaluation, and prioritization of compatible and feasible (contextually) first order (correction within existing system) and second order (correction by system redesign) initiatives to advance governance of the metasystem • Definition of capabilities and competencies (individual and organizational) necessary to engage systemic innovations to advance the state of CSG and evolve context • Definition of system capacity (resources, infrastructure) and compatibility for engagement of systemic innovation initiatives

(continued)

Table 13.2 (continued)

Governance development activity	Purpose	Objectives
Transformation	Implementation of systemic metasystem governance developmental strategy, decisions, actions, and initiatives to influence system trajectory, GRL advancement, and contextual development	• Holistic deployment planning and resource allocation for initiatives in support of metasystem governance development • Assignment of responsibilities and accountabilities for achievement of transformation initiatives • Exploration of the potential failure modes and mitigation actions necessary to increase probability of success of launched initiatives • Launching of selected initiatives to enhance the metasystem • Integration and assessment of ongoing initiatives in relationship to CSG development strategy and priorities • Identification and assessment of emerging 'rogue' initiatives against the deficiencies, pathologies, and priorities (blueprint) for strategic CSG development
Evaluation	Assessment of the effectiveness of metasystem initiatives, ongoing strategic performance of the metasystem, and development of the metasystem	• Identification of the minimal set of indicators (measures) that serve to show progression of the metasystem development efforts and shifting state of the metasystem • Assess effectiveness of initiatives undertaken for systemic metasystem transformation • Provide feedback for continuing relevance of transformation strategy in light of new system knowledge, understanding, and contextual changes

(continued)

Table 13.2 (continued)

Governance development activity	Purpose	Objectives
Evolution	Setting and monitoring the trajectory and maturation of metasystem governance and system identity	• Monitor the long-range purposeful trajectory of the system in response to internal and external shifts • Enhance the continuing maturity (GRL advancement) of the system of interest, taking the long view, uncorrupted by short term aberrations • Ensure continuity, sustainability, and viability of the system in relationship to changes in the system, context, and environment • Prevent system erosion through methodical development consistent with shifting demands on the system of interest within the context and environment

of initiatives to CSG development. If initiatives are truly targeted to improving the system, their utility with respect to addressing priorities, deficiencies, and identified needs should be capable of withstanding scrutiny. Thus, decision makers are provided actionable intelligence concerning the contribution of different 'well meaning' activities currently underway or being contemplated to improve CSG. If development initiatives, either ongoing or being considered, cannot be 'justified' as to their relevance to the most pressing needs for improving the state of CSG and context, they should be called into question.

CSG development is not envisioned as an easy approach to system improvement. On the contrary, it is viewed as a difficult development path. This path is fraught with potential obstacles that should be considered by individuals or entities contemplating engaging the approach presented for CSG development.

13.5 Exercises

The following exercise provide an opportunity to examine the concepts presented in this chapter through several questions.

1. For a situation of your choosing, identify elements from the depiction of the complex system problem domain identified in Fig. 13.1. What is the significance of the nature of this domain for practitioners? Future systems? What about critical space 'infrastructures?

2. Identify a 'system superhero' that you have come across. Identify why this system superhero might be detrimental to the long-range survivability of the system(s) they continually save. What can be done in the case of a system superhero unwilling to 'relinquish' their super powers over a system?
3. What do each of the three fields supporting CSG (Governance, Systems Theory, and Management Cybernetics) bring to CSG development? What about the development of your critical space infrastructure?
4. Succinctly explain the essence of the CSG Paradigm. What difficulties might be encountered in the deployment of this paradigm in Critical space infrastructure domain?
5. What guidance and cautions might you suggest for practitioners who might be considering initiation of a CSG based initiative?

References

Ackoff, R. L. (1997). Systems, messes and interactive planning. *The Societal Engagement of Social Science, 3*(1997), 417–438.

Adams, K. M., Hester, P. T., Bradley, J. M., Meyers, T. J., & Keating, C. B. (2014). Systems theory as the foundation for understanding systems. *Systems Engineering, 17*(1), 112–123. https://doi.org/10.1002/sys.21255.

Beer, S. (1979). *The heart of the enterprise*. New York: Wiley.

Casti, J. (2012). *X-Events: Complexity overload and the collapse of everything*. New York: William Morrow.

Davidz, H. L. (2018). Systems engineering pathology: Leveraging science to characterize dysfunction. In *Disciplinary convergence in systems engineering research* (pp. 683–696). Cham, Switzerland: Springer. https://doi.org/10.1007/978-3-319-62217-0_47.

Katina, P. F. (2015). Emerging systems theory-based pathologies for governance of complex systems. *International Journal of System of Systems Engineering, 6*(1/2), 144–159. https://doi.org/10.1504/IJSSE.2015.068806.

Katina, P. F. (2016). Metasystem pathologies (M-Path) method: Phases and procedures. *Journal of Management Development, 35*(10), 1287–1301. https://doi.org/10.1108/JMD-02-2016-0024.

Keating, C. B. (2014). Governance implications for meeting challenges in the system of systems engineering field. In *2014 9th International Conference on System of Systems Engineering (SOSE)* (pp. 154–159). Adelaide: IEEE. https://doi.org/10.1109/sysose.

Keating, C. B. (2015). Complex system governance: Theory to practice challenges for system of systems engineering. In *2015 10th System of Systems Engineering Conference (SoSE)* (pp. 226–231). https://doi.org/10.1109/SYSOSE.2015.7151955.

Keating, C. B., & Katina, P. F. (2011). System of systems engineering: Prospects and challenges for the emerging field. *International Journal of System of Systems Engineering, 2*(2/3), 234–256.

Keating, C. B., & Katina, P. F. (2012). Prevalence of pathologies in systems of systems. *International Journal of System of Systems Engineering, 3*(3/4), 243–267. https://doi.org/10.1504/IJSSE.2012.052688.

Keating, C. B., Katina, P. F., & Bradley, J. M. (2014a). Complex system governance: Concept, challenges, and emerging research. *International Journal of System of Systems Engineering, 5*(3), 263–288.

Keating, C. B., Katina, P. F., & Bradley, J. M. (2014b). Complex system governance: Concept, challenges, and emerging research. *International Journal of System of Systems Engineering, 5*(3), 263–288.

Keating, C. B., & Morin, M. (2001). An approach for systems analysis of patient care operations. *The Journal of Nursing Administration, 31*(7–8), 355–363.

Rittel, H. W. J., & Webber, M. M. (1973). Dilemmas in a general theory of planning. *Policy Sciences, 4*(2), 155–169. https://doi.org/10.1007/BF01405730.

Skyttner, L. (2005). *General systems theory: Problems, perspectives, practice* (2nd ed.). Singapore: World Scientific Publishing Co., Pte. Ltd.

Troncale, L. (2013). Systems processes and pathologies: Creating an integrated framework for systems science. *INCOSE International Symposium, 23*(1), 1330–1353. https://doi.org/10.1002/j.2334-5837.2013.tb03091.x.

von Bertalanffy, L. (1968). *General system theory: Foundations, developments, applications*. New York: George Braziller.

von Foerster, H., Mead, M., & Teuber, H. L. (1953). *Cybernetics: circular causal and feedback mechanisms in biological and social systems*. New York: Josiah Macy, Jr Foundation.

Postface

The book "Critical Space Infrastructures" provided a detailed overview of an issue which is garnering more and more attention not only from the authorities and various direct stakeholders, but increasingly from the wider public—our growing dependence on space systems which provide critical services enabling a wide variety of applications. Every developed and sophisticated nation is dependent on space services in the operation of some or all of its critical infrastructure systems, and the book emphasizes the fact that these space systems are also critical components of the wider infrastructure system-of-system. It is this relationship which should drive further research into the topic and steer the attention of the various authorities towards the problem of collective management of space and collective action to address serious issues, such as the various specific threats.

Even smaller countries, lacking in space systems of their own, are dependent on the timely provisioning of space services of a certain quality and quantity, which turns them into stakeholders of an evolving space governance system. As the book underlines, space is a very international environment, and unilateral action is insufficient to mitigate the effects of natural and artificial phenomena, of accidental and deliberate threats, on collective security.

This means that all countries, regardless of apparent stake in space, must make themselves heard to argue for the accessibility, affordability and sustainability of access to critical space services, and to bring their own contribution to space governance. This is a priority for the space agencies of Europe, which have, through their membership in the European Space Agency (ESA) and other bodies, through consultations in the United Nations Committee on the Peaceful Use of Outer Space (UNCOPUOS), and through exchanges with private companies and other stakeholders, emphasized the risks and opportunities of space reliance. Just as has been promoted in other fields, such as regional security and stability, there should be a focus on as many countries as possible becoming providers and not just

© Springer Nature Switzerland AG 2019

A. Georgescu et al., *Critical Space Infrastructures*,
Topics in Safety, Risk, Reliability and Quality 36,
https://doi.org/10.1007/978-3-030-12604-9

beneficiaries of space security governance. A lesser known example, which is familiar to some of the authors, is the case of Romania, the second Eastern European ESA Member to join and one of only four members at the time of completion for this work.[1]

The present book advances this issue and offers a basis for future development of a useful concept not only for understanding the security issues inherent in our growing dependence on space systems, but also for creating the capabilities and knowledge required for security governance efforts to safeguard these critical systems against disruption or destruction. The book explored various facets of critical space infrastructure study and protection, delving into various side issues such as their role in crisis and emergency situation management, the possible developments to cope with a global orbital "commons" whose inventory of assets and resources is more and more controlled by private entities, and what multi-stakeholder governance models look like and the tools that must be developed for them to function adequately.

Ultimately, for such a speculative and emerging topic, the authors agree and have found agreement in interactions with other stakeholders that the capabilities of space systems will grow in quantity and quality and will be the basis of new applications with astonishing impact on Earth. But our dependence will also grow and collective action becomes not just a recommendation, but an obligation if we are to avoid severe disruptions affecting business continuity, quality of life and the confidence of citizens, investors and consumers.

We have much work ahead of us not only to set up the means by which critical space infrastructure protection can be performed, such as Space Situational Awareness, or the technologies that increase resilience (such as shielding or repair capabilities), but also to align the various spacefaring states, international institutions and other active stakeholders into a coherent and functional whole that can marshal resources and provide standards and good practices. These are some of the main directions for research and policy debates. As the authors of this book also note, it is this very high threshold of coordination and cooperation that must be met

[1]The Romanian efforts are spearheaded by the Romanian Space Agency (RoSA), which recognized early on the cross-cutting issues related to the study of space systems and the challenges to their security. To that end, RoSA undertook several studies on these issues, including with partners such as the Ministry of Defense of Romania, state research institutions and various private research entities, such as the EURISC Foundation. ROSA took the lead on an International Academy of Astronautics Study Group on "Space Systems as Critical Infrastructure" and, under the aegis of this prestigious institution, started organizing a yearly international conference on "Space Systems as Critical Infrastructures" in Mamaia, a Romanian resort town on the Black Sea. The year 2018 saw the seventh edition of this event, bringing together experts from several countries, from private and public entities and from among the Romanian decision makers, to discuss issues related to space development and space-related security. These efforts have been invaluable in the development of the field. Under a project financed by the Romanian Government, RoSA will host the SCIPRO (Space Critical Infrastructure at RoSA) Center of Excellence, which will conduct research on the topic and develop innovative instruments and even products which are related to this field, as well as advising Romanian and global decision makers on matters pertaining to this issue. The project is led by Dr. Adrian V. Gheorghe, one of the authors of this book.

for effective critical space infrastructure protection which distinguishes space infrastructures from most terrestrial counterparts and provides the main challenge to their protection.

November 2018
Bucharest, Romania — Alexandru Georgescu, Ph.D.
Norfolk, USA — Adrian V. Gheorghe, Ph.D.
Bucharest, Romania — Marius-Ioan Piso, Ph.D.
Spartanburg, USA — Polinpapilinho F. Katina, Ph.D.

Index

© Springer Nature Switzerland AG 2019

A. Georgescu et al., *Critical Space Infrastructures*, Topics in Safety, Risk, Reliability and Quality 36,
https://doi.org/10.1007/978-3-030-12604-9

Printed by Printforce, the Netherlands